Law as if Earth Really M

CW01509970

This book is a collection of judgments drawn from the innovative Wild Law Judgment Project. In participating in the Wild Law Judgment Project, which was inspired by various feminist judgment projects, contributors have creatively reinterpreted judicial decisions from an Earth-centred point of view by rewriting existing judgments, or creating fictional judgments, as wild law. Authors have confronted the specific challenges of aligning existing Western legal systems with Thomas Berry's philosophy of Earth jurisprudence through judgment writing and rewriting. This book thus opens up judicial decision-making and the common law to critical scrutiny from a wild law or Earth-centred perspective.

Based upon ecocentric rather than human-centred or anthropocentric principles, Earth jurisprudence poses a unique critical challenge to the dominant anthropocentric or human-centred focus and orientation of the common law. The authors interrogate the anthropocentric and property rights assumptions embedded in existing common law by placing Earth and the greater community of life at the centre of their rewritten and hypothetical judgments. Covering areas as diverse as tort law, intellectual property law, criminal law, environmental law, administrative law, international law, native title law and constitutional law, this unique collection provides a valuable tool for practitioners and students who are interested in learning more about the emerging ecological jurisprudence movement. It helps us to see more clearly what a new system of law might look like: one in which Earth really matters.

Nicole Rogers is based in the School of Law and Justice at Southern Cross University, Australia.

Michelle Maloney is the National Convenor of the Australian Earth Laws Alliance, and teaches Earth jurisprudence at Griffith University, Brisbane, Australia.

Law, Justice And Ecology
Series editor: Anna Grear
Cardiff Law School, Cardiff University, UK

In an age of climate change, scarcity of resources, and the deployment of new technologies that put into question the very idea of the 'natural', this book series offers a cross-disciplinary, novel engagement with the connections between law and ecology. The fundamental challenge taken up by the series concerns the pressing need to interrogate and to re-imagine prevailing conceptions of legal responsibility, legal community and legal subjectivity, by embracing the wider recognition that human existence is materially embedded in living systems and shared with multiple networks of non-humans.

Encouraging cross-disciplinary engagement and reflection upon relevant empirical, policy and theoretical issues, the series pursues a thoroughgoing, radical and timely exploration of the multiple relationships between law, justice and ecology.

Titles in the series:

Law and Ecology
New environmental foundations
Andreas Philippopoulos-Mihalopoulos

Law and the Question of the Animal
Edited by Yoriko Otomo and Edward Mussawir

Wild Law – In Practice
Edited by Michelle Maloney and Peter Burdon

Earth Jurisprudence
Peter Burdon

Contributions to Law, Philosophy and Ecology
Exploring Re-Embodiments
Edited by Ruth Thomas-Pellicer, Vito De Lucia and Sian Sullivan

Law as if Earth Really Mattered

The Wild Law Judgment Project

Edited by
Nicole Rogers and Michelle Maloney

Routledge
Taylor & Francis Group

LONDON AND NEW YORK

First published 2017 by Routledge

2 Park Square, Milton Park, Abingdon, Oxfordshire OX14 4RN
711 Third Avenue, New York, NY 10017

Routledge is an imprint of the Taylor & Francis Group, an informa business

First issued in paperback 2018

Copyright © 2017 selection and editorial matter, Nicole Rogers and Michelle Maloney; individual chapters, the contributors.

The right of Nicole Rogers and Michelle Maloney to be identified as the authors of the editorial material, and of the authors for their individual chapters, has been asserted in accordance with sections 77 and 78 of the Copyright, Designs and Patents Act 1988.

All rights reserved. No part of this book may be reprinted or reproduced or utilised in any form or by any electronic, mechanical, or other means, now known or hereafter invented, including photocopying and recording, or in any information storage or retrieval system, without permission in writing from the publishers.

Notice:
Product or corporate names may be trademarks or registered trademarks, and are used only for identification and explanation without intent to infringe.

British Library Cataloguing in Publication Data
A catalogue record for this book is available from the British Library

Library of Congress Cataloging-in-Publication Data
Names: Rogers, Nicole, editor. | Maloney, Michelle M., editor.
Title: Law as if earth really mattered : the wild law judgment
    project / Edited by Nicole Rogers and Michelle Maloney.
Description: New York, NY : Routledge, 2017. | Includes index.
Identifiers: LCCN 2016046198 | ISBN 9781138669086 (hardback) |
    ISBN 9781315618319 (ebook)
Subjects: LCSH: Environmental law. | Environmental law—Philosophy. |
    Environmental law, International.
Classification: LCC K3585 .L397 2017 | DDC 344.04/6—dc23
LC record available at https://lccn.loc.gov/2016046198

ISBN: 978-1-138-66908-6 (hbk)
ISBN: 978-0-367-02419-2 (pbk)

Typeset in Sabon
by Apex CoVantage, LLC

MIX
Paper from
responsible sources
FSC
www.fsc.org   FSC™ C013985

Printed in the United Kingdom
by Henry Ling Limited

Contents

Notes on contributors

Afshin Akhtar-Khavari is an Associate Professor at the Griffith Law School, Griffith University, Australia. He is a former Deputy Head and (acting) Head of the Griffith Law School. He is currently a member of the World Commission on Environmental Law for the International Union for the Conservation of Nature and a lead author for the Intergovernmental Platform on Biodiversity and Ecosystem Services on a report dealing with land degradation and restoration. He is on the editorial board of the Griffith Law Review. He publishes in the general area of international law and environmental humanities and recently completed a co-authored monograph with Routledge on *Ecological Restoration in International Environmental Law*, and is currently working on a co-edited collection in the Griffith Law Review dealing with law and ecological recovery. He is also writing his next book, which deals with the subject of fear in the law and ecological recovery discourse.

Greta Bird has long been an activist, scholar and advocate in the area of justice and Australia's First Nations peoples. Her education about the injustice of colonisation and the 'terra nullius' fiction began in stories shared in bush camps, prisons and court houses during ethnographic research. Books such as *The Civilising Mission: Race and the Construction of Crime* and *The Process of Law in Australia* contributed to bringing issues such as the 'invasion' of Australia, genocidal white practices and the wrongful denial of First Nations sovereignty to a broad readership. Greta was an Associate Professor at Southern Cross University for 18 years and currently teaches at the University of South Australia.

Jo Bird is interested in the intersections between human rights, environmental law and bioethics, both in Australia and internationally. She has an undergraduate background in arts and law, which included studies in environmental law. She then obtained an Australian Postgraduate Award to complete a PhD in law at the University of Melbourne. She was awarded the Vallejo Ganter scholarship to conduct research in Malaysia and has also travelled to Mexico, working with *No Mas Muertes*, a human rights

NGO. Jo currently works with Professor Irene Watson on an ARC grant on Indigenous Knowledges and Law.

Susan Bird is currently a lecturer at Central Queensland University Law School, Australia. Her approach to research, while philosophically focused, is not located purely in the imaginary. She is an active participant in the topics of her research, which involves interacting with those that resist official governance structures. Her chapter is based in part on interviews conducted with Kyle Magee, the defendant in this wild law judgment. Like Magee, Susan believes that urgent action on environmental destruction will involve restructuring market-driven political systems that currently place too much power in the hands of multinational corporations. Some of her current research interests are law and geography, anarchy and property, and free speech in neo-colonial spaces. Her doctoral thesis involved a critical ethnographic methodology which invited participants to contribute to the voice and direction of the research.

Cristy Clark is a lecturer at the Southern Cross University School of Law and Justice, Australia, where she teaches human rights, competition and consumer law, and equity. She has a BA/LLB from ANU and a Masters in International Social Development from UNSW, and did her PhD with the Australian Human Rights Centre at UNSW on the emerging human right to water, with a focus on the Mazibuko water rights case in South Africa and the privatisation of Manila's water system in the Philippines. Cristy writes about the human right to water, water governance, the right to a healthy environment and the right to participation. She is also the co-founder of the Feminist Writers Festival.

Benedict Coyne is a human rights lawyer and advocate at Brisbane law firm Anderson Fredericks Turner. He is also the national President of Australian Lawyers for Human Rights (ALHR); a national network of Australian solicitors, barristers, academics, judicial officers and law students who practise and promote international human rights law in Australia. He has a background in environmental advocacy, activism and campaigning, and is a board member of the Australian Earth Laws Alliance (AELA). Benedict recently graduated with Distinction from a Master of Studies in International Human Rights Law at the University of Oxford. In 2009 he completed a graduate law degree at Southern Cross University and was awarded the university medal for outstanding academic achievement. Benedict is a passionate advocate and has received numerous awards for his work including the Australian Lawyers Alliance/ Amnesty International 2014 National Emerging Lawyer of the Year Award, and the 2015 Queensland Civil Justice Award.

Cormac Cullinan is a director of Cullinan & Associates, a specialist environmental law firm based in Cape Town, South Africa and past recipient of the South African Environmentalist of the Year Award. He has worked on environmental governance issues in more than 20 countries around the world. His groundbreaking book *Wild Law: A Manifesto for Earth Justice* has been translated into several languages and has helped inform and inspire the growing international Rights of Nature movement. In 2010 Cormac led the drafting of the *Universal Declaration of the Rights of Mother Earth*, which was proclaimed on 22 April 2010 by the People's World Conference in Bolivia. He is a member of the Executive Committee of the Global Alliance for Rights of Nature, drafted the Peoples' Convention that establishes the International Tribunal on the Rights of Nature and was the presiding judge at the Tribunal hearings in December 2015 in Paris.

Robert Cunningham is an Associate Professor at Curtin Law School, Curtin University, Perth, Australia. Throughout his professional career, he has engaged with the law in his capacity as both legal practitioner and academic. As a legal practitioner his efforts have concentrated on the provision of legal information, court advocacy and education within Community Legal Centres and the not-for-profit sector. He currently practices as a barrister, specialising in corporate contracting and public interest advocacy. In academia his focus has been on how the law interfaces with international trade, sustainability, corporate accountability, and information regulation. His latest book, published by Edward Elgar, is entitled *Information Environmentalism: A Governance Framework for Intellectual Property Rights*. Along with a PhD from the Australian National University, Robert holds a Bachelor of Business (Accounting), Bachelor of Laws (Hons), Master of Laws (Hons), and a Graduate Certificate of Legal Practice from the University of Technology, Sydney. He currently teaches Corporations Law, Corporate Governance, and Professional Responsibility.

Felicity Deane is a lecturer at the Faculty of Law, Queensland University of Technology (QUT), Brisbane, Australia. She completed a Bachelor of Law and a Bachelor of Commerce at the University of Queensland in 1999. Immediately following graduation, she commenced work and study in the United States in the disciplines of accounting and law. She returned to Australia to complete a Postgraduate Diploma at Monash University. Before returning to studies, Felicity worked as a Senior Policy Officer with the Queensland Police Service. She commenced PhD studies in December 2009 at Queensland University of Technology. Her PhD, entitled 'The Clean Energy Package and WTO Law: An Analysis of Compliance Issues', was subsequently completed in August 2013. In January 2014 she commenced her time as a lecturer within the QUT Law School on the early career academic programme. She has published a number of articles on the topic of emissions trading, market based mechanisms and the WTO Law. Her book *Emissions Trading* *and WTO Law: A Global Analysis* was published in March 2015. Her recent research has focused on climate change and transparency, in particular in regards to emissions trading in China.

Julia Dehm is a postdoctoral fellow at the Rapoport Center for Human Rights and Justice at the University of Texas School of Law, USA, working on a multi-year project rethinking human rights for the twenty-first century. Julia was previously a Resident Fellow at the Institute for Global Law and Policy (IGLP) at Harvard Law School. Her doctoral research at the Melbourne Law School examined the social implications of a specific carbon offset scheme under the United Nations Framework Convention on Climate Change umbrella, called Reducing Emissions from Deforestation and Forest Degradation (REDD+). She has published academic articles in the *Journal of Human Rights and the Environment*, the *London Review of International Law*, the *Macquarie Journal of International and Comparative Environmental Law* and special climate justice-themed editions of the *Journal of Australian Political Economy* and *Local-Global Journal*. She holds a BA, LLB (Hons), and PhD from the University of Melbourne.

Kate Galloway is an Assistant Professor of Law at Bond University, Gold Coast, Australia. Her principal field of research is property law with a particular interest in the intersection of real property and sustainability, and the implications for the environment, and for justice for Aboriginal and Torres Strait Islander Australians. She has published and presented internationally on issues of sustainability in legal education. As well as teaching property law at Bond University, Kate teaches mining and resources law and governance at James Cook University. She serves on the editorial committees of the *Alternative Law Journal* and the *Legal Education Review* and blogs at kategalloway.net.

Bee Chen Goh is Professor of Law, Director of Research and Research Training, and Director of the Centre for Peace and Social Justice at the School of Law and Justice, Southern Cross University (Gold Coast Campus), Australia. She is a Director and Fellow of the Australian Academy of Law, Fellow of the Cambridge Commonwealth Society, and Fellow of the Society of Advanced Legal Studies in London. Her teaching and research interests include mediation and ADR, especially on cross-cultural (Sino-Western) dispute resolution. She also has a special interest in environmental dispute resolution. Her publications include *Negotiating with the Chinese* (Dartmouth, 1996), *Law Without Lawyers, Justice Without Courts: On Traditional Chinese Mediation* (Ashgate, 2002) and *Activating Human Rights and Peace: Theories, Practices and Contexts* (Ashgate, 2012).

Hope Johnson is a research associate at the School of Law, Queensland University of Technology (QUT), Australia, where she is a member of the

International Law and Global Governance research programme. Recently, she submitted her PhD thesis on international agricultural law and food policy with the support of the Queensland University of Technology and the Institute for Future Environments. Hope's thesis critically analysed the international regulation of agriculture across international trade, environmental and human rights law. Her principal research interests relate to sustainable natural resource management and the regulation of food and agriculture at national and international levels. She has published in areas such as international investment law, food labelling and green criminology.

Bridget Lewis is a senior lecturer at the School of Law, Queensland University of Technology (QUT), Australia, where she is the co-leader of the Rights and Regulation research theme within the International Law and Global Governance Research Programme. Her principal research interests relate to international human rights law, particularly the area of environmental human rights. She has also published in the areas of environmental and disaster displacement, climate justice and indigenous rights. Bridget is on the editorial board of the *Asia-Pacific Journal of Environmental Law*. She completed her PhD at Monash University on the topic of 'The human right to a good environment in international law and the implications of climate change'.

Rowena Maguire is a senior lecturer at the School of Law, Queensland University of Technology (QUT), Australia. She is the Co-Chair of the International Law and Global Governance Research programme and programme theme leader of the Climate and Environmental Governance programme. She predominantly researches in the area of international climate and forestry law with a focus on equitable design and implementation. She has published a solo monograph on international forest governance (Edward Elgar) and has co-edited four collections with Routledge examining international climate and human rights governance. Rowena is currently a Chief Investigator on an Australian Research Council Discovery Project examining the integrity of the international climate regime. She teaches in the areas of environmental law, international law, refugee and immigration law and law in context.

Michelle Maloney holds a Bachelor of Arts (political science and history) and Laws (Honours) from the Australian National University and a PhD in Law from Griffith University, Australia. She is the co-founder and National Convenor of the Australian Earth Laws Alliance (AELA), a national not-for-profit organisation whose aim is to increase understanding and practical implementation of Earth-centred law, governance and ethics. She is on the Executive Committee for the Global Alliance for the Rights of Nature, a member of IUCN World Commission on Environmental Law and is a member of the UN Harmony with Nature Knowledge Network.

She is co-editor of *Wild Law in Practice* (Routledge, 2014) and has written about the powerful alternative jurisprudence being created by the International Rights of Nature Tribunal.

Edward Mussawir is a lecturer at the Griffith Law School, Griffith University, Australia. He is the Managing Editor of the *Griffith Law Review: Law, Theory, Society* and teaches legal theory and civil procedure. He has various research interests in jurisprudence including jurisdiction, legal personality, the work of Gilles Deleuze and the legal status of animals. The latter has given rise to some recent studies on the technical meaning of animals in jurisprudence, focusing particularly on how different juridical forms of civil liability have constructed the animal as an object or subject of right.

The Honourable Justice Brian J Preston SC is the Chief Judge of the Land and Environment Court of New South Wales. Prior to being appointed in November 2005, he was a senior counsel practising primarily in New South Wales in environmental, planning, administrative and property law. He has lectured in postgraduate environmental law for over 23 years. He is the author of Australia's first book on environmental litigation and 97 articles, book chapters and reviews on environmental law, administrative and criminal law. He holds numerous editorial positions in environmental law publications and has been involved in a number of international environmental consultancies and capacity-building programmes, including for judiciaries throughout Asia. Justice Preston is an Official Member of the Judicial Commission of NSW, Fellow of the Australian Academy of Law and is Co-Chair of the IBA Working Group on a Model Statute on Climate Change Claims and Remedies.

Aidan Ricketts is an experienced social and environmental activist, social change trainer, academic and published author. Aidan has over 25 years hands-on experience as a campaigner and trainer and continues to provide strategic workshops, advice and training for emerging social movements and campaigns. Aidan has recently published *The Activists' Handbook: A Step-By-Step Guide to Participatory Democracy* which is published internationally by Zed Books, London. Aidan is currently employed as an academic at Southern Cross University School of Law and Justice, Lismore, Australia. Aidan's academic qualifications include undergraduate and postgraduate degrees in law and a postgraduate degree in education. He is currently completing a PhD examining the application of complexity theory to social movement organisation and mobilisation.

Matthew Rimmer is a Professor of Intellectual Property and Innovation Law at the Faculty of Law, at the Queensland University of Technology (QUT), Australia. He is a leader of the QUT Intellectual Property and Innovation Law research programme, and a member of the QUT Digital Media Research Centre (QUT DMRC), the QUT Australian Centre for

Health Law Research (QUT ACHLR), and the QUT International Law and Global Governance Research Programme. Matthew has published widely on copyright law and information technology, patent law and biotechnology, access to medicines, plain packaging of tobacco products, intellectual property and climate change, and Indigenous intellectual property. He is currently working on research on intellectual property, the creative industries, and 3D printing; intellectual property and public health; and intellectual property and trade, looking at the *Trans-Pacific Partnership*, the *Trans-Atlantic Trade and Investment Partnership*, and the *Trade in Services Agreement*. His work is archived at QUT ePrints, SSRN Abstracts and Bepress Selected Works.

Nicole Rogers is a senior lecturer at the School of Law and Justice, Southern Cross University, Australia. She has two first class honours degrees in arts and law, a Masters degree in natural resources law and a PhD in law. She was a founding member of the School and has developed and taught Environmental Law as a core unit in its LLB programme since the programme was first introduced in 1993. Her current research is in the areas of wild law, climate change litigation and climate change activism. In various publications and in her PhD, she has applied the 'interdiscipline' of performance studies theory to legal and activist performances. Nicole is a member of the IUCN World Commission on Environmental Law and lawyer on the Southern Cross University Human Research Ethics Committee. She conceived of a Wild Law Judgment Project in 2012 and, together with Michelle Maloney, she has led the Wild Law Judgment Project since it was developed, introduced and launched in 2014.

Tom Round has degrees in arts (with first-class honours in government) and law from the University of Queensland, Australia, and a PhD (in politics and public policy) from Griffith University. He has lectured and tutored in the Schools of Criminology (formerly Justice Administration) and Politics at Griffith University, and the School of Government at UQ; he also worked in the federal Department of Immigration and Multicultural Affairs (as it then was). Before teaching at SCU, he was a Research Fellow at Griffith's Key Centre for Ethics, Law, Justice, and Governance (KCELJAG). Tom has co-edited two books, *Beyond the Republic: Meeting the Global Challenges to Constitutionalism* (2000) and *Asia-Pacific Governance: From Crisis to Reform* (2003), and published several book chapters and journal articles. He also co-wrote *Retrospectivity and the Rule of Law* (Oxford University Press, 2006) with Professor Charles Sampford and others.

In his first career, **Stephen Summerhayes** practised law for approximately 20 years in both Australia and England. In 2003, he established his own legal firm to support studies in environmental science and international studies and, thereafter, a postgraduate research Master of Environmental Science. His international studies took him to Latin America for two years, learning and working for community organisations in Ecuador and Guatemala. He is currently the executive officer of a regional organisation of councils advancing sustainable urban water management. He tutors adult literacy and numeracy and has taught philosophical ethics to primary school children.

Irene Watson belongs to the Tanganekald, Meintangk-Bunganditj First Nations Peoples and their territories which include the Coorong and the south-east of South Australia, and is Pro Vice Chancellor, Aboriginal Leadership and Strategy and a Research Professor of Law with the University of South Australia. Irene was awarded an ARC Indigenous Discovery Award in 2013, and has recently completed the project, titled *Indigenous Knowledge: Law, Society and the State*. Irene has published extensively in law and Indigenous Knowledges and in 2015 she published *Aboriginal Peoples, Colonialism and International Law: Raw Law* (Routledge, 2015), and more recently has completed *Indigenous Peoples as Subjects in International Law*, (Routledge, 2016).

Katie Woolaston is a PhD candidate at Griffith University, Queensland, Australia. After completing her undergraduate law degree with Honours in 2006, she commenced practising as a corporate lawyer in Sydney, Australia. Katie now works as a research assistant across universities in Queensland, whilst teaching and completing her PhD. Her recent research involves wildlife law and human–wildlife interactions. Her most recent article, co-authored with Evan Hamman, appeared recently in the *Environmental Planning and Law Journal* and was entitled *The operation of the precautionary principle in Australian environmental law: An examination of the Western Australian white shark drum line programme*. She is also an ardent environmentalist, feminist and wildlife warrior.

Preface

Imagine thinking of law from the perspective of nature and not the common anthropocentric approach. In a nutshell this is 'Wild Law'. Nicole Rogers and Michelle Maloney are the co-editors of a ground-breaking book that is part of the 'Wild Law Judgment Project', an innovative initiative that looks to developing wild law or Earth laws jurisprudence. Inspired by feminist judgment projects, a number of highly regarded legal thinkers have re-analysed and re-written well known judgments from national and international cases, placing nature squarely at the centre of the case. The results are thought-provoking and possibly could lead to transformational jurisprudence.

This book is well-timed as greater attention globally is focusing on ecological jurisprudence. There is a growing movement for national environmental courts as reflected in the establishment of a Judicial Institute for Environmental Law by the IUCN World Commission on Environmental Law. Likewise, international tribunals, courts and adjudicative bodies are taking bolder decisions for the protection of nature. However, the existing model that is based on sustainable development requires a counter-balance that is provided by wild law.

This book will likely, and should, provoke similar rewritings for cases as an important intellectual rethinking of how we view law in face of the impending consequences of climate change, loss of biodiversity, ocean acidification and sadly more. This project and others that will surely follow in other jurisdictions will stimulate a much-needed discussion in relation to alternative ways of thinking and judging applicable to different systems of law around the world.

This book will help to reinvigorate innovative and ethical legal thought, reminding us that law is not simply for protection of human property rights, relations and security but must be integrated into the holistic approach that recognises the central place of nature.

Wild law's time has come.

Professor Nilufer Oral
Istanbul Bilgi University, Turkey
Chair, IUCN Academy of Environmental Law
Member of United Nations International Law Commission
6 September 2016

Introduction

Chapter 1

The Wild Law
Judgment Project

Nicole Rogers and Michelle Maloney

The Wild Law Judgment Project was launched at a workshop in 2014 by the editors of this collection, Dr Nicole Rogers from the School of Law and Justice, Southern Cross University and Dr Michelle Maloney from the Australian Earth Laws Alliance. The project forms part of a developing wild law or earth laws jurisprudence. In engaging with existing judicial reasoning from a wild law perspective and exploring the ways in which wild changes to existing laws change the nature of judicial reasoning, the project highlights opportunities for achieving wild law outcomes in the courtroom. At this critical juncture in Earth's history, when we confront an ecological crisis precipitated by human activity and accelerated by an ongoing adherence to the 'business as usual' mantra, the task of 'wilding' law is an urgent one.

The Wild Law Judgment Project has been inspired by various feminist judgment writing projects and in particular by the Australian feminist judgment project, in which judgments from all areas of Australian law have been rewritten from a feminist perspective. The Wild Law Judgment Project, however, instigated a different form of rewriting. While the contributors to the feminist judgment projects have interrogated masculinist assumptions and rewritten judgments from a feminist perspective, contributors to the Wild Law Judgment Project have selected judgments and rewritten them from a wild law viewpoint, interrogating the dominant anthropocentric principles which underpin existing law, and incorporating into their judgments some of the key principles and values of the burgeoning wild law/earth laws movement. In a departure from the methodology devised for the feminist judgment projects, a number of contributors have written hypothetical judgments and relied upon hypothetical legislative developments, both domestic and international. The collection of rewritten and hypothetical judgments and the accompanying commentaries is the end result of a collaborative ideological challenge to the hegemony of anthropocentrism in the common law and in all contemporary western-style legal systems.

In recent years, there have been a number of feminist judgment writing projects undertaken in different jurisdictions. In both the United Kingdom and Australia, two separate rewriting projects have culminated in edited collections of rewritten judgments.[1] In Canada, a 'virtual court' comprised of academics, activists and litigators, the Women's Court of Canada, has embarked upon the lengthy exercise of rewriting Canadian equality jurisprudence from a feminist perspective. Contributors to the Wild Law Judgment Project have confronted quite a different challenge. Although there are many feminisms, women judges, or women writing as judges,[2] can bring their own perspectives as women, albeit women of a particular nationality, class and race, into the process of judging. It is, arguably, a more challenging enterprise for human 'judges' to move beyond the confines of our own human-centred concerns and to try to imagine and read law from the perspective of other non-human species or even from the perspective of Earth. This is quite a different undertaking from the anthropocentric task of judging in accordance with principles of sustainable development,[3] or even in accordance with sustainability principles.

There are, of course, considerable philosophical difficulties in attempting to speak for Earth or non-human species and, in fact, considerable philosophical difficulty in imagining Earth or non-human species interested in or convinced of the importance of judging. As Derrida points out, animals other than the human animal lack the consciousness of good and evil,[4] and this distinction underlies the human habit of judging. Nevertheless, let us not underestimate what might be involved in attempting to judge or even to see human activities with the gaze of the non-human animal. Derrida has written that 'as with every bottomless gaze, as with the eyes of the other, the gaze called animal offers to my sight the abyssal limit of the human'.[5] We can extrapolate from this statement: the possibility of judging through non-human eyes offers up the 'abyssal limit(s)' of human-centred law.

1 The United Kingdom Feminist Judgment project produced the edited collection of judgments in R Hunter, C McGlynn and E Rackley (eds), *Feminist judgments: from theory to practice* (Hart Publishing, 2010) and the Australian Feminist Judgment project produced the edited collection of judgments in H Douglas, F Bartlett, T Luker and R Hunter (eds), *Australian feminist judgments: Righting and re-writing law* (Hart Publishing, 2014).
2 However not all contributors to the Australian Feminist Judgment project were female; see J Crowe, '*U v U* [2002] HCA 36' in H Douglas, F Bartlett, T Luker and R Hunter (eds), *Australian feminist judgments: Righting and re-writing law* (Hart Publishing, 2014) 365.
3 Justice Brian Preston discusses some examples of judging in accordance with principles of sustainable development in 'The art of judging environmental disputes' (2008) 12 *Southern Cross University Law Review* 103, 115–121.
4 J Derrida and D Wills, 'The animal that therefore I am (more to follow)' (2002) 28 *Critical Inquiry* 369, 373.
5 Ibid., 381.

1 Writing judgments wildly

The idea behind the Wild Law Judgment Project was that judgments, selected from any area of law, would be viewed and rewritten through a wild law lens. Wild law is a term devised by South African lawyer Cormac Cullinan, in his 2002 book of that title.[6] Cullinan perceives wild law as an expression of Thomas Berry's philosophy of Earth jurisprudence;[7] wild laws, he has written, 'are laws that regulate humans in a manner that creates the freedom for all the members of the Earth Community to play a role in the continuing co-evolution of the planet'.[8] Existing legal systems and existing laws are profoundly anthropocentric in their orientation. As we have previously observed in an earlier article on the project, the rewriting project was 'intended to disrupt and unsettle the established human and property-centred practices of the common law', by placing 'all life, and all of life's support systems, at the centre of judgments', and 'contesting the place of humanity at the centre of existing notions of justice'.[9]

In the ongoing discussions about the importance of wild law principles, academics and other theorists tend to focus on legislative and/or constitutional change. The recognition and protection of rights of nature in Bolivian law[10] and the Ecuadorian Constitution[11] are cited as international examples of the practical implementation of wild law principles. Less attention is directed towards the flexibility inherent in the common law and the wild possibilities in the art of judging. Yet, even within the constraints of our existing legal structure and our existing regulatory regimes, there is scope for the expression and implementation of wild law principles in individual judgments. This involves the recognition and careful nurturing of what Cormac Cullinan has described as 'the flashes' of wild law[12] in existing legal doctrines, the re-examining of 'all legal rules and conventions with wide open, wild eyes'.[13] This sort of creative exercise reveals the wild potential even in existing, flawed, property-centred, anthropocentric legal doctrines and, indeed, the wild potential in judging. As every law student quickly comes to realise, while the process of judging is a rigorous one and one in

6 C Cullinan, *Wild law* (Siber Ink, 2002).
7 Ibid., 10.
8 Ibid.
9 N Rogers and M Maloney, 'The Australian wild law judgment project' (2014) 39 *Alternative Law Journal* 172, 173.
10 *Law of the Rights of Mother Earth* (Bolivia) (2010).
11 *Constitution of the Republic of Ecuador* (Ecuador) (2008) arts 71–74.
12 Cullinan, above n 6, 10.
13 N Rogers, 'Where the wild things are: finding the wild in law' in P Burdon (ed.), *Exploring wild law. The philosophy of Earth Jurisprudence* (Wakefield Press, 2011) 183, 188.

which certain steps must be followed,[14] the conclusions reached by judges are never foregone conclusions.

The Wild Law Judgment Project, while influenced by the methodology of the feminist judgment projects, evolved as a different kind of foray into judgment writing. At the project's launch at the University of New South Wales in November 2014, Professor Heather Douglas, one of the project leaders of the Australian Feminist Judgment project, outlined the parameters of that project. One important parameter, which was also a parameter in the United Kingdom Feminist Judgment project,[15] is that the rewritten judgments had to be structured around existing legal principles[16]; they contained *possible* or *plausible* judicial outcomes. At the launch, Justice Brian Preston, Chief Judge of the New South Wales Land and Environment Court, presented a paper in which he acknowledged that 'accept(ing) the law as it currently exists, but explor(ing) where there is scope for finding, interpreting and applying the law to best meet the justice – including the ecological justice – of the situation' is indeed one approach to writing judgments wildly.[17] However he identified another possible approach. Justice Preston stated that:

> The other approach is to challenge the existing law and to mould it to fit the earth's demands. This falls outside the orthodox technique and logic of judging disputes. Judges are not permitted to be legislators. However the object of engaging in this second approach is to highlight the inadequacies of the existing law. Judgments would be rewritten to identify the reformed laws and show how the application would affect the outcome of the case.

Justice Preston argued that 'both approaches have utility and could be employed productively in the Wild Law Judgments Project'.

The participants in the first workshop reached an agreement that the Wild Law Judgment Project should draw on both these approaches. Contributors were thus empowered, if they wished, to redraft existing laws and to rewrite

14 Justice Preston discusses these steps in his chapter 'Writing judgments wildly'.
15 R Hunter, C McGlynn and E Rackley, 'Feminist judgments: an introduction' in R Hunter, C McGlynn and E Rackley, *Feminist judgments: from theory to practice* (Bloomsbury publishing, 2010) 6.
16 H Douglas, F Bartlett, T Luker and R Hunter, 'Introduction: righting Australian law' in H Douglas, F Bartlett, T Luker and R Hunter, (eds), *Australian feminist judgments: righting and re-writing law* (Hart Publishing, 2014) 1, 13.
17 Justice Preston's address, entitled 'Writing judgments wildly', is reproduced as the second chapter in the collection. The title to this collection was inspired by Justice Preston's suggestion, in this essay, that wild law can be described as 'wild is beautiful: a study of law as if nature mattered'.

judgments based on these rewritten laws. Some have availed themselves of this opportunity. Others have elected to rewrite judgments based on existing legal principles. In addition, some of the judgments in the collection are hypothetical and/or futuristic judgments rather than rewritten judgments. Two such hypothetical judgments included in the collection, written by Justice Preston and Cormac Cullinan, were devised independently of the project and delivered, respectively, at a mock trial in Australia in 2012 and at a civil society International Rights of Nature Tribunal in 2014.

Part of the innovative approach in this project has involved the critique and reshaping of principles within a number of different areas of law from a wild law perspective. Traditionally, ecocentric concerns have been corralled within the limited and permissive framework of environmental law.[18] There is clearly, however, potential for a wild infiltration of other legal categories. In a recent development, an international group of legal scholars has adopted the 2016 *Oslo Manifesto for Ecological Law and Governance*, in which the inadequacies of current environmental law are highlighted and the need to replace this 'anthropocentric, fragmented and reductionist' framework for a new framework of ecological law is asserted. In the manifesto, it is stated that:

> We do not need more laws, but different laws from which no area of the legal system is exempted. The ecological approach to law is based on ecocentrism, holism, and intra-/intergenerational and interspecies justice.[19]

In the Wild Law Judgment Project, wild law seeps into other areas of the legal system including constitutional law, international law, torts, criminal law, administrative law, intellectual property law and native title law.

II: The judgments

In structuring the collection of judgments, we have drawn out common themes in clusters of judgments. Themes which appear in a number of judgments include: standing for non-human entities, the wellbeing of other entities as a foremost consideration, the need for an equitable and precautionary

---

18 See M C Wood, *Nature's Trust. Environmental law for a new ecological age* (Cambridge University Press, 2014) Part I, for a critique of the limitations of existing environmental law.

19 *Oslo Manifesto for Ecological Law and Governance*, adopted at the IUCN WCEL Ethics Specialist Group Workshop, IUCN Academy of Environmental Law Colloquium, University of Oslo, 21 June 2016.

approach in decision-making processes on mining activities, the flexibility of statutory interpretation principles in the context of mining rights and mining activities, points of intersection between First Nations law and wild law, limitations in existing international law, the application of criminal law to environmental activists, and future visioning in the wild law context.

A: Standing for all entities

In his influential essay in 1972,[20] Christopher Stone posed the tantalising question of whether trees and other 'natural' objects should have standing to bring legal actions to protect their own interests and rights. The issue of standing for non-human entities remains an important one, particularly in light of existing obstacles to standing in environmental cases and ongoing practical difficulties experienced by activists in attempting to mount legal challenges on behalf of other species and the environment. Contributors have addressed the issue of legal standing of non-human entities both directly and indirectly in a number of the rewritten judgments.

In his fictitious and futuristic judgment *Green sea turtles by the representative, Meryl Streef v the State of Queensland and the Commonwealth of Australia*, originally delivered at a mock trial organised by the Environmental Defender's Office, Victoria, in Melbourne in 2012, Justice Preston confers standing on green sea turtles. Placing a novel twist on the doctrine of nuisance in its application to non-human entities, he finds in favour of the turtles against the Australian Commonwealth and Queensland governments on the basis that the governments' approval of eight coalmines in the Galilee Basin in 2012 had contributed to anthropocentric climate change and hence to the destruction of the turtles' habitat, the Great Barrier Reef. The public nuisance is also created by the governments' general failure to reduce greenhouse gas emissions and/or take ameliorating action in relation to the impacts of climate change on the Reef.

In the judgment that follows, the entire marine community of the Great Barrier Reef has become the litigant. *Great Barrier Reef v Australian Federal and State governments and others* is the first judgment handed down by the International Rights of Nature Tribunal.[21] The Tribunal was created in 2014 by the Global Alliance for the Rights of Nature, as a permanent forum for people around the world to give a voice to the natural world and

---

20 C D Stone, 'Should trees have standing? – Toward legal rights for natural objects' (1972) *Southern California Law Review* 450.
21 See <http://therightsofnature.org/rights-of-nature-tribunal>.

to protest the destruction of the Earth.[22] The International Tribunal offers a powerful civil society 'alternative' to state centred international law and aims to build a body of Earth-centred jurisprudence for the wider Earth community. The Tribunal has convened three times and has in turn fostered the creation of regional chambers in the United States and Australia.[23] The *Great Barrier Reef* case articulates what law would look like if the *Universal Declaration for the Rights of Mother Earth*, a document created at the World People's Conference on Climate Change and the Rights of Mother Earth at Cochabamba Bolivia on 22 April 2010, becomes the foundation of our legal systems.

Benedict Coyne revisits a judgment which signified an existential crisis for the Queensland lungfish: *Wide Bay Conservation Council Inc v Burnett Water Pty Ltd (No 8)*.[24] In that case, the challenge to the Federal Minister's conditional approval of the Paradise Dam was unsuccessful. Coyne modifies the application of the existing Australian federal environmental regulatory regime, the *Environment Protection and Biodiversity Conservation Act 1999* (Cth), by envisaging a number of unashamedly 'utopian' future legislative developments both at international and national levels;[25] these enable him to reason his way to a conclusion better equipped to protect this ancient endangered species. His fictitious *Rights of Nature and Mother Earth Act 2018* (Cth) confers standing upon the seven lungfish, which are bringing this particular action in trespass and also seeking injunctive relief under the *Environment Protection and Biodiversity Conservation Act* and this later Act.

Standing for non-human species is also addressed in Hope Johnson, Bridget Lewis and Rowena Maguire's revised and improved version of the *International Convention on Whaling*, in which there is provision for a Special Representative for Whales.

Two other contributors review the legal and political status of non-human species and other entities in their judgments. Stephen Summerhayes is prepared to find that Country is a legal entity with a unique 'ecological title' in his rewriting of the groundbreaking Australian High Court decision in the *Mabo* case.[26] Hence he concludes that the Meriam people, rather than

---

22 See M Maloney, 'Building an alternative jurisprudence for the Earth: the International Rights of Nature Tribunal', *Vermont Law Review* (forthcoming 2017).

23 See, e.g., the Rights of Nature Tribunal Australia <www.earthlaws.org.au/events/tribunal2016>.

24 *Wide Bay Conservation Council Inc v Burnett Water Pty Ltd (No 8)* [2011] FCA 175.

25 These are 'the adoption by the UN General Assembly of the Universal Declaration on the Rights of Mother Earth; the drafting, signing and ratification on the International Convention on the Rights of Mother Earth 2018 and the subsequent passage of the Rights of Nature and Mother Earth Act 2018 in Australia'.

26 *Mabo v Queensland (No 2)* (1992) 175 CLR 1.

possessing native title rights as was found in the original decision, are rather custodians and constructive trustees for Country with rights under this trust as well as an obligation to maintain the land's ecological integrity.

Tom Round considers the related question of political representation of other species in his recalibration of what democracy means in a wild law context. In rewriting *Attorney-General (Cth); Ex Rel McKinlay v The Commonwealth*,[27] Round contends that 'legislators should represent all living things' and sets out a number of possible ways in which this could be achieved.

B: *Wellbeing of non-human species*

In other contributions, the wellbeing of other species influences the authors' reasoning and shapes their conclusions. By venturing into the field of wild negligence with her rewriting of the seminal English torts decision *Donoghue v Stevenson*,[28] Bee Chen Goh has provided us with valuable insights into the synergy between Buddhist principles and the precepts of wild law. Cullinan has observed that environmental destruction is also rife in countries in which a large number of Buddhists and Taoists reside.[29] Nevertheless, as Goh points out in her judgment and Round discusses in his accompanying commentary, the karmic consequences of the way in which we treat non-human and human individuals underpin Buddhist thought and can contribute to judgments which vindicate non-human entities. In Goh's judgment, the decomposing snail infamously found in a partially consumed bottle of ginger beer moves to the foreground; ultimately, she proposes that the doctrine of negligence should be extended to all sentient beings and finds for the snail.

In Edward Mussawir's rewritten Canadian judgment of *Shaw v McCreary*,[30] however, the wellbeing and rights of a captive bear remain part of the subtext. Mussawir refrains from a rewriting which emphasises and protects the bear's rights, whether these are rights not to be held in captivity or rights to behave 'wildly' when the bear escapes captivity. He identifies his approach as one in which a space might be opened 'within the language and logic of law itself' such that 'the juridical outline of an animal might take clearer shape'. Thus Mussawir, while seeking to expose the anthropocentric bias inherent in the Court's 'juridical erasure' of the bear, nevertheless

---

27 *Attorney-General (Cth); Ex Rel McKinlay v Commonwealth* (1975) 135 CLR 1.
28 *Donoghue v Stevenson* [1932] AC 562.
29 Cullinan, above n 6, 31.
30 *Shaw v McCreary* [1890] 19 *Ontario Reports* 39.

retains the focus of the original judgment on the rights and duties of the humans implicated in the bear's captivity and actions.

C: Wild law, climate change and coalmines

The looming crisis of climate change has influenced three rewritten Australian environmental law judgments, which deal with challenges to proposed coalmine developments. It is worthy of note that in this particular cluster of rewritten judgments we find a point of intersection with the Australian feminist judgment collection. Lee Godden and Jacqueline Peel selected the same judgment from the fertile field of Australian coalmine litigation as that selected by Felicity Deane and Katie Woolaston in this collection: namely, *Wildlife Preservation Society of Queensland and Proserpine/Whitsunday Branch Inc v Minister for the Environment and Heritage and others*.[31] The decision to rewrite this particular judgment in both projects is to some extent unsurprising, given the original judge's contentious conclusion that there was no obligation on the Australian government under the *Environment Protection and Biodiversity Conservation Act* to consider the impact of the considerable greenhouse gas emissions from the proposed mine on protected areas in Australia.

The two rewritten judgments reveal some strong similarities between a feminist and wild law rewriting; in both, a quite different outcome is reached once principles of sustainable development and in particular the precautionary principle are factored into the decision-making process. Peel points out in her commentary that a feminist perspective 'allows for the adoption of more holistic conceptions of the environment that emphasise the interconnectedness of all life forms and the need for a nurturing approach that creates a bond between people and nature'. There is a 'duty of care' for human and non-human creatures.[32] Godden therefore concludes in her rewritten judgment that decision-makers under the Act should apply a precautionary and 'widely-framed' holistic approach, factoring in the cumulative and indirect impacts of the burning of coal from the mine.[33] However this precautionary approach does not necessarily require decision-makers to refuse

31 *Wildlife Preservation Society of Queensland and Proserpine/Whitsunday Branch Inc v Minister for the Environment and Heritage and others* (2006) 232 ALR 510.
32 J Peel, 'Addressing climate change inequities: the contribution of a feminist judgment' in H Douglas, F Bartlett, T Luker and R Hunter (eds), *Australian Feminist Judgments. Righting and rewriting law* (Hart publishing, 2014) 133, 136.
33 L Godden, '*Wildlife Preservation Society of Queensland and Proserpine/Whitsunday Branch Inc v Minister for the Environment and Heritage and others*' in H Douglas, F Bartlett, T Luker and R Hunter (eds), *Australian Feminist Judgments. Righting and rewriting law* (Hart publishing, 2014) 138, 149.

to issue the approval. By way of contrast, Deane and Woolaston, applying a wild law interpretation, find that the precautionary principle demands such a refusal. They state in their judgment that:

> A wild law interpretation of the precautionary principle would require caution at a much higher level than that which is currently accepted in the application of this international requirement. . . . In the event that an action may have a possibility of irreversible damage, any application for its approval must necessarily be refused.

Julia Dehm rewrites a later Australian coalmine case: *Xstrata Coal Queensland Pty Ltd & Ors v Friends of the Earth – Brisbane Co-Op Ltd & Ors, and Department of Environment and Resource Management*.[34] By re-interpreting the meaning of 'environmental impact' to include the burning of coal from the proposed mine, re-assessing the scientific evidence presented, and factoring in the equitable concerns raised by the concept of a global climate budget, she reaches a different conclusion about the proposed Wandoan mine to that reached by the Queensland Land Court in the original decision.

D: Statutory interpretation principles and mining

Another proposed Australian coalmine development is rejected in Kate Galloway's rewriting of *Hancock Coal v Kelly and Department of Environment and Heritage Protection (No 4)*.[35] As stated above, a key question to consider in the context of this project is whether there are doctrines already underpinning the existing legal system, and principles already enshrined within it, which can be drawn upon in wild rewritings to produce different modes of reasoning and/or different outcomes. Challenging the avowed neutrality of statutory interpretation principles, Galloway makes creative use of such principles in considering the statutory regime applicable to the proposed Alpha mine in the Galilee Basin in Queensland. She draws upon a seminal text, the 1217 *Forest Charter*, as well as a number of international and other sources, in support of an approach which priorities ecological integrity and environmental interconnectedness.

Aidan Ricketts also demonstrates that existing principles of statutory interpretation can be effectively wielded to achieve a wild law outcome.

---

34 *Xstrata Coal Queensland Pty Ltd & Ors v Friends of the Earth – Brisbane Co-Op Ltd & Ors, and Department of Environment and Resource Management* [2012] QLC 013.
35 *Hancock Coal v Kelly and Department of Environment and Heritage Protection (No 4)* [2014] QLC 12.

In rewriting *Newcrest Mining (WA) Ltd v Commonwealth*,[36] the author considers the Australian government's efforts to prevent mining and other activities in protected areas in the context of what has been construed by the High Court as a property rights guarantee in the Australian Constitution: section 51(xxxi). Ricketts, masquerading as Justice Gaia of the Australian High Court, reasons his way through the logic of statutory interpretation to a conclusion far less palatable than the original judicial outcome for a mining company deprived of the benefits of its mining leases.

E: *Rights of local communities and mining*

Coal seam gas mining is another contentious form of mining which has been vigorously opposed by local communities confronted with the prospect of contaminated water sources and catchment areas, and many other groups including farmers and environmentalists. It is a community's right of consultation in relation to its local environment which forms the basis of Cristy Clark's rewriting of an administrative law challenge to a decision to suspend operations under a coal seam gas mining exploration licence. In rewriting *Metgasco Limited v Minister for Resources and Energy*,[37] she turns her attention to the community opposition to proposed exploratory activities at Bentley in northern New South Wales and the political and judicial consequences of a concerted community campaign to prevent these from taking place. As she explains, her emphasis on the rights of local communities resonates with wild law principles. Cormac Cullinan has observed that 'one of the things that we can begin to do in order to express Earth jurisprudence is to consciously open spaces within the dominant legal systems within which communities can begin to give expression to an Earth-centred worldview'.[38]

F: *The intersection of First Nations law and wild law*

Irene Watson rejected the methodology of judgment rewriting in the Australian Feminist judgment project. In her chapter in that collection, she wrote that:

> the rewriting of the judgment of *Kartinyeri* in accordance with the methodology of this project would not prise open places for Nunga women because the rewriting needs to be done from 'another place', outside

---

36 *Newcrest Mining (WA) Ltd v Commonwealth* (1997) 190 CLR 513.
37 *Metgasco Limited v Minister for Resources and Energy* [2015] NSWSC 453.
38 Cullinan, above n 6, 207.

the jurisdiction of the Australian common law and the sovereignty of the Australian state.[39]

As Kathy Bowrey has pointed out, Watson's chapter is written in 'an Aboriginal women's voice' and 'in her own terms'.[40] Similarly her chapter in this collection is written as a critique 'of the colonial foundation and practices of Australian law' rather than as a judgment, and her focus is on 'the authority held by First Nations to speak on the ongoing sustainability and health of our natural world'. She queries the authority of the colonial legal system to approve the controversial developments currently proposed on the land of her ancestors and affirms the continuing authority of First Nations laws, which are the laws of the land and impose an obligation to care for country. She writes that:

> what is really essential to the survival of all species is to progress a horizontal dialogue between colonialist interests and First Nations-centred epistemologies.

Central to this dialogue is the inclusion of 'First Nations' perspectives on authority and power' and 'the understanding that the natural world holds authority'.

Watson notes that the *Mabo* case,[41] in which the Australian High Court found that First Nations people had common law native title rights in land, had significant limitations in that it 'failed to secure our obligations to land and law'. The deficiencies of *Mabo* are also addressed in two other chapters in the collection. Stephen Summerhayes rewrites *Mabo* in light of the First Nations' beliefs and traditions of care for country, so eloquently discussed by Irene Watson in her chapter. He attempts to align 'land law with Aboriginal traditional laws and customs rather than trying to mould or interpret Aboriginal rights within inappropriate Western legal models' and finds that rights in land are vested in the Meriam people as custodians of and constructive trustees for Country, a newly recognised legal entity which possesses an 'ecological title'.

Mabo is also revisited by Greta and Jo Bird, in their rewriting of a judgment which was never handed down: the judgment on the proposed Muckaty

---

39 I Watson, 'First Nations stories, grandmother's law' in H Douglas, F Bartlett, T Luker and R Hunter (eds), *Australian feminist judgments: Righting and re-writing law* (Hart Publishing, 2014) 46, 53.
40 K Bowrey, 'Commentary on *Kartinyeri v the Commonwealth*' in H Douglas, F Bartlett, T Luker and R Hunter (eds), *Australian feminist judgments: Righting and re-writing law* (Hart Publishing, 2014) 41, 45.
41 *Mabo v Queensland* (No 2) (1992) 175 CLR 1.

nuclear waste dump in the Northern Territory. The Commonwealth government withdrew from the Federal Court action brought by traditional owners concerned about the nomination of a site on their land. Greta and Jo Bird assert that the Commonwealth may well have withdrawn from the hearing to avoid legal arguments and a decision based on the 'inherent sovereignty of First Nations people'. In their imaginary judgment, delivered as a High Court decision while sitting on country at Warlmanpa, they tackle the 'glaring lacuna at the heart of Australian jurisprudence': the issue of the continuing sovereignty of First Nations people. They revisit the decision in *Mabo* that sovereignty is not a justiceable issue and find that the First Nations people of Warlmanpa are 'a sovereign peoples, living their law', which includes care for and custodianship of country.

G: International law and its limitations

While most judgments selected by participants are judgments originating in various Australian jurisdictions, two rewritten judgments in the collection are based on recent international law decisions: the *Whaling in the Antarctic* decision handed down by the International Court of Justice in 2014 and the decision in the *San Juan River* case handed down by the International Court of Justice in 2015.

In their rewriting of the *Whaling in the Antarctic* decision, Hope Johnson, Bridget Lewis and Rowena Maguire have adopted principles and reasoning drawn from wild law in reframing the 1946 *International Convention for the Regulation of Whaling*[42] to remove its problematic exception for 'scientific whaling'. Although the original decision led to a positive outcome for Australia, the Court also acknowledged that 'scientific whaling' remained an option for signatory nations under the Convention. The authors have opted to rewrite the Convention in accordance with wild law principles and eliminate this possibility before revisiting the judgment.

In rewriting the *San Juan River* case, Afshin Akhtar-Khavari departs from the Court's limited interpretation of the principle of transboundary harm in international law to incorporate considerations about the impact of a proposed road along the San Juan River and associated urban development on Planet Earth itself. In his separate opinion, the capacity of ecosystems to restore themselves is also a fundamental consideration in the application of this principle.

---

42 Opened for signature 2 December 1946, 161 UNTS 72 (entered into force 10 November 1948).

H: Prosecuting environmental activists

How does a wild law perspective alter the interpretation and implementation of criminal law? In three chapters, contributors consider the plight of activists prosecuted for altruistic acts intended to protect Earth and both human and non-human species. Matthew Rimmer analyses the sentencing judgment of Justice Davies after Australian climate change prankster Jonathon Moylan pleaded guilty to an offence under the *Corporations Act 2001* (Cth). In January 2013, Moylan issued a press release on ANZ letterhead, stating that the bank had withdrawn its financial support for Whitehaven's Maules Creek coal project. In the wake of Moylan's hoax, a cacophony of different voices was heard as academic and mainstream commentators, members of the public, corporate interest groups and those charged with the protection of 'market integrity' aired their views on the most appropriate means to deal with this high profile political activist. In his chapter, Matthew Rimmer expresses reservations about 'engag[ing] in judicial impersonation in respect of a case of impersonation'. Thus, rather than rewriting the judgment, he 'chart[s] the polyvocal debate of the "Stand with Jono" controversy'. Ultimately, it is not only the judge who passes judgment on, and sentences such, activists.

Susan Bird and Nicole Rogers also address the criminalisation of environmental activists and the extent to which an acknowledged freedom of expression, whether statutory or constitutional, should protect activists from prosecution. In rewriting *Magee v Wallace*,[43] Bird looks at whether the freedom of expression in the Victorian *Charter of Human Rights and Responsibilities Act 2006* should have served as a defence to the actions of activist Kyle Magee, who habitually paints and posts over corporate billboards. In finding for Magee, Bird points out that Magee 'is engaging in a public service in his mission to raise awareness of environmental concerns'. Her conclusions, she writes, 'arise from the principle that environmental rights should be given precedence over the rights of the corporation'.

The plight of another environmental activist, Lawrence Levy, is considered in Nicole Rogers's rewriting of *Levy v Victoria*.[44] This case involved a constitutional challenge to a Victorian Regulation designed to prevent activists such as Levy from retrieving dead and wounded birds during duck shooting season, in order to administer aid and attract media attention. Nicole Rogers departs from the reasoning in the original judgment in finding that a constitutional freedom of implied political communication operated to invalidate the Regulation. Her reasoning is based on the recognition of

43 *Magee v Wallace* [2014] VSC 643.
44 *Levy v Victoria* (1997) 189 CLR 579.

preservation of ecological integrity as an important component of public safety, arguably the purpose served by the Regulation, and on the need to acknowledge the rights of individuals to render assistance and bear witness to mass murder and massacres, irrespective of whether the victims are human or non-human beings.

I: Looking ahead

We have already discussed two futuristic judgments, contributed by Justice Preston and Benedict Coyne. These contributors are exploring future possibilities in the form of hypothetical, holistic, less anthropocentric statutes and international conventions and providing visionary content on what law could be. A third such judgment has been contributed by Robert Cunningham. In 2037, he hands down a judgment in which he reviews the issue of a 'biological mining data licence' by the Australian government. In so doing, he considers the application of two futuristic legislative initiatives. The first of these permits the collecting, owning and selling of biological data within the physical environment; the second enacts the concept of an 'information commons' and creates information commons rights. In his judgment Cunningham refers to appropriately futuristic human-nature interface technology and a seminal case which will be decided in 2035. He concludes that the Commonwealth of Australia cannot facilitate the private appropriation of biological data that falls within the domain of the information commons'.

III: Conclusion

The judgments from the Wild Law Judgment Project constitute a coherent body of wild law judgments which may lack the 'force of law'[45] or conventional legitimacy but which have a powerful political and educative value. This collection of judgments is complemented and augmented by the increasing number of ecocentric judgments handed down by the Rights of Nature Tribunals all over the world.[46]

The feminist judgment projects have led to a number of reflective pieces on the value of alternative forms of judging as an extra-judicial tool to shed light on the potential and limitations of orthodox judging.[47] Rosemary

---

45 This reference is from J Derrida, 'Force of law: The "mystical foundation of authority"' (1990) 11 *Cardozo Law Review* 920.
46 See e.g., M Maloney, 'Finally being heard: the Great Barrier Reef and the International Rights of Nature Tribunal', (2015) 3 *Griffith Journal of Law and Human Dignity* 40.
47 See e.g., H Roberts and L Sweeney, 'Review essay: why (re)write judgments?' (2015) 37 *Sydney Law Review* 457.

Hunter's observations on the value of the feminist judgments as a teaching tool, as a tangible demonstration of how feminist ideas can influence judicial decision-making and as a trigger for critical thinking about the nature and outcomes of judicial decision-making,[48] are useful in reflecting on the educational value of the wild law judgments. Similarly, legal scholars, practitioners, students and the wider community can draw on the wild law judgment collection to reflect critically on the anthropocentric bias inherent in all judicial decision-making and on how this bias shapes both judicial reasoning and case outcomes.

In the face of the ecological crisis, we must change the way we think and act and judge as a matter of urgency. The growing body of wild law judgments, whether handed down by Rights of Nature Tribunals, crafted by rewriting existing judgments or envisaged as a consequence of future legislative change, highlight the transformative potential of the common law and its inherent flexibility in the context of this crisis.

48 Discussed in Douglas et al., above n 16, 17.

Chapter 2

Writing judgments 'wildly'

Justice Brian Preston

Michelle Maloney and Nicole Rogers have asked me, as a current judge, to make some introductory comments on 'writing judgments "wildly"'. This choice of words made me wonder what inside knowledge they possessed about how I write my judgments.

'Wildly' as an adverb means 'in a wild manner'. 'Wild', as an adjective, has many meanings. One of these meanings is: 'of unrestrained violence, fury, and intensity'.[1] Now I know that occasionally the nefarious conduct of particular parties or their lawyers in a few environmental cases does raise my blood pressure to unhealthy levels, but I have not yet resorted to unrestrained violence in the courtroom and I will never do so. I hope the organisers did not have this meaning in mind in selecting the topic of my introductory comments.

Another meaning of 'wild' is 'frantic, distracted, crazy or mad'. Whilst I am resigned to the fact that, in response to some of my judgments, at least one of the parties may have thought that 'he must be mad to have made the decision he did', I do try not to be crazy in my judgments. But then again I used to think that *One Flew Over the Cuckoo's Nest* was a nature film about avian fauna. I will therefore pass over this meaning.

A third meaning of 'wild' is 'undisciplined, unruly, flawless or turbulent'. This meaning would be more of a problem for a judge who is meant to uphold the law and the rule of law. Judges are not meant to be undisciplined and lawless in their judgments. One of Lord Denning's books is entitled *The Discipline of Law*.[2] Lord Denning speaks of the discipline of the law in the sense of imparting instruction in the principles of the law as they have been, as they are, and as they should be. The theme of his book is that:

> the principles of law laid down by the Judges in the 19th century – however suited to social conditions of that time – are not suited to the

1 S Butler (ed.), *Macquarie Dictionary* (online edition at 8 November 2014) 'Wild'.
2 A Denning, *The Discipline of Law* (Butterworths, 1979).

social necessities and social opinion of the 20th century. They should be moulded and shaped to meet the needs and opinion of today.[3]

Lord Denning believed that judges can play a role in making the law correspond with the justice that the case requires. In the book, Lord Denning discusses how he did this in the various cases he decided over his long career as a judge. In doing so, Lord Denning did not believe that he was behaving in an undisciplined or lawless manner; to the contrary, he was acting within the law in a disciplined manner to develop legal principles. Here is an approach to writing judgments to which I will return.

A fourth meaning of 'wild' is 'unrestrained by reason or prudence'. Again, this meaning poses a problem for judges. The essential characteristic of adjudication – the act of judging is the application of reason to reach a decision. Courts are institutionally committed to acting on the basis of reasoned argument.[4]

A fifth and more general meaning is 'to behave in an unrestrained or uncontrolled manner'. Judges are constrained by the law and the act of judging. Their task is to adjudicate the dispute before the court applying the relevant law to the facts found on the evidence before the court. Again, it is antithetical to the act of judging for a judge to behave in a manner unrestrained or uncontrolled by the law and the evidence in the case to be judged. This much can be accepted. But that still leaves unanswered the questions of:

- What is the law to be applied?
- How is it to be interpreted?
- What facts should be found and inferences should be drawn on the evidence before the court?
- How does the law, properly interpreted, apply to the facts as found?
- What remedies and relief should be granted if breach of the law be found?

These questions help frame what are the restraints and the controls on the judge who is judging the dispute before the court. As Lord Denning has observed, there is scope within these restraints and controls to make the law correspond with the justice that the case requires.

Of course, my remarks so far are a play on the word 'wildly'. I really do know that this workshop is to discuss the Australian Wild Law Judgments Project. But as you will soon hear, some of my word play remarks

---

3 Ibid., v.
4 As Lon Fuller explained in his classic article 'The forms and limits of adjudication' (1978) 92 *Harvard Law Review* 353, 371.

actually do have relevance to the topic of writing judgments from a wild law perspective.

Wild law is a theory of earth-centred law and governance. It takes an eco-centric or nature-centred approach rather than a human-centred approach. It values the earth not merely instrumentally as a commodity belonging to humans but also intrinsically as a community to which we belong. If I can appropriate and adapt the title of Edward Schumacher's famous book on economics, wild law can be described as: 'Wild is beautiful: a study of law as if nature mattered'.

If we were to apply this approach, if nature really did matter, what would the law be and what would judgments of the court applying the law be? This is the central concern of the Wild Law Judgments Project. The brief I have been given, therefore, is to begin the discussion on how would judges decide cases and how would judgments be written if nature really did matter.

I would suggest that there are at least two approaches that could be taken. One approach is to examine where in the judging process there are opportunities for adopting a wild law perspective. This first approach accepts the law as it currently exists, but explores where there is scope for finding, interpreting and applying the law to best meet the justice – including the ecological justice – of the situation. It is the more orthodox approach, applying the same technique and logic as is used in judging other disputes. I discussed this approach in an article I wrote on 'The Art of Judging Environmental Disputes'.[5] In large part it accords with Lord Denning's approach that I have summarised earlier. Judgments would be rewritten to identify where and how the opportunities have been taken to prefer a wild law perspective and how doing so would affect the outcome of the case.

The other approach is to challenge the existing law and mould it to fit the earth's demands. This falls outside the orthodox technique and logic of judging disputes. Judges are not permitted to be legislators. However, the object of engaging in this second approach is to highlight the inadequacies of the existing law. Judgments would be rewritten to identify the reformed laws and show how the application would affect the outcome of the case.

Both approaches have utility and could be employed productively in the Wild Law Judgments Project.

Let me now explain in a little more detail the opportunities available under each approach for rewriting judgments from a wild law perspective.

Under the first approach, the opportunities arise in finding, interpreting and applying the law. The first step of judging is finding the law – ascertaining which of the many rules in the legal system is to be applied, or, if none

5 B Preston, 'The art of judging environmental disputes' (2008) 12 *Southern Cross University Law Review* 103.

is applicable, reaching a rule for the particular case. This step of supplying a new rule is to be undertaken by a principled and rational process. Different but equally legitimate methods may be used. They include the method of analogical reasoning following the line of logical progression from similar cases; the method of evolution along the line of historical development of a principle; the method of tradition along the line of customs of the community; and the method of sociology along the lines of justice, morals and social welfare and the mores of the day.[6] A wild law perspective may inform and supply a different rule than was selected by the court in the judgment being considered.

Having found the applicable rule of law, the second step of judging is interpreting the rule of law. Interpretation of the law is always required, for a variety of reasons. First, all legal rules involve classifying particular cases as instances of general terms. For any rule, it is possible to distinguish clear central cases, where the rule certainly applies, and cases where there is doubt as to whether the rule applies, there being reasons both asserting and denying that it applies. It is in this penumbra of doubt that judicial interpretation of the rule and its application in the case at hand is required.

Second, indeterminacy arises from the need to use ordinary English words. The English language is indeterminate and irreducibly open textured. Just like the rules, words used to formulate the rules can be seen to contain a core of certainty and a penumbra of doubt. Judicial interpretation is needed to ascertain the appropriate meaning of words in the applicable rule.

Third, there is indeterminacy in the rules in legislation. Legislators can have no knowledge of all the possible combinations of circumstances which the future may bring. This inability to anticipate brings with it a relative indeterminacy of aim. It is impossible to have a complete legislative provision in advance covering every case. Judicial interpretation is required to determine whether rules in legislation apply to the particular circumstances of the case at hand.

Fourth, the rules may use very general standards, such as reasonableness, fairness, or what is just and equitable, or general concepts such as the community or the public interest. By using these general standards or concepts, the rules incorporate extra legal norms into the law.[7] The use of these standards and concepts enables changes in the community's values to be taken into the law.[8] The standards and concepts are relative to time and place. They will, therefore, vary in content from time to time and place to

---

6 B Cardozo, *Nature of the Judicial Process* (Yale University Press, 1949) 30–31.
7 O W Holmes, Jr, *The Common Law* (Dover Publications, 1991) 1.
8 Ibid.

place. Courts need to interpret what is the current meaning and content of the standards and concepts in deciding the case.

Finally, there is indeterminacy inherent in the common law system of precedent. The effect of the court exercising its interpretive role is to create the law, even though this may be interstitial and subject to many constraints.[9]

A wild law perspective may inform and lead to the adoption of a different interpretation of the rule of law than that adopted by the court in the judgment being considered.

The third step of judging is to apply the law as found and interpreted. This third step involves two stages. The first stage is to find the facts relevant to the identified rule of law. Fact finding and inference drawing requires the court to exercise its intellectual judgment on the evidence submitted to it in order to ascertain the truth. The facts able to be found will be constrained by the evidence before the court. However, the available evidence may nevertheless provide a foundation for finding more facts than the court chose to find in the judgment being considered. Adopting a wild law perspective may warrant further fact finding than would be required by adopting an anthropocentric perspective.

For example, if fact finding of the impact of conduct on the environment is required, perhaps in assessing the impacts of a proposed development or in assessing the objective harmfulness of criminal conduct for sentencing, factual findings could be made not just on the direct impacts on specific biota, such as threatened species of plants or animals, but also on the indirect impacts on ecological functioning and services and the ecological relationships between that biota and its biotic and abiotic environment.

The second stage is to apply the identified rule of law to the facts as found. In this second stage, consideration needs to be given to the types and content of the remedies or relief to be granted if a breach of law be found. Adopting a wild law perspective may expand the choice and result in different remedies being granted to those granted by the court in the judgment being considered. For example, remedial orders could be made to restore ecosystem functioning and services or ecological interrelationships, or compensation could be ordered for the affected environment and not merely for affected humans.

Consideration also needs to be given to whether the applicable law accords a judicial discretion as to the remedy, relief or punishment to be granted if a breach of law were to be found. The duty of the court in matters of judicial discretion is to exercise its moral judgment as to what is right, just, equitable or reasonable in the case. Again, this provides an opportunity for a court to take a range of considerations into account, pertaining to both the private

---

9 H L A Hart, *The Concept of Law* (Oxford University Press, 2nd ed, 1994) 273.

interests of the parties and third parties, as well as the public interest. The public interest could include environmental considerations such as upholding conservation of biological diversity and ecological integrity. Adopting a wild law perspective may result in a different exercise of the discretion to grant or withhold relief to that originally exercised by the court.

This first approach can be pursued for many different types of judgments, not only those in environmental and planning law cases, but also cases in resource law, administrative law, constitutional law, property law, agricultural law, tort law, contract law, commercial law, corporations law, intellectual property law, trade practices law, consumer law and criminal law. There is also scope for applying this approach to judgments on procedural law to highlight how a wild law perspective could improve access to justice, including environmental justice.

The second approach to writing judgments from a wild law perspective is more involved. It requires identifying and reforming the aspects of existing laws that prevent or impede achieving earth justice. This step of reformulation of the law is in substitution of the first step of finding the law and the second step of interpreting the law under the orthodox method of judging. The reformed law is then applied in the third step of judging. This may involve new fact finding to meet the needs of the reformed law. The application of the reformed law to the facts as newly found would be expected to lead to a different outcome to that held by the court in the judgment under consideration. This, of course, is one of the principal objects of the exercise – to highlight how the inadequate law leads to non-earth-centred outcomes.

My identification and brief explanation of these two approaches is not intended to constrain how the Wild Law Judgments Project could be undertaken. Other approaches can legitimately be adopted to analysing and rewriting judgments. Indeed, the purpose of this workshop is to encourage a wild germination of ideas and approaches. One topic for discussion is: 'What do we mean by writing from a wild law perspective?'. My remarks are not intended to be, and should not be taken to be, a complete answer to that question. Rather, my remarks are intended to stimulate discussion, to trigger the germination process.

I have been asked to illustrate what might be a wild law judgment by reference to the hypothetical judgment I gave in a mock trial concerning green sea turtles. This was a case where the green sea turtles brought proceedings in public nuisance against the Federal and State government who, the turtles claimed, caused a public nuisance by both their actions and omissions to act. The governments' actions comprised granting various approvals to coal mines in Queensland. Coal mining results in onshore and offshore greenhouse gas emissions. The governments' omissions to act included their failure to take action to reduce Australia's greenhouse gas emissions and to mitigate the impacts of climate change on the Great Barrier Reef. The turtles claimed that greenhouse gas emissions from mining and burning of coal have contributed to climate change. Climate change caused damage to the Great Barrier Reef. The Great Barrier Reef is the habitat of the turtles. Hence, the climate change induced damage to the Great Barrier Reef impacted directly on the turtles.

At the outset, I have to say it was much easier and much more enjoyable giving judgment in a hypothetical case rather than in a real case. There are no disappointed litigants to appeal the judgment, no rebukes from appellate courts and no censure in the media.

In writing the hypothetical judgment, I adopted both of the two approaches I have outlined. I worked within the constraints of the existing law, but I found opportunities to take a wild law perspective. I also reformed aspects of the law where they were barriers.

The turtles' cause of action was in public nuisance. I applied the established elements of that cause of action but I interpreted key concepts from a wild law perspective.

In the first element, concerned with whether there was interference with a public rather than a private right, I construed the concept of the public to include not only humans but also the turtles. This step was, however, made possible by my adopting the second approach of reforming the law. I assumed the law had been amended to extend the class of rights holders from humans only so as to include non-human life, such as the turtles. Once that step was taken, I was readily able to find that there was an interference with the rights of this wider public.

The second element was whether the interference was unreasonable. I was able to interpret and apply existing legal rules but I also reformed the law in one respect. This was to assume legislative change to adopt the public trust doctrine. This assisted in finding that the governments' conduct was contrary to law, by failing to discharge their public trust duties to take action to protect common natural resources. I also found that the governments' defence to the public nuisance was ineffectual.

The third element concerned the remedies that should be granted for the established public nuisance. Here, the law of public nuisance required that the plaintiffs – the turtles – suffer special injury different to the ordinary member of the public. Once turtles were accepted as being members of the public, and having standing to sue, it was easy to conclude that they would, on the facts found, suffer special injury. The turtles' lives and health would be significantly adversely affected. Such harm was different in kind to the harm suffered by human members of the public.

The issue of the type of remedy was important. Clearly, the equitable remedy of an injunction was required, firstly, to restrain the governments from granting further coal mine approvals so as to mitigate future greenhouse gas emissions and, secondly, to compel the governments to take action to mitigate the adverse effects on the turtles' habitat and to enable the turtles to better adapt to the affected environment. That left unabated and unaddressed the damage already caused to the turtles. That harm could be compensated by an award of damages. I was not in a position to deal with that remedy and I reserved that claim to another day. But I had in mind that the assessment of the damages would need to take a turtle-centric perspective – what monetary amount would compensate for the loss of the turtles that had already died, the loss of their habitat and the loss that would continue even with mitigation of greenhouse gas emissions until viable populations of turtles and their habitat could be restored? This would involve a different method of assessing damages to conventional methods.

There is one other aspect of the sea turtles judgment that I would like to mention. This concerns the language and style of the judgment. A wild law perspective might suggest that the language, structure and style of traditional judgments ought to be readdressed. My sea turtles judgment did not radically depart from the traditional judgment language, structure and style. But I did commence the judgment in a less conventional and more poetic manner. I hope you will forgive me if I read that opening as it might inspire your creativity:

> In 1949, the foresighted forester, Aldo Leopold, in his famous book, *A Sand County Almanac*, challenged society to reflect on the unsustainable exploitation of the environment and its consequences, including loss of animals and birds. In his essay 'Goose Music', Leopold asked:
>
>> And when the dawn-wind stirs through the ancient cottonwoods, and the grey light steels down from the hill over the old river sliding softly past its wide, brown sand bars – what if there be no more goose music?
>
> In 1962, the eminent zoologist and biologist, Rachel Carson, in her classic book, *Silent Spring*, described the chronic, bioaccumulative effects pesticides, including DDT, can have up the food chain, particularly on birds, and called for governmental action. The title 'Silent Spring', referred to Carson's fear that uncontrolled use of pesticides would eventually result in a season in which no birds could be heard because they had all died from exposure to pesticides. The title was inspired by the John Keats poem, 'La Belle Dame Sans Merci' which contains the lines:
>
>> 'The sedge was wither'd from the lake
>> And no birds sing'.
>
> Today, in 2032, a group of green sea turtles have raised alarm and made a clarion call for governmental action to address a pernicious, pervasive and pressing threat to the environment – climate change's devastation of

the habitat of the turtles. They ask us to reflect on a future where there are no more turtles.

The turtles have issued their call for action in these legal proceedings.

This little vignette about the endangered green sea turtles may offer ideas for writing judgments from a wild law perspective.

By this time in my remarks, you are no doubt starting to think again about those different meanings I proffered about the word 'wildly'. Some of you may be thinking 'he really is mad', but a few more may be thinking 'he is behaving in an unrestrained or uncontrolled manner'. So, let me take control and restrain myself from speaking any further on 'Writing Judgments "Wildly"'.

Part I

Standing and wellbeing of non-human species

Chapter 3

Green sea turtles by their representative, Meryl Streef v The State of Queensland and the Commonwealth of Australia

Justice Brian Preston

Judgment 18 February 2032

A: Turtles bring public nuisance proceedings

In 1949, the foresighted forester, Aldo Leopold, in his famous book, *A Sand County Almanac*, challenged society to reflect on the unsustainable exploitation of the environment and its consequences, including loss of animals and birds. In his essay 'Goose Music', Leopold asked:

> And when the dawn-wind stirs through the ancient cottonwoods, and the grey light steels down from the hill over the old river sliding softly past its wide, brown sand bars – what if there be no more goose music?[1]

In 1962, the eminent zoologist and biologist, Rachel Carson, in her classic book, *Silent Spring*, described the chronic, bioaccumulative effects pesticides, including DDT, can have up the food chain, particularly on birds, and called for governmental action.[2] The title 'Silent Spring', referred to Carson's fear that uncontrolled use of pesticides would eventually result in a season in which no birds could be heard because they had all died from exposure to pesticides. The title was inspired by the John Keats poem, 'La Belle Dame Sans Merci' which contains the lines:

> The sedge was wither'd from the lake
> And no birds sing.

Today, in 2032, a group of green sea turtles have raised alarm and made a clarion call for governmental action to address a pernicious, pervasive and pressing threat to the environment – climate change's devastation of the

---

1 Aldo Leopold, *A Sand Country Almanac* (Oxford University Press, 1949).
2 Rachel Carson, *Silent Spring* (Houghton Mifflin, 1962).

habitat of the turtles. They ask us to reflect on a future where there are no more turtles.

The turtles have issued their call for action in these legal proceedings. They have invoked what the law terms a cause of action in public nuisance against the Commonwealth Government of Australia and the State Government of Queensland. The turtles have brought proceedings by their representative, Meryl Streef. The turtles claim that the governments, by both their actions and omissions to act, have caused a public nuisance. The governments' actions comprise granting approval, 20 years ago, in 2012, to eight major coal mines in Queensland's Galilee Basin. These mines together have produced and will produce onshore and offshore greenhouse gas emissions of 29.1 gigatonnes of CO_2 equivalent over the life of the mines from 2012 to 2050. The governments' omissions to act include their failure to take action to reduce Australia's greenhouse gas emissions and to mitigate the impacts of climate change on the Great Barrier Reef.

The turtles claim that greenhouse gas emissions from human activities, including the mining and burning of coal, have induced climate change. The evidence before the Court of Dr Jones, a climate scientist, establishes that climate change has caused damage to the Great Barrier Reef in a number of ways: first, a rise in sea levels leading to increased water depth (and associated decreased photosynthesis) and changes in tidal variation and water movements; second, a rise in sea temperatures, leading to coral bleaching and changes in the distribution of sea grasses; third, an increase in severe weather events such as storms and cyclones; and fourth, an increase in CO_2 in coastal waters, causing an increase in carbonic acid in the ocean which has interfered with the coral's ability to build skeletons.

The turtles depend on the Great Barrier Reef to live, feed and breed. The evidence before the Court of Dr Fuentes, a marine biologist, establishes that the climate change-induced damage to the Great Barrier Reef has impacted directly on the turtles. Sea level rise has affected the significant nesting areas on low-level sand beaches along the coastline adjoining the Great Barrier Reef. Rising temperatures have caused sand temperatures to exceed the upper limit for egg incubation (34°C), leading to reduced numbers of sea turtles hatching. In addition, because sand temperature determines the sex of the hatching turtle, increased sand temperatures have also caused a bias in the sex ratio towards more female turtles. Finally, because coral and sea grasses form the core of the turtles' diet, damage to these grasses has disturbed the turtles' feeding patterns, including by forcing them to travel further distances in search of food.

The turtles claim that the governments' actions and omissions have unreasonably interfered with the interests of the community at large, interests recognised as rights of the general public entitled to protection. The turtles seek the equitable remedy of an injunction. They seek a negative injunction prohibiting further approval of coal mines in the future without imposition of conditions requiring mitigation and offsetting of greenhouse gas emissions, and a positive injunction that the governments offset past and future greenhouse gas emissions from coal mines. They also seek a positive injunction that the governments mitigate the adverse effects of climate change on the Great Barrier Reef which is the habitat of the turtles, in particular, by implementing a relocation programme to ensure the ongoing survival of the turtles.

The governments have defended the turtles' claim.

For reasons I will explain, I find for the turtles. They have established their claim that the governments have caused a public nuisance. It will be necessary to hold a further hearing to determine the appropriate injunctive relief.

B: *Nature of the public nuisance claim*

A public nuisance is an unreasonable interference with a right common to the general public. The origins of the public nuisance doctrine are found in interferences and infringements of the rights of the British Crown, which were applied to any actions that produced an inconvenience or some kind of harm to members of the public. A public nuisance was regarded as a crime, and it came to be defined as:

> an unlawful act, or omission to discharge a legal duty, which act or omission endangers the lives, safety, health, property or comfort of the public or by which the public are obstructed in the exercise or enjoyment of any right common to all Her Majesty's subjects.[3]

There are three primary questions to be answered in resolving the turtles' public nuisance claim:

- Have the governments caused an interference with a public, as opposed to a solely private, right?
- Was the interference unreasonable?
- Have the turtles sustained the kind of unique or special injury that differentiates their harm from that suffered by the general public so as to be entitled to remedies for the interference?

I: *Interference with a public right*

I will start with the first question. This involves two aspects: first, has there been an interference with a public right, and second, did the governments cause that interference?

---

3 *Kent v Minister of State for Works* (1973) 2 ACTR 1, 25–26.

The essence of the tort of public nuisance is interference with interests of the community at large – interests that are recognised as rights of the general public entitled to protection. The class that is the general public has traditionally been comprised only of humans, in particular the citizens of the country concerned. The rights, therefore, have been rights of humans. However, the legislature has the power to extend the class of rights holders. In this case, sometime prior to 2032, the Federal and State legislatures have enacted environmental protection legislation extending the rules of standing to allow non-human members of the community of life on earth, including the turtles, to bring proceedings to enforce public rights, including to arrest and abate public nuisances. This statutory extension of standing has effected an enlargement of membership of the class of the general public and consequently of the content of the rights of the general public which are protected.

Traditionally, the rights of the public entitled to protection from interference include the life, health, safety, morals, peace, comfort and convenience of the public and the public's rights to use, enjoy and preserve the aesthetic and ecological values of public places and common natural resources. Interferences with these public rights can constitute public nuisances.

The public also has an interest in certain common natural resources such as the air, waterways and forests, which are held in trust by the government for the benefit and use of the general public. This is the concept of the public trust that the earth's natural resources are held in trust by the present generation for future generations. The government, as trustee, is under a fiduciary duty to deal with the trust property, being the common natural resources, in a manner that is in the interests of the general public. Thus, the government should not alienate or harm the trust property unless the public benefit that would result outweighs the loss of the public use or social wealth derived from the natural resources.

In this case, the climate change-induced effects have significantly interfered with the lives, health, safety, comfort and convenience of the turtles who, by the statute, have become members of the public. The evidence of Dr Jones and Dr Fuentes establishes that the turtles' lives and health are threatened. But there are also effects on the public rights of humans. Turtles are of cultural significance for Aboriginal and Torres Strait Islander people. Turtles also have high value to the tourism industry. Interference with the turtles diminishes public use, enjoyment and recreation of the public area and the natural resources of the Great Barrier Reef.

Having established interference with public rights, the next step is to determine whether the governments' conduct (their actions and omissions) has caused this interference. The governments' conduct is a link in the chain of causation resulting in the interference with the public rights.

The governments approved the Galilee Basin coal mines in 2012. The approvals made the coal mining lawful. The coal mining results in onshore and offshore greenhouse gas emissions of considerable magnitude. Altogether, 29.1 gigatonnes of carbon dioxide equivalent is estimated to have been and will be emitted between 2013 and 2050.

The governments have also failed to take action to reduce or mitigate Australia's greenhouse gas emissions. Together, the greenhouse gas emissions from the mining and burning of coal and from Australia generally are contributors to climate change. Dr Jones estimates that the combined emissions from the Galilee Basin coal mines will raise global temperature by 0.012°C. This comprises 0.4 per cent of the global mean warming anticipated by 2100 and 0.5 per cent of the 2.2°C temperature increase anticipated by 2050. This incremental warming is contributing to the adverse effects on the Great Barrier Reef. These include contributing to the estimated 30km^2 of Great Barrier Reef coral reef habitat being critically bleached now, increasing to 60km^2 by 2050 and 90km^2 by 2100. Thus, the emissions are having a material and measurable impact on the ecology of the region, and the habitat of the turtles.

A similar causal chain connecting the plaintiff's injuries and the defendant government's conduct was accepted by the United States Supreme Court in *Massachusetts v Environmental Protection Agency*[4] (concerning lack of governmental regulation of greenhouse gas emissions from motor vehicles) and by the United States Court of Appeals for the 5th Circuit in *Comer v Murphy Oil USA*[5] (concerning lack of governmental regulation of greenhouse gas emissions contributing to global warming, including the ocean's temperature, which caused sea level rise and increased ferocity of Hurricane Katrina, damaging the plaintiff's coastal Mississippi property).

It is true that the governments' conduct is not the sole cause of greenhouse gas emissions globally or of climate change. There are, of course, other causes. However, the fact that there are other causes does not negate the governments' conduct as being *a* cause. As with pollution of public waterways, it is no defence to prove that the polluting act of one person by itself is not a nuisance and only becomes a nuisance when combined with the polluting acts of others. Each of two or more persons whose conduct is a contributing cause of a single and indivisible harm to an injured party is liable to the injured party for the entire harm.

It is also not an answer to prove that the taking of the governmental actions urged by the turtles (namely, not approving the coal mines and taking steps to reduce Australia's greenhouse gas emissions) would not, by themselves, have mitigated global climate change and the interferences with the public rights in this case. As the United States Supreme Court observed in *Massachusetts v EPA*,

---

4 549 US 497 (2007).
5 607 F 3d 1049 (5th Cir 2010).

massive problems are not solved in one fell swoop but rather are whittled away over time by a series of incremental steps.[6] There, the Court recognised that governmental regulation of motor vehicle emissions would not by itself reverse global warming but that it was sufficient to show that it would slow or reduce it. The governmental actions urged by the turtles were hardly tentative steps or so insignificant as not to have been worth taking. A reduction in the greenhouse gas emissions from mining and burning coal from the Galilee Basin coal mines, and more generally a reduction in Australia's domestic greenhouse gas emissions, would have slowed the pace of global emissions, no matter what happened elsewhere. The magnitude of injury would therefore have been less.

I find, therefore, that the governments' conduct was a material contributor to the ultimate interference with public rights in this case.

2: Interference was unreasonable

The second question is whether the interference with the public rights was unreasonable. I find that it was, in two ways.

First, the governments' conduct was contrary to law. I find the governments misexercised their statutory powers by granting approvals for the coal mines without imposing effective conditions mitigating and offsetting the greenhouse gas emissions produced by the mines. I find that the governments failed, when granting the approvals, to consider the mandatory relevant matters of the principles of ecologically sustainable development, including the precautionary principle, intergenerational equity and the conservation of biological diversity and ecological integrity. The governments failed to give proper, genuine and realistic consideration to, first, the material contribution the greenhouse gas emissions produced by the mines would be likely to have on climate change and, second, the effects climate change would have on common natural resources such as the Great Barrier Reef and its biota. The failure to consider relevant matters is evidenced by the governments' failure to impose any conditions of approval mitigating or offsetting the greenhouse gas emissions.

The governments' conduct in failing to take action to reduce Australia's greenhouse gas emissions also involves a failure to discharge the legal duty imposed on the governments by the recent environmental protection legislation that is now in force in 2032. Under this legislation, the governments have a statutory duty, according with the public trust doctrine, to take action to protect common natural resources. Greenhouse gas emissions cause climate change and climate change results in adverse natural weather

6 549 US 497, 524 (2007).

events, such as have occurred in relation to the Great Barrier Reef. The Great Barrier Reef is part of the public trust. The governments have a duty to protect it from harm.

Second, the governments knew or had reason to know that their conduct in approving the coal mines would lead to mining, which produces greenhouse gas emissions, which causes climate change, which leads to environmental harm, which results in the interference with public rights. All of these links in the chain of causation were reasonably foreseeable. The governments had actual or constructive knowledge that their conduct would have significant and long lasting adverse effects on the public rights.

In these two ways, the governments' conduct caused an unreasonable interference with the public rights.

For these reasons, I find that the turtles have established their claim that the governments' conduct has caused and is causing a public nuisance. The onus then shifts to the governments to justify or excuse their conduct.

3: Defences to public nuisance are ineffectual

I find that the governments have not established any defences to the public nuisance. The governments' conduct was not authorised by statute. It is true that a governmental body is not liable for a nuisance which is attributable to the exercise by it of either a duty imposed on it by a statute or a power conferred by a statute, in the absence of negligence. However, in this case, the governments misexercised their statutory powers and failed to exercise statutory duties imposed on them. Their conduct cannot be said, therefore, to be the exercise of those duties or powers. Moreover, the government's conduct did involve negligence, in the special sense that the governments did not exercise the duties and powers with all reasonable regard and care for the interests of the general public.

As to the other defence raised by the governments, that the governments' conduct is not the only cause of the interference which constitutes the public nuisance, this is an ineffectual defence for the reason I have earlier given. The presence of other causes does not relieve the governments of responsibility for their material contribution to the interference.

4: Remedies to address public nuisance

The third question is what remedies should be granted for the established public nuisance.

Where a plaintiff is seeking monetary damages, the plaintiff must satisfy the special injury rule, that is, to establish that they have suffered special injury different to the ordinary member of the public. In this case, the turtles have suffered special injury different to the ordinary member of the public. The turtles' lives and health are significantly adversely affected by the interference. Their injuries are different in kind from the harms suffered by human members of the public. If the turtles seek damages, it will be necessary to hold a further hearing to determine the appropriate measure of damages for the special injury suffered by the turtles.

The turtles seek equitable remedies in the form of injunctions. The turtles seek prohibitory injunctions to restrain the governments from granting further approvals to coal mining without imposition of conditions mitigating and offsetting greenhouse gas emissions and to require the governments to take action to offset greenhouse gas emissions from the coal mines. The turtles also seek mandatory injunctions compelling the governments to take action to mitigate the adverse effects on their habitat and to enable the turtles to better adapt to the affected environment. This includes a relocation programme for the turtles to suitable habitats. It will be necessary to hold a further hearing to determine the appropriate terms of the prohibitory and mandatory injunctions.

For these reasons, I find for the turtles that the governments have committed a public nuisance and I adjourn the proceedings for further hearing at a date to be fixed.

The Hon Justice Brian J Preston
18 February 2012
Mock Trial, Melbourne

Chapter 4

Great Barrier Reef v Australian Federal and State governments and others

Cormac Cullinan

Commentary

This judgment was delivered by the International Rights of Nature Tribunal[1] in December 2014. The tribunal is a permanent people's tribunal initiated by an international civil society network, the Global Alliance for the Rights of Nature and formally established on 4 December 2015 by the entry into force of the *Peoples' Convention for the Establishment of the International Rights of Nature Tribunal*.[2] The initial hearing in the case of *Great Barrier Reef v Australian Federal and State governments and others* took place in January 2014, at Quito in Ecuador and was brought to the Tribunal by the Australian Earth Laws Alliance (AELA). A regional tribunal hearing then took place on 15 October 2014, in Brisbane, Australia. The members of the regional tribunal were Professor Brendan Mackey (President), Emeritus Professor Ian Lower, Sam Watson, indigenous community leader from South East Queensland and Janelle Fabio, youth representative. Benedict Coyne appeared as Mother Earth Defender and Abraham O'Neill was the volunteer defence counsel for the Australian and Queensland governments, with witnesses including Joanne Bragg, Glen Holmes, Sean Ryan, Brynn Matthews and Dr Michelle Maloney.

The hearing before the International Rights of Nature Tribunal took place on 5th and 6th December 2014, in Lima, Perú. The members of the Tribunal were: Alberto Acosta, Tribunal President, economist and President of the Constituent Assembly from Quito, Ecuador; Raúl Prada Alcoreza, philosopher, sociologist, author and former member of the Bolivian Constituent Assembly of 2006–2007, from Bolivia; Hugo Blanco, political leader, leader of the Confederación Campesina del Perú, from Perú; Tantoo

---

1 For further information about the International Tribunal see <http://therightsofnature. org/rights-of-nature-tribunal>. For details about the Global Alliance for the Rights of Nature, see <http://therightsofnature.org>.
2 The text of the People's Convention can be found at <http://therightsofnature.org/ convention-rights-of-nature-tribunal>.

Cardinal, actress, Métis peoples, from Canada; Blanca Chancoso, Kichwa leader and educator from Cotacachi, Imbabura, Ecuador; Tom Goldtooth, director of Indigenous Environmental Network, from Turtle Island, United States; Francios Houtart, professor, philosopher, theologian, from Belgium; Osprey Orielle Lake, co-Founder and Executive Director, Women's Earth and Climate Network (WECAN), from the United States; Edgardo Lander, sociologist, professor, from Venezuela; Veronika Mendoza, congresswoman, representing the region of Cusco, from Perú; Rocío Silva Santiesteban, National Human Rights Coordinator, author and professor, from Perú; Atossa Soltani, founder and Executive Director of Amazon Watch, from the United States; and Terisa Turner, professor of sociology and anthropology, former United Nations Energy Specialist, from Canada. At the hearing, Ramiro Avila was the Mother Earth Defender.

The defendants, who were not represented at the hearing, were the Australian and Queensland governments; elected representatives and officials of those governments who performed any of the political, legislative, executive or administrative acts that legitimated or facilitated activities destructive to the Great Barrier Reef, or who omitted to take the actions necessary to prevent that destruction; companies and people involved in the construction of ports, roads and other infrastructure to facilitate the transportation of coal across the Great Barrier Reef, the mining, transportation or purchase of coal to be transported across or in the vicinity of the Great Barrier reef or other associated activities harmful to the Great Barrier Reef; and organisations and people engaged in other activities harmful to the Great Barrier Reef such as industrial developments near the reef, dredging, and agricultural activities that increase the levels of sediment in coastal waters. The complainant, the Australian Earth Laws Alliance, was represented by Dr Michelle Maloney (via video) and Erin Fitz-Henry. Indigenous leader Gudju Gudju and Senator Larissa Waters (via video) represented the affected parties. Expert witnesses were Professor Brendan Mackey (via video) and Dr Glen Holmes. Judge Tantoo Cardinal gave the unanimous judgement of the Tribunal.

Judgment

This case was heard by the Tribunal presided over by Alberto Acosta during the second session of the Tribunal held from 5 to 6 December in Lima, Peru. The hearing was dedicated to the memory of José Tendetza who was murdered shortly before he was due to give evidence before the Tribunal. This is the unanimous judgment of the Tribunal.

A: Proceedings

The Australian Earth Laws Alliance (AELA) submitted the application for the case of the Great Barrier Reef to the Tribunal in Quito in January 2014.

This was followed by a hearing of a Regional Chamber of the Tribunal on 15 October 2014 in Brisbane, Australia which preceded the hearing of the case by the Tribunal which sat in Lima, Peru from 5 to 6 December 2014.

In finalising this judgement the Tribunal also took account of testimony from Gudju Gudju, an elder of the Gimul people, one of the many groups of indigenous peoples who have lived adjacent to the Great Barrier for approximately 70,000 years or 2,250 generations and have a particularly intimate relationship with it.

B: Initial hearing (Quito, January 2014)

The Australian Earth Laws Alliance (AELA)[3] submitted the application for the case of the Great Barrier Reef to the Tribunal in Quito in January 2014. Dr Michelle Maloney, the National Convenor of the AELA, presented the case to the Tribunal and requested that it be accepted for hearing by the Tribunal at a future date.

The Tribunal (Judge Vandana Shiva presiding) decided that the Tribunal should hear the matter on a future date because the complainant (AELA) had provided credible evidence that, as a direct consequence of the development of coal ports and the proposed shipping of large volumes of coal across the Great Barrier Reef:

(a) serious violation of the rights, and breaches of the duties, established by the *Universal Declaration of the Rights of Mother Earth* ('the Declaration')[4] are occurring and are likely to continue; and

(b) there is a present and very real threat to the Great Barrier Reef community.

In announcing the Tribunal's decision to hear the case, Judge Cormac Cullinan noted that the submissions made to the Tribunal provided evidence of potentially serious violations of specific rights of the Great Barrier Reef and the members of that community,[5] and of potentially serious failures on the part of human beings and the public and private institutions established by

---

3 See <www.earthlaws.org.au>.
4 The *Universal Declaration of the Rights of Mother Earth* ('the Declaration') was adopted by the World People's Conference on Climate Change and the Rights of Mother Earth at Cochabamba, Bolivia on 22 April 2010. Its text can be found at <http://therightsofnature.org/universal-declaration>.
5 These include the rights recognised in the Declaration: to continue their vital cycles and processes free from human disruptions (Declaration art 2(1)(c)); to integral health (art 2(1)(g)); to be free from contamination and pollution (art 2(1)(h)); and to wellbeing (art 2(3)).

humans to respect and live in harmony with Mother Earth.[6] He also pointed out that the facts of this case raised other important issues worthy of consideration by the Tribunal. For example, if current concentrations of greenhouse gases in the atmosphere are already causing significant climate change, is any significant increase in the rate of production of hydrocarbons (such as the opening of the vast new coal mines that are to be the source of the coal shipped across the Great Barrier Reef) a violation of the rights of Nature?

C: Hearing of the Australian Regional Chamber of the Tribunal

On 15 October 2014, an Australian Regional Chamber of the Tribunal conducted hearings in Brisbane, Queensland, to hear evidence from local witnesses for the Great Barrier Reef case. The evidence presented at that hearing proved that human activities were causing significant damage to the Great Barrier Reef.

After deliberation, the Regional Chamber made the following findings based on the evidence presented to it:

(a) The Great Barrier Reef's rights to exist, thrive and evolve have been, and are being, violated. In particular, the following three articles of the Declaration are being violated: the right to exist,[7] the right to continue its vital cycles and processes free from human disruptions,[8] and the right to maintain its identity and integrity as a distinct, self-regulating and interrelated being.[9]

(b) Intervention by the Australian Federal Government and Queensland State Government to prevent further industrial development along the coastline adjacent to the Reef would assist the Great Barrier Reef community to restore itself to integral health and it was the responsibilities of those governments to do so.

(c) The Regional Chamber noted that the significant body of domestic and international law that already exists to protect the Great Barrier Reef had failed to prevent the ongoing damage to the Reef.[10] It concluded that although appropriate enforcement of existing environmental law in

---

6 These include the duties of all human beings, all States and all public and private institutions, to ensure that the pursuit of human wellbeing contributes to the wellbeing of Mother Earth now and in the future (art 3(2)(d)), and to establish precautionary and restrictive measures to prevent human activities from causing species extinction, the destruction of ecosystems or the destruction of ecological cycles (art 3(2)(i)).

7 Art 2(1)(a).

8 Art 2(1)(c).

9 Art 2(1)(d).

10 These are listed in Sch 3.

Australia would assist in protecting the health of the Reef, this would be insufficient. The overwhelming impacts of the ongoing growth in current modes of production and consumption, require a new ecocentric framing of ethical and legal systems such as that given legal expression in the Declaration.

The Regional Chamber suggested that testimony should be received from the indigenous communities who are custodians of the land and sea country linked to the Great Barrier Reef.

In 2015 the Australian Earth Laws Alliance held discussions with indigenous communities in Cairns, and testimony from a small selection of indigenous people was received. One indigenous representative, Gudju Gudju of the Gimul people, noted that his people had lived in Australia for at least 70,000 years or 2,250 generations. He and his people are concerned about the health of land and sea country of the Great Barrier Reef. One of the reasons they are advocating for their own sovereignty, is so that they can continue to apply the traditional laws which they used for millennia to protect the Great Barrier Reef. He noted that he and his people are very concerned that the organisations currently responsible for managing the Reef don't truly understand the interconnectedness of all life on the land and sea, and consequently the importance of protecting it all. His people want to ensure this is better understood and taken into account in management and decision-making.[11] He also stated that his people do not want more coal mines, coal ports or coal seam gas developments.

D: Issues for determination by the Tribunal

The complainants requested this Tribunal:

(a) to determine whether or not the current and proposed shipping of coal across the Great Barrier Reef in Australia, the construction of ports and other infrastructure to facilitate that process, and the consequential increases in global combustion of coal, violates the Declaration;

(b) to determine which parties should be held accountable for failing to fulfil their duties as recognised in the Declaration;

(c) to determine what restorative measures should be taken;[12] and

---

11 A video of the testimony presented by Gudju Gudju, may be viewed at <https://youtube/z-ocsRYPCzo>.

12 The Declaration provides that if human activities violate any of the rights recognised in the Declaration then the injured party (for example, the Great Barrier Reef community and its members) has a right to full and prompt restoration for the violation of those

(d) to determine what preventive or precautionary measures should be taken to prevent future violations of the Declaration being caused by the extraction and transportation of coal.[13]

E: Evidence before the Tribunal

The Tribunal, after hearing the testimony and considering evidence submitted to it and analysing it and supplementing it with its own research, accepted the following as factually correct.

1: Value of the Great Barrier Reef community

The Great Barrier Reef is the world's largest coral system and one of the greatest wonders of Earth. It is the largest structure made by living organisms and can be seen from space. The reef extends for 2,300 km, includes 9,000 tropical islands covering an area of 344,400 km^2 and has more than 3,000 individual coral systems. It is home to at least 1,625 species of fish, 600 types of coral, 133 species of sharks, 30 species of whales and dolphins, and 3,000 species of molluscs.

The Defender of Mother Earth drew the Tribunal's attention to the fact that the Great Barrier Reef is a community consisting of an enormous number and diversity of beings whose existence and wellbeing is dependent on that of the Reef. Consequently references in this judgement to the rights of the 'Great Barrier Reef' or of the 'Great Barrier Reef community' must be understood as references not only to the collective rights of that community but also to the rights of its members.

2: Threats to the Reef caused by humans (anthropogenic threats)

In 1981 the United Nations Educational, Scientific and Cultural Organisation (UNESCO) recognised the value of the Great Barrier Reef to humanity by declaring the Reef to be a World Heritage Site. However, in June 2011,

---

rights (art 2(1)(j)) and human beings, all States, and all public and private institutions must: respect, protect, conserve and where necessary, restore the integrity, of the vital ecological cycles, processes and balances of Mother Earth (art 3(2)(f)); and guarantee that the damages caused by human violations of the inherent rights recognized in the Declaration are rectified and that those responsible are held accountable for restoring the integrity and health of Mother Earth (art 3(2)(g)).

13 The Declaration recognises that human beings, all States, and all public and private institutions have a duty to establish precautionary and restrictive measures to prevent human activities from causing species extinction, the destruction of ecosystems or the disruption of ecological cycles (art 3(2)(i)).

UNESCO issued a warning to the Australian government about threats to the life of the Great Barrier Reef. Among the threats noted by UNESCO are:

(a) the impacts of 1.99 million tourists a year;
(b) damage caused by thousands of freight ships through the ports on the northern coast of Australia, which create risks of pollution from fuel and coal dust (particularly during transhipping operations) and from cargo losses; and
(c) damage to the Reef caused by sediments agitated in the construction of new ports.

Dredging and sea dumping of dredge material have caused sedimentation across the Reef to a far greater extent than had been predicted. This reduces water quality which affects corals and other marine life.

The phenomenon of climate change is also having a negative effect on the Great Barrier Reef. A 0.4°C rise in average water temperature is already causing severe bleaching on much of the Reef and the increased frequency of severe weather events, ocean acidification and sea level risk caused by climate change also put the Reef at risk. The combustion of coal mined in Australia (approximately 50.8 million tons of coal per year) makes a significant contribution to accelerating climate change and consequently to accelerating harm to the Reef.

UNESCO made 14 recommendations to the Australian Government, which have not been addressed to date.

The damage arising from these activities (which are well documented) include: marine pollution, the death and displacement of fish, ocean acidification, destruction of the coral (it is estimated that up to 48 per cent of the coral has already been lost), and significant impacts on the livelihood and health of more than 50 indigenous communities. If destructive human activities continue, the Great Barrier Reef will die along with the marine life it supports and this will have severe impacts on many species and people (particularly indigenous peoples).

Instead of the Federal Government and the Queensland government responding by stopping or restricting the human activities that are causing this harm, they are permitting the intensification of these activities. For example, these governments have authorised the opening of vast new coal mines that will double the annual coal production of Queensland and dramatically increase the volume of coal exported via the waters of the Great Barrier Reef. Three of the largest coal mines are the Carmichael Mine, proposed by Adani, and Kevin's Corner and Alpha Coal Mine, proposed by GVK Hancock.

The expansion of the ports along the Queensland coast and the increase in shipping that is occurring to accommodate the escalating coal exports is increasing the risks to the Great Barrier Reef community. Dredging to expand Gladstone harbour is releasing toxins in the sediment and silt which remains in suspension in harbour waters and inhibits the ability of fish to extract oxygen from the water. The offshore disposal of dredge spoil has increased the turbidity of the water, which has a negative effect on sea grass meadows and on the marine life which depends on them.

3: The oceanic context

Oceans constitute a single interconnected system that covers 71 per cent of the surface of Earth and contains 97 per cent of Earth's water.

Humans are changing the fundamental chemical properties of the oceans by increasing the carbon dioxide concentrations in the atmosphere and by undertaking a wide range of activities that pollute marine waters (such as those occurring around the Great Barrier Reef). The ocean is absorbing so much carbon dioxide that the PH level is being reduced. The ocean is 30 per cent more acidic today that it was before humans began burning fossil fuels and is more acidic than it has been at any time in approximately 55 million years. The increase in acidification is having widespread impacts, particularly on oceanic beings that require calcium. Reefs are corroding, shells are becoming thinner and if in acidification continues it may become impossible for marine beings to form shells. If that occurs it will result in the destruction of many species of plankton, with devastating impacts on the functioning of ecosystems and on the wellbeing and existence of many populations of species, including whale-kind.

Scientists predict that coral reef ecosystems may become extinct within the next 50 years because they will not be able to regenerate their calcium carbonate structures.

F: The rights of the Great Barrier Reef

The Declaration recognises in Article 2(1) that the Great Barrier Reef and myriads of beings that form part of that community have the following rights, among others:

- the right to life and to exist (article 2(1)(a));
- the right to wellbeing (article 2(3));
- the right to a place and to play its role in Mother Earth for her harmonious functioning (article 2(2));
- the right to continue their vital cycles and processes free from human disruptions (article 2(1)(c));
- the right to integral health (article 2(1)(g));
- the right to be free from contamination, pollution and toxic or radioactive waste (article 2(1)(h)); and
- the right to full and prompt restoration for the violation of the rights recognized in this Declaration caused by human activities; (article 2(1)(j)).

On the basis of the evidence considered by the Tribunal it is clear that a range of activities being carried out by humans in Australia (particularly within the State of Queensland) are polluting the waters of the Great Barrier Reef with sediments and toxic chemicals, and causing direct physical damage to the Reef to such an extent that the integrity and health of the various communities of which the Reef is comprised have been deteriorating for several decades (if not longer) and that degradation is accelerating. These activities include:

(a) the mining, export and combustion of coal, which contribute to climate change and the warming and acidification of the oceans;
(b) the discharge of pollutants from land-based sources, such as agricultural runoff, into the waters of the Great Barrier Reef;
(c) shipping, with the risk of shipping incidents such as oil spills and coal dust pollution;
d) the construction, expansion and dredging of ports adjacent to the Great Barrier Reef; and
e) the impact of tourism on the Reef, including high level of boats travelling through the waters, large numbers of tourists entering the waters and the consequent pollution and impacts on coral through tourist interactions with the Reef.

The Declaration specifically recognises that the Great Barrier Reef and the beings of that community have the right to integral health (article 2(1)(g)). In referring to 'integral health' the Declaration reminds us of the essential interrelatedness of every aspect of the Earth community and that the health of the whole system affects the health of any aspect or being within it, and *vice versa*. As the preamble to the Declaration notes: 'we are all part of Mother Earth, an indivisible, living community of interrelated and interdependent beings with a common destiny'. This understanding is also reflected in Article 1 which records that 'Each being is defined by its relationships as an integral part of Mother Earth' (Article 1(3)). In this case it is clear that the health of the great oceanic systems of Earth and that of the Great Barrier Reef are interrelated and inseparable.

There is no doubt that the rapid acidification and warming of the oceans that is currently occurring pose grave dangers to most fundamental rights of the Great Barrier Reef: its right to exist and to continue its vital cycles that contribute to the health of Earth. These threats arise from global warming caused primarily by the excessive combustion of hydrocarbons, including coal, by human beings. In this case many of the direct assaults on the Great Barrier Reef arise from the construction, expansion and operation of coal ports that are intended to dramatically increase the volumes of coal exported from Australia. This coal will be combusted, which will in turn increase global warming and ocean warming and acidification to the detriment of the Great Barrier Reef. This is a clear example of how these deliberate human activities are violating the integral health of the Great Barrier Reef community.

Human activities are undoubtedly violating the rights of the Great Barrier Community, recognised in the Declaration. These human activities include both activities that have a direct impact on the Great Barrier Reef such as those referred to above, and activities that contribute to climate change and ocean warming and acidification which have indirect but potentially even more severe impacts.

Extremely serious and ongoing violations of the rights of the Great Barrier Reef are being caused both by human activities that have direct destructive impacts on the Great Barrier Reef (such as dredging and the discharge of pollutants into coastal waters) and by those that have indirect, but no less serious, impacts (such as activities that increase global warming, and particularly the increase of carbon dioxide in the atmosphere). Such severe harm to the Great Barrier Reef community has already been caused by these activities, which continue to intensify, that it is now necessary to recognise the reality that this is currently a global emergency which requires urgent and decisive responses.

We find that these deliberate human activities are violating the right of the Great Barrier Reef community to integral health (article 2(1)(g)) and its related rights to wellbeing (article 2(3)); to continue its vital cycles and processes free from human disruptions (article 2(1)(c)); to be free from contamination, pollution and toxic or radioactive waste (article 2(1)(h)); to play its role in Mother Earth for her harmonious functioning (article 2(2)); and ultimately to exist (article 2(1)(a)).

G: Accountability

In order to establish whether or not the defendants must be held accountable, we must identify whether or not they (or others) have breached any of the duties recognised by the Declaration, and if so, to what extent.

The parties that are most responsible for serious violations of the Declaration, often do not suffer the harmful consequences of their actions. In this case those that will suffer most include other species, ocean-dependent communities, future generations, and the Earth community as a whole. This is unjust and perpetuates anti-social decision-making. Consequently the Declaration imposes a duty on all human beings, all States, and all public and private institutions to recognize and promote the full implementation and enforcement of the rights and obligations recognized in the Declaration (article 3(2)(b)), and to guarantee that those responsible for violations of the inherent rights recognized in this Declaration are held accountable for restoring the integrity and health of Mother Earth (article 3(2)(g)).

1: Accountability of mining companies, banks and other private sector organisations

The Declaration recognises that, among other obligations, human beings, all States, and all public and private institutions must:

(a) act in accordance with the rights and obligations recognized in the Declaration;
(b) ensure that the pursuit of human wellbeing contributes to the wellbeing of Mother Earth, now and in the future; and
(c) recognize and promote the full implementation and enforcement of the rights and obligations recognized in the Declaration.

These duties must rest particularly heavily upon those organisations which intentionally promote increased extraction and combustion of fossil fuels, including by funding these activities. The fact that these activities undermine, rather than contribute to, the wellbeing of Mother Earth, renders them illegitimate and unlawful. In fact from the perspective of the Earth community as a whole, the continuation of such activities is profoundly anti-social and must be discontinued as soon as possible.

2: Accountability of individuals

The Declaration records that 'Every human being is responsible for respecting and living in harmony with Mother Earth' (article 3(1)). Consequently no-one is exempt from ensuring that they do not contribute to climate change and thereby to the warming and acidification of the oceans.

The responsibilities of the people who invest in, fund, promote, manage or undertake these harmful activities is not lessened by the legal fiction that these anti-social activities are carried out by a juristic person such as a company. Each human being is responsible for the consequences of their actions – particularly when they are foreseeable.

Although there is some merit in the argument that responsibility must also be borne by everyone who uses the energy generated by the burning of the coal exported via the ports adjacent to the Great Barrier Reef, their responsibility is minor in comparison with that of those directly involved in the harmful activities or with the power to regulate and prevent the harm.

3: Accountability of federal and state governments and public institutions

Governments and public institutions, and the people who work in them bear a particular responsibility to act and must meet a higher standard by virtue of the regulatory powers and responsibilities vested in them. For example, the Declaration requires that States and public institutions must:

(a) establish and apply effective norms and laws for the defence, protection and conservation of the rights of Mother Earth (article 3(2)(e));
(b) guarantee that the damages caused by human violations of the inherent rights recognized in this Declaration are rectified and that those responsible are held accountable for restoring the integrity and health of Mother Earth (article 3(2)(g));
(c) empower human beings and institutions to defend the rights of Mother Earth and of all beings (article 3(2)(h)); and
(d) establish precautionary and restrictive measures to prevent human activities from causing species extinction, the destruction of ecosystems or the disruption of ecological cycles (article 3(2)(i)).

The evidence placed before the Tribunal indicates that there has been a catastrophic failure on the part of the Australian Federal Government and the Government of Queensland to take the necessary action to comply with these duties in relation to the Great Barrier Reef. Instead of strengthening the protection of the Great Barrier Reef in response to incontrovertible evidence that it is being damaged, and taking decisive action to limit Australia's growing contribution to climate change, they have done the reverse. This is a clear contravention of the Declaration and is consequently unlawful.

A government that establishes a power supply system to deliver electricity generated from the combustion of fossil fuels to people, and subsidises the price of that power (for example by not requiring polluting power companies to pay for the elimination or reduction of emissions or to remedy the negative consequences) is promoting global warming. This culpability is exacerbated if the government is not actively promoting a rapid transition to renewable sources of energy. In comparison, the culpability of a person who uses electricity generated from fossil fuels to meet basic human needs is negligible in comparison to those who perpetuate a system for providing that energy to the exclusion of less harmful sources.

4: Accountability of public officials and voters

The primary responsibility for this contravention lies not only with the governments (that is with intangible juristic persons created by legal fictions) but also (to varying degrees) with the decision-makers in political parties who formulated and promoted the policies that legitimised these actions, the elected public representatives and public officials who authorised these actions, and the people who voted for those political parties and elected officials knowing that they intended to take these actions. Each one of them is personally responsible for the part which they have played in creating the immense harm to Mother Earth and the members of the Earth community (including humans) caused by their deliberate actions and inaction.

The Federal Government of Australia and the Government of Queensland and the people who work in them bear a particular responsibility to act and must meet a higher standard by virtue of the regulatory powers and responsibilities vested in them.

H: Are these limitations to the rights of the Reef necessary to maintain integrity, balance and health?

Since the rights recognised by the Declaration are not absolute and the rights of any being must of necessity be limited in the interests of maintaining the health and integrity of the whole Earth community, it is also necessary to consider the extent to which these violations of the rights of the Great Barrier Reef may be considered to be limitations on those rights which are justifiable in the light of the interests of the Earth community as a whole.

The Declaration recognises that there may be situations in which the rights of specific beings conflict with (or appear to conflict with) the rights of other beings. Article 1(7) states that:

> The rights of each being are limited by the rights of other beings and any conflict between their rights must be resolved in a way that maintains the integrity, balance and health of Mother Earth.

In this case the human activities that give rise to the violations of the rights of the Great Barrier Reef (for example the mining, export and combustion of coal) are primarily commercial activities that are undertaken to enrich juristic entities (companies) and the people that own shares in them.

The revenues from these activities and the resources that they make available to human beings (for example, electricity from coal-fired power stations) contribute to increasing the standard of life and in some cases, the wellbeing of some human beings. However, these apparent benefits to some humans are being achieved at a considerable cost to the Earth community as a whole (Mother Earth). The costs to all beings, of humans continuing to burn fossil fuels, is increasing dramatically as we approach and exceed the limits of Earth's capacity to reabsorb and metabolise the greenhouse gasses which these processes emit.

The right of human beings to pursue their own wellbeing is not unqualified and is limited by the duty in article 3(2)(d) of the Declaration to ensure that the pursuit of human wellbeing contributes to the wellbeing of Mother

Earth, now and in the future. In other words, the Declaration does not recognise that human beings have a right to seek to enhance their wellbeing at the expense of the wellbeing of the Earth community. Whatever the short-term gains which humans may make in this way must inevitably be overshadowed by the long-term losses which humanity will suffer as a consequence of the deterioration of their habitat.

A complete cessation of coal exports from Australia via the waters of the Great Barrier Reef would obviously have a significant impact on the Australian economy and may result in certain corporations ceasing to exist. However, it is unlikely to result in the death of any humans. Certainly it is improbable to suggest that the effects would be sufficient to constitute a serious violation of those rights of human beings that are recognised in the Declaration (for example, the right to life and to exist), or indeed under international human rights law. On the contrary, a significant reduction in the mining, export and combustion of coal is likely to enhance the rights of human beings to clean air, to integral health, and to be free from contamination and pollution (article 2(1) paragraphs (f), (g) and (h)). The evidence before the Tribunal also suggest that such a cessation would enhance the wellbeing of indigenous and local human communities who depend on the Reef for sustenance. Consequently we find that the violations of the fundamental rights of the Reef cannot be justified on the basis that these constitute acceptable limitations on the rights of the Reef that are necessary to protect fundamental human rights.

Even if we were wrong in this regard and there is a conflict between the rights of human beings (i.e., what is best for humans) and the rights of the Great Barrier Reef (i.e., what is best for the Reef) the role of the Tribunal is to identify how such a conflict can be resolved in a way that maintains the integrity, balance and health of Mother Earth (article 1(7)). The evidence shows that the integrity, balance and health of Mother Earth is being seriously compromised by global warming, ocean warming and acidification, and direct assaults on the Reef. Consequently it is not an option to resolve any conflicts between the rights of humans and the rights of the Reef in a way that legitimises the continuation of the human activities that create these consequences.

I: *Restoration measures*

According to scientific evidence presented at the Regional Tribunal, the Reef has a tremendous capacity to heal itself and if the many human activities that are causing ongoing harm were stopped, a significant amount of natural regeneration could be expected. However, it may no longer be possible to rectify all the harm that human activities have caused to the Great Barrier Reef, and to achieve a full and prompt restoration of these violations of the rights recognized in the Declaration as required by article 1(j) as read with articles 3(2)(g)). However, even if this were proved, it cannot justify a failure to take measures to mitigate such harm. The duty to act does not depend on the likelihood of success.

In this regard the primary duty of the Tribunal is to identify what is necessary:

(a) to give effect to the right of the Great Barrier Reef to full and prompt restoration for the violation of the rights recognized in the Declaration caused by human activities (article 2(1)(j)); and
(b) to guarantee that the damages caused by human violations of the inherent rights recognized in the Declaration are rectified, and that those responsible are held accountable for restoring the integrity and health of Mother Earth (article 3(2)(g)).

An enormous range of actions could be taken to protect the rights of the Great Barrier Reef and to ensure that it is restored to health. Consequently this judgement only focuses on a few measures which the Tribunal considers to be essential in this regard.

First, it will be necessary to do everything humanly possible to stop and reverse the acidification of the oceans and the direct destructive impacts on the Reef. This will require the following, among other measures:

(a) a complete cessation of coal exports from Queensland; and
(b) a dramatic global reduction in the emission of greenhouse gasses.

Given the fact that concentrations of greenhouses gasses in the atmosphere are already dangerously high and continued acidification of the oceans can only be prevented by rapidly reducing concentrations of CO_2 in the atmosphere, these measures are urgent and must be commenced immediately. A failure to act with urgency, particularly with knowledge of the consequences of delay and the real likelihood that delay itself may render any response ineffective, is itself a culpable violation of the Declaration and of the rights of all beings. Whether or not nation States have agreed to an international treaty that imposes binding obligations to reduce emissions of greenhouse gasses does not affect the existence of the duty to take effective preventive and restorative measures urgently, nor diminish the culpability of those who fail to act.

J: Conclusions

This Tribunal makes the following findings.

Deliberate human activities are directly and indirectly violating the right of the Great Barrier Reef community to integral health (article 2(1)(g)) and its related rights to wellbeing (article 2(3)); to continue its vital cycles and processes free from human disruptions (article 2(1)(c)); to be free from contamination, pollution and toxic or radioactive waste (article 2(1)(h); to play its role in Mother Earth for her harmonious functioning (article 2(2)); and ultimately to exist (article 2(1)(a)).

These violations are not necessary to maintain the integrity, balance and health of Mother Earth, and are consequently unjustifiable, illegitimate and unlawful.

The Australian Federal Government and the Government of Queensland have failed to comply with their obligations under the Declaration, and in particular, have failed to establish and effectively implement laws and other measures that:

(a) conserve the Great Barrier Reef community and protect its inherent rights (in contravention of article 3(2)(e));
(b) prevent human activities from causing species extinction, the destruction of ecosystems or the disruption of ecological cycles (article 3(2)(i)); and
(c) ensure those responsible for violating the inherent rights of the Great Barrier Reef community restore its integrity and health (article 3(2)(g)).

In order to begin the process of restoring the integrity and health of the Great Barrier Reef community, the Australian Federal Government and the Government of Queensland must immediately:

(a) comply with the UNESCO recommendations;
(b) prohibit the expansion of the coal ports adjacent to the Great Barrier Reef;
(c) prohibit any activities that pollute the waters of the Great Barrier Reef, or harm it directly or indirectly;
(d) enforce a rapid reduction in the mining and combustion of coal within, and the export of coal from, the areas under their jurisdiction; and
(e) ensure that human beings that have benefitted from the activities that have harmed the Great Barrier Reef contribute (financially and otherwise) to measures to restore it to integral health, and that the degree to which a person is culpable and has benefitted from those activities be taken into account in apportioning responsibility for contributing to those measures.

Any person who directly or indirectly funds, invests in, participates in, authorises or facilitates the process of mining and exporting coal via the Great Barrier Reef ceases doing so and immediately begins taking or contributing to the taking of effective measures to restore the integrity and health of the Great Barrier Reef.

The following activities constitute unlawful violations of the rights of Mother Earth and must cease immediately:

(a) the granting of any authorisations for the export of coal via the waters of the Great Barrier Reef;

(b) providing funding for, or facilitating, the export of coal via the Great Barrier Reef;

(c) investing in the companies that are developing and operating coal mines that are or will be exporting coal via the Great Barrier Reef, including Adani Enterprises Limited (developers of the Carmichael Coal Mine with an estimated annual output of 60 million tonnes); GVK Coal Developers, the parent company of Hancock Galilee (Pty) Ltd (developers of the Kevin's Corner mine) and Hancock Coal Pty Ltd, (developers of the Alpha coal project) which each have an estimated annual output of 30 million tonnes; Waratah Coal Pty Ltd (developers of the Galilee Coal Project with an estimated annual output of 40 million tonnes), and Macmines Austasia Pty Ltd (developers of the China Stone Coal Project with an estimated annual output of 38 million tonnes);

(d) the use of agricultural fertilisers in a manner that pollutes the sea surrounding the Reef.

The Tribunal calls upon all people of good conscience who recognise the value of their membership of the magnificent community of life that we call Earth or Mother Earth to take appropriate action to ensure the implementation of this judgment and to defend the rights of the Great Barrier Reef.

Chapter 5

The fraught and fishy tale of *Lungfish v The State of Queensland*

Benedict Coyne

Commentary

The rewritten judgment follows a case[1] in 2011, which concerned the survival of the endangered Queensland lungfish on the Burnett River. The matter was brought by the Wide Bay Burnett Conservation Council (WBBCC) represented by the Environmental Defenders Office Queensland.

The Burnett River is home to the Queensland lungfish; one of only two endemic populations of the Australian lungfish. The applicant sought to compel the owner and operator of the Paradise Dam on the Burnett River (Burnett Water Pty Ltd) to comply with a condition subsequent to the Dam's approval (made by the federal Minister for the Environment pursuant to the *Environment Protection and Biodiversity Conservation Act 1999* (Cth) (EPBC Act)). That condition was relevant to the Queensland lungfish which inhabited the part of the Burnett River affected by the Paradise Dam, including allowing lungfish to traverse the dam wall for the purposes of breeding.

This was a public interest test case as it was the first case to focus on compliance with Ministerial approval conditions under the EPBC Act. In recognition of the public interest importance of this case, the applicant received a grant of Commonwealth public interest funding. The Federal Environment Minister had approved the Paradise Dam under the EPBC Act in 2002. Construction of the Dam did not begin until 2006. However, subsequent to the 2002 approval the Queensland lungfish was listed as vulnerable under the EPBC Act and in 2003 a further condition was added requiring Burnett Water to install a fish transfer device or 'fishway' on the dam 'suitable for lungfish'. The fishway was mandated to commence operation when the dam

---

1 *Wide Bay Conservation Council Inc v Burnett Water Pty Ltd (No 8)* [2011] FCA 175.

became operational. Burnett Water Pty Ltd had installed both upstream and downstream fishways on the dam, however, the applicant alleged they were not suitable for lungfish. The matter was listed for a two-week hearing, one week in November 2009 and one week in February 2010. On 4 March 2011, Justice Logan dismissed WBBCC's application, finding that the fishway was suitable for lungfish.

The judgment was selected for rewriting because of the author's experience of being involved in the litigation and because the judgment is eminently emblematic of an anthropocentric legal structure and legal system. That is, there were very clear contentions on the expert evidence that the fishways were not suitable for the lungfish, and were having a significant and detrimental impact on the health and wellbeing of the Burnett River lungfish population. This was more pressing, as the species had recently been added to the vulnerable species list. Yet, the tone of the judgment appears dismissive of any problems of the more-than-human-world with a human development, because it is convenient to do so. To the author's mind, this accords with an underlying premise of Earth jurisprudence, that the legal system, anthropocentric by genesis and nature, inevitably seeks to serve its masters and facilitate the broad justification of human development at the expense of the environment via an authoritative anthropocentric framework. The lungfish was not able to give evidence in the case and nor were there any measures by which an advocate for the lungfish was able to give evidence in court. The original case illuminates the multiple and various shortcomings of our legal system in dealing with loss and damage in other than pecuniary terms.

This version differs from the original judgment because it is set in an eco-utopian Australian future of sorts, when a raft of international earth laws have been implemented and a future Australian government has properly ratified and domestically implemented these laws. This judgment is set in the not-too-distant-future, in which I envision certain unprecedented developments have occurred, including the adoption by the UN General Assembly of the *Universal Declaration on the Rights of Mother Earth*; the drafting, signing and ratification of the *International Convention on the Rights of Mother Earth 2018* and the subsequent passage of the *Rights of Nature and Mother Earth Act 2018* in Australia.

In this context, the rewritten judgment allows the concerns of the lungfish to be heard, and is a representative action brought by a group of Queensland lungfish in the tort of trespass against the operators of the Paradise Dam on the Burnett River for which the group members seek damages and injunctive relief. It both reflects and progresses an earth laws jurisprudential analysis, and a positive and optimistic illumination of how things could be in eco-harmony for the lungfish, for humans, and for the more-than-human world.

IN THE FEDERAL COURT OF AUSTRALIA
QUEENSLAND DISTRICT REGISTRY
GENERAL DIVISION   QUD 034 of 2020

BETWEEN: LOUIE THE LUNGFISH and ORS
                  Applicant

AND   MINISTER FOR THE ENVIRONMENT, CLIMATE
                  CHANGE and FUTURE GENERATIONS
                  First Respondent

                  STATE OF QUEENSLAND
                  Second Respondent

                  COMMONWEALTH OF AUSTRALIA
                  Third Respondent

JUDGE   COYNE J

DATE: 20 NOVEMBER 2020

PLACE: BRISBANE

Reasons for judgment

A: Introduction

This is a representative action brought by a group of Queensland lungfish in the tort of trespass against the operators of the Paradise Dam on the Burnett River for which the group members seek damages and injunctive relief.

The group also seeks injunctive relief for breaches of various sections of the *Rights of Nature and Mother Earth Act 2018* (Cth) (Rights of Nature and Mother Earth Act) and the *Environment Protection and Biodiversity Conservation Act 1999* (Cth) (EPBC Act).

Due to the significant global impact of over two centuries of industrialisation, the planet and humanity are on a collision course of considerable degree and a somewhat immediate trajectory. Increasing global temperatures as a result of excessive greenhouse gas emissions has created a dire, urgent situation for the continued survival of many species with which we, humanity, co-inhabit the Earth. According to the leading scientists, the Earth is in the midst of the sixth mass extinction of life.[2] Scientists estimate that

2 G Ceballos et al., 'Accelerated modern human-induced species losses: entering the sixth mass extinction', *Science Advances* vol 1, no 5, 19 June 2015 e1400253 <http://advances. sciencemag.org/content/1/5/e1400253>.

every 24 hours 150–200 species of plant, insect, bird and mammal become extinct.[3] This is nearly 1,000 times the 'natural' or 'background' rate and scientists say it is greater than anything the world has experienced since the vanishing of the dinosaurs nearly 65 million years ago. Around 15 per cent of mammal species and 11 per cent of bird species are classified as threatened with extinction.[4]

Recent scientific studies estimate the natural world contains some 8.7 million species; three-quarters are land-bound, comprising mostly insects, and one-quarter are found in the oceans which comprise 70 per cent of the Earth's surface.[5] However, the vast majority of species have not been identified.[6] Scientists estimate that 86 per cent of existing land-bound species and 91 per cent of ocean species still await description, which has been estimated at taking more than 1,000 years.[7]

The Australian continent hosts numerous, globally significant biodiversity hotspots and, due to the fragile environment of the world's driest inhabited continent, many species are listed as threatened, vulnerable and endangered. This list includes the Queensland lungfish (*Neoceratodus forsteri*), which is the subject of this judgment.

B: The Queensland lungfish

David Attenborough once described the Queensland lungfish as a 'remarkable, ancient and extraordinary fish'.[8] Lungfish first appeared in the fossil record approximately 380 million years ago. They are often termed a 'living fossil' and are a unique, extraordinary and incredibly significant species in the context of physiology, anatomy and evolutionary biology.

Lungfish are the only species of fish on planet earth that have proper lungs homologous with other tetrapod species including mammalian human beings.

They are relics of ancient fish groups related to the ancestors of amphibians, reptiles, birds and mammals. Globally, there are six different species of

---

3 World Conservation Union, United Nations Environment Programme and World Wide Fund for Nature, *Caring for the Earth – A strategy for Sustainable Living*, 1991 <https://portals.iucn.org/library/efiles/edocs/CFE-003.pdf>.
4 Ibid.
5 C Mora et al, 'How Many Species are there on Earth and in the Ocean?' (2011) 9(8) *PLoS Biol* e1001127 <http://journals.plos.org/plosbiology/article?id=10.1371/journal.pbio.1001127>.
6 Ibid.
7 Ibid.
8 British Broadcasting Corporation, 'Land Invaders', *Life in Cold Blood*, 2008 (D Attenborough).

lungfish: four African species in the genus *Protopterus* (Family Protopteridae); one South American species *Lepidosiren paradoxa* (Family Lepidosirenidae); and the Queensland lungfish, which is the only species in the Family Ceratodontidae. There are approximately 400 species of fish which breathe air, but without proper functioning lungs. The lungfish is a truly remarkable relic of the long evolutionary journey from the amphibious ocean depths to the shoreline of land-dwelling biped-ability, which allowed for the genesis of Homo sapiens.

All lungfish must breathe air to survive, but the Queensland lungfish is further unique in that it can rely on its gills solely when at rest in aerated water. It remains the only existing member of an extensive array of lungfish species in Australia dating as far back as the Triassic, Cretaceous and Tertiary periods.[9]

Dr Slim Phishlipz, a Swedish fish biologist and expert witness in this case has described their evolutionary significance as follows:

> This ancient fish is the biz! It has profound lessons for us to learn about our own evolutionary journey and yet currently this fish is fighting off extinction. . . . A recent scientific discovery evidences that this intriguing creature is an indispensable link in the transformative sequence of events that saw amphibious creatures crawl out of the world's waters and take one small step toward becoming land dwelling upright backboned mammals like you and me. Crikey, it's a fascinating fish! But just as we are learning about this, its breeding grounds and habitats are disappearing! And like their former colleagues, the dinosaurs, it's frightening to think that they may soon become extinct – on our watch. If that was so unfortunate as to happen we would lose a very precious lesson and opportunity to fully understand our own deep ancestral history.

Whilst historically lungfish have enjoyed limited protection under Queensland law including from fishing (*Fish and Oyster Act 1914* (Qld) (rep) and the *Fisheries Act 1994* (Qld)), in 2002 it was officially declared a 'vulnerable species' under the EPBC Act and today the species numbers less than 10,000.

The problem is that the preferred breeding habitats of the lungfish are disappearing due to structures like the Paradise Dam. Lungfish require shallow, vegetated and quiet pools in which to spawn and feed. They are very

9 Australian Broadcasting Corporation, 'Lungfish', *Catalyst*, 20 April 2006 <www.abc.net. au/catalyst/stories/s1620397.htm>.

particular about and loyal to their spawning sites. If a site has been interfered with, lungfish will not spawn, but return the next year. So reproduction of the species will be lost for a year. If this continues to happen at current rates, then their population will become extinct.

C: The Paradise Dam on the Burnett River

The Burnett River is located on the southern Queensland coast with a total catchment area of approximately 33,000 square kilometres. Its mouth opens north of the City of Bundaberg. Its source arises eastward in the Dawes Range, and soon flows south through Eidsvold and Mundubbera, meandering along a trajectory soon joined by the Nogo and Auburn Rivers, both of which swallow large areas of the Burnett River's western catchment. The main river then enters a north-easterly flow and journeys toward the coast.

The Paradise Dam is located on the lower portion of the Burnett River, approximately 80 km southwest of the city of Bundaberg. It is situated at approximately 131.2 km Average Middle Thread Distance (AMTD).

The Dam was constructed between 2003 and 2006 by Burnett Water Pty Ltd (ACN 097 206 614) (Burnett Water), a wholly owned subsidiary of a Queensland Government owned corporation, SunWater Limited. Due to the 2016 global financial crisis and the subsequent global depression, coupled with the long harsh drought of recent years, in November 2016 the two companies were wound up and their liabilities now fall squarely within the remit of the First Respondent. Relevantly, the Dam's construction and its continued operation constitutes a 'controlled action' which required the written approval under section 133 of the EPBC Act by the Minister administering that Act. On 25 January 2002 the requisite ministerial approval was given by the then Minister for the Environment, the Hon DA Kemp and provided as follows:

> to construct and operate the Burnett River Dam with a capacity of 300,000 megalitres, on the lower Burnett River at 131.2 km Average Middle Thread Distance, and make controlled discharges of water for agricultural, commercial, domestic and environmental uses (EPBC 2001/422) by Burnett Water Pty Ltd subject to the conditions set out in annexure 1.

At that particular time the Queensland lungfish was not listed as a threatened species under the EPBC Act and therefore it is not mentioned in the approval. However, on 8 August 2003 the original approval was amended by the Minister thereby adding seven further conditions to the original two conditions. The additional conditions were concerned with facilitating and monitoring the welfare and continued survival of the lungfish populations proximate to the Paradise Dam, and included Condition 3 which is relevant to this judgment:

> Additions
>   3. Burnett Water Pty Ltd must install a fish transfer device on the Burnett River Dam suitable for the lungfish. The fishway will commence when the dam becomes operational.

The collective conditions were deemed valid until 1 January 2052.

A court action was taken by the Wide Bay Burnett Conservation Council for judicial review of Condition 3 in 2008 and 2009, asserting that the fishway was not a suitable mechanism to facilitate lungfish movements above and below the Paradise Dam. However, that legal action failed. Subsequently, the population of lungfish continued plummeting and in 2017 the lungfish was listed as threatened and in 2018 as endangered under the EPBC Act.

D: Australia's international obligations to nature and Mother Earth

The continued operation of the Paradise Dam is also subject to Australia's international obligations as a signatory to both the *Universal Declaration on the Rights of Mother Earth 2016* and the *International Convention of the Rights of Mother Earth 2017*, the obligations of which have been distilled domestically into the *Rights of Nature and Mother Earth Act 2018* (Cth).

Following increasing, incontrovertible scientific evidence of climate turmoil in increasing extreme weather events throughout the world, an unprecedented paradigm shift occurred at the COP 21 United Nations Framework Convention on Climate Change conference in Paris in December 2015 with a global agreement on emissions cuts in the Paris Climate Accord. The nations of the world moved to usher in a new era of ecologically sustainable trajectory for human development.

By late 2016, the United Nations General Assembly had unanimously adopted the *Universal Declaration on the Rights of Mother Earth*, previously submitted to the UN General Assembly for adoption in 2010 by Bolivian President Evo Morales. Bolivia had then hosted the World People's Conference on Climate Change and the Rights of Mother Earth on Climate Change, a response to the failure of the COP15 Copenhagen negotiations in December 2009. The preambular paragraph of the founding Declaration states:

> We, the peoples and nations of Earth:
>   considering that we are all part of Mother Earth, an indivisible, living community of interrelated and interdependent beings with a common destiny; gratefully acknowledging that Mother Earth is the source of

life, nourishment and learning and provides everything we need to live well; recognizing that the capitalist system and all forms of depredation, exploitation, abuse and contamination have caused great destruction, degradation and disruption of Mother Earth, putting life as we know it today at risk through phenomena such as climate change; convinced that in an interdependent living community it is not possible to recognize the rights of only human beings without causing an imbalance within Mother Earth.

It is unusual for international legal instruments to be drafted external to the United Nations and submitted to the General Assembly for adoption. However, a consortia of developing nations including a coalition of small island States, international non-government organisations including the Global Alliance for the Rights of Nature and essentially the citizens of the world, pulled the leaders of developed nations up by their bootstraps to insist that they must urgently transition to a fossil fuel free global economy. On Earth Day, 22 April 2017, the UN General Assembly unanimously adopted the binding *Convention on the Rights of Mother Earth*.

Amidst this positive action, Australia signed the Convention on 22 May 2017 and subsequently ratified it on 22 April 2018, and then set about making plans to legislate rights of nature domestically, inspired by countries like Ecuador and Bolivia. Ecuador had amended rights for nature into its Constitution in September 2008 and in December 2010 Bolivia's Plurinational Legislative Assembly passed legislation recognising Rights of Nature as the Law of the Rights of Mother Earth (*Ley de Derechos de la Madre Tierra* – Law 071 of the Plurinational State). The Maori peoples of Aoteoroa have also utilised the Treaty of Waitangi to recognize the rights of nature, including providing legal standing to river systems such as the Whanganui River in court.

During the second reading speech Prime Minister Tanya Plibersek, also Minister for the Environment, Climate Change and Future Generations stated that:

> The global community has spoken loud and clear against environmental devastation with the adoption of the *Universal Declaration of Rights of Mother Earth* and its international convention. Those documents declare a global consensus that the ongoing destruction of our fragile planet for short-term profiteering is no longer acceptable nor viable for a functional planet, for an effective and egalitarian economic system and for the welfare of future generations. The subsequent binding incarnation in the Convention further hardens humanity's righteous resolve to protect this precious planetary jewel for the benefit of present generations and those well into the distant future. We have moral responsibilities to our children, and their children, and so on, ad infinitum.

The *Rights of Nature and Mother Earth Bill 2018* (Cth), was passed by both houses of Parliament in record time on 22 July 2018. The preambular paragraph of the Explanatory Memorandum states:

> Australia has proudly ratified the Convention, the adoption of which marks a common standard of achievement for all peoples and all nations of the world whereby every individual and institution takes responsibility for promoting through teaching, education, and consciousness raising, respect for the rights of the more-than-human world recognised in the Convention and ensure through prompt and progressive measures and mechanisms, national and international, their universal and effective recognition and observance among all peoples and States in the world. This is an urgently necessary and long overdue, powerful and positive paradigm shift toward a healthier and more equitable planet for all.

As the Australian constitutional referendum for indigenous recognition and a constitutional bill of rights was a success in 2017, the Australian government, determined to take leadership in a time of climate crisis, promised to hold a further referendum to include rights of nature into the Australian Constitution in the near future. This has not yet occurred but remains on the agenda by the three major parties in the next election in 2021.

E: Standing generally

With regards to the cause of action in the tort trespass, and for the injunctive relief sought, section 14 of the *Rights of Nature and Mother Earth Act 2018* (Cth) provides that:

> Every living species shall be entitled to, in a court of law, pursue the protection of their rights as enshrined in this Act and in any relevant international covenant including the *2017 International Covenant of the Rights of Mother Earth*.

During the second reading speech, the Minister for the Environment, Climate Change and Future Generations emphasised that:

> The Explanatory Memorandum to the Bill contemplates a new era of scientific understanding and systems theory whereby humanity becomes fully cognisant of its place in an interconnected world wide web of life and where the health of the whole is interdependent upon the health of regional and local ecosystems right down to the welfare of individual species. In that regard, and following the brave new path forged by numerous world leaders, the proposed Act will bestow litigious rights upon the more-than-human world who will have the ability to bring

grievances to an independent arbiter of law and seek redress for violations of their individual and collective rights.

Prior to the passage of the *Rights of Nature and Mother Earth Act*, the rules for standing to sue for damages and for injunctive relief were constituted by existing common law authorities on standing: *Australian Conservation Foundation Inc v Commonwealth of Australia*[10]; *Bateman's Bay Local Aboriginal Land Council v Aboriginal Community Benefit Fund Pty Ltd*[11] and *Truth About Motorways Pty Ltd v Macquarie Infrastructure Investment Management Ltd*.[12] These authorities had been developed from Lord Denning's observations that standing to claim prerogative relief is denied 'to a mere busybody who is interfering in things which do not concern him'.[13] The law has since evolved much further beyond mere masculine pronouns and anthropocentric norms.

Section 475 of the EPBC Act codified the common law rules of standing for injunctive relief for contraventions of the same act. Standing was conferred on:

a. the Minister; or
b. an interested person (other than an unincorporated organisation); or
c. a person acting on behalf of an unincorporated organisation that is an interested person.

However, the rules of standing including to claim injunctive relief under the EPBC Act were further expanded by the passage of the *Rights of Nature and Mother Earth Act*. Section 54 of Chapter II of that Act provides as follows:

A species listed on the Commonwealth threatened species register and endangered species register is an 'interested person' for the purpose of seeking standing to claim the following forms of relief:

(a) Declaratory Relief;
(b) Injunctive Relief;
(c) Damages.

---

10 *Australian Conservation Foundation Inc v Commonwealth of Australia* (1980) 146 CLR 493.
11 *Bateman's Bay Local Aboriginal Land Council v Aboriginal Community Benefit Fund Pty Ltd* (1998) 194 CLR 247.
12 *Truth About Motorways Pty Ltd v Macquarie Infrastructure Investment Management Ltd* (2000) 200 CLR 591.
13 *R v Greater London Council; Ex parte Blackburn* [1976] 1 WLR 550, 559.

Nothing more need be said on standing generally as, first, it is not opposed, and second the legislation makes the point very clear: the lungfish group members have standing to bring their case for a fair trial for all the forms of relief sought.

F: The class action claim

The claim is also brought to this court in the vehicle of a representative or 'class action' by the seven endangered Queensland lungfish named in the schedule pursuant to Part IVA of the *Federal Court of Australia Act 1976* (Cth). Section 33C of that Act provides as follows:

(1) Subject to this Part, where:

  (a) seven or more persons have claims against the same person; and

  (b) the claims of all those persons are in respect of, or arise out of, the same, similar or related circumstances; and

  (c) the claims of all those persons give rise to a substantial common issue of law or fact.

The lead Plaintiff, 'Louie the Lungfish', at all material times was a permanent resident of the Burnett River below Paradise dam. His date of birth was 11 September 1948 and he is 68 years old. The group definition for the class action is articulated by the plaintiffs' legal representatives in the Further Amended Statement of Claim as follows:

A Queensland lungfish who resides on the Burnett River either above and/or below Paradise Dam and whose use of the river and ability to traverse upriver and downriver, including for breeding purposes, has been affected and obstructed by the construction and ongoing operation of the Paradise Dam.

The specific relief sought by the lungfish is as follows:

  i. Declaratory relief that the construction and continuing operation of the Paradise Dam is a trespass and an unlawful imposition upon the territory and biological integrity of the Queensland lungfish and must be demolished in an ecologically sensitive and responsible manner;

  ii. Injunctive relief that the Second Respondent cease all operations at the Paradise Dam and prepare for the demolishing of the Paradise Dam in an ecologically sensitive and responsible manner;

  iii. Injunctive relief that the Second Respondent prepare to rehabilitate and regenerate the Dam site and river system to its condition prior to construction of the Paradise Dam;

iv. Damages as detailed below;

v. Costs.

G: The nature of the trespass claim

The tort of trespass is a civil and private wrongdoing or an intentional infringement of a private right by which a wrongdoer causes a direct injury to a person or to their property without any lawful justification. 'Trespass' in common parlance is the intentional unauthorised invasion or interference of property or chattel including one's right to use and enjoyment of land whether above, below or on the surface: *Stoneman v Lyons*,[14] *Lawlor v Johnston*.[15]

Trespass to land is thus not limited to the land's surface and it includes the subsoil, airspace, watercourses and anything that comes with the land. In modern times courts have limited the ancient property law principle of absolute dominion, '*cuius est solum eius est usque ad coelum et ad inferos*' meaning 'for whoever owns the soil, it is theirs up to Heaven and down to Hell'.

The tort of trespass is sometimes confused with the tort of nuisance, but the two are quite distinct. A trespass action protects against an invasion of one's right to exclusive possession of land. For example, if a leaseholder builds an unauthorised structure upon another's land they have committed a trespass. However, if a person builds a coal-fired power station then they may be liable in nuisance for noise and pollution that impacts the neighbours. Typical remedies for trespass include damages and injunctive relief. The tort of trespass, whether to person or property, requires only proof of injury and not actual damage. At common law, each and every unauthorised entry onto a person's property is considered a separate trespass.

Lawful justification of an invasion of person or property may be satisfied where consent is provided for the transgression of the person or property to occur.

In this case we are concerned with trespass to property, the lungfish group members' habitat and breeding sites on the Burnett River in the vicinity of the Paradise Dam.

H: Consideration

It is well known that dams can be disruptive and destructive forces upon riverine ecosystems. It is therefore of utmost importance that the development and operations of dams are treated with significant care by the regulating authorities.

---

14 *Stoneman v Lyons* (1975) 133 CLR 550; 8 ALR 173.
15 *Lawlor v Johnston* [1905] VLR 714.

The claim brought is, in the words of the Honourable Justice Brian Preston, 'a clarion call for governmental action to address a pernicious, pervasive and pressing threat to the environment',[16] which in this case is the Paradise Dam. The lungfish, like the Green Sea Turtles (*Green Sea Turtles By Their Representative, Meryl Streef v The State Of Queensland And The Commonwealth Of Australia*), also ask us, humanity, to reflect upon a possible and even probable future where there are no more lungfish amongst the many thousands of species being extinguished forever every single day as a result of human activity.

I accept the expert evidence of Dr Phishlipz in this case that the extinguishment of the lungfish is a very real possibility in circumstances where their fragile habitat is significantly impacted upon by the ongoing existence and operation of the Paradise Dam. The expert evidence of the group members demonstrated clearly that the event of the construction of the Paradise Dam and its ongoing operation have had a significant detrimental and destructive impact upon their property, being their habitat upon the Burnett River.

The seven group members bringing this claim vary in ages but are all permanent residents of the Burnett River below Paradise Dam and their lives and breeding cycles and breeding abilities were at all material times significantly affected by the de facto existence of the Paradise Dam in concert with the Minister's wholly inadequate de jure conditions for the dam to attempt to cater for the breeding requirements of the lungfish by requiring a 'fish-lift' to purportedly act as a transport device by which the group members and members of the class are to traverse the dam wall and move upstream to breed.

By consent of the parties and in order to narrow down the time at trial it was agreed that three of the group members provide viva voce evidence at trial via a professional interpreter or 'fish whisperer'. However, all seven group members submitted comprehensive statements including the particularities of the damage that they directly attribute to the construction and ongoing operation of the Paradise Dam.

I accept the evidence of each of the group members that they have never used the fish lift because it was dysfunctional, wholly inadequate, and impossible to do so. The expert evidence of Professor Bildit was also to the same effect.

On the expert evidence of Dr Willis, I am satisfied that the construction and ongoing operation of the Paradise Dam has caused a direct and ongoing

---

16 *Green Sea Turtles by their Representative, Meryl Streef v The State of Queensland and The Commonwealth of Australia* (Chapter 2).

interference with the earth property rights and common property rights of the group members.

The Respondent parties both accepted that a trespass onto the habitat of the lungfish had occurred but they both argued they had legal justification for such trespass, being the consent of the group members. They argued that the class of lungfish provided an implied consent by not disputing the original construction and operation of the Dam.

I cannot accept such a submission on the evidence. It is clear that the lungfish were never adequately consulted, nor consulted at all on the proposed construction and operation of the Dam. I am satisfied by the evidence of the group members that no proper impact study was ever conducted in regards to the impact the Dam was to have on the lungfish specifically. I accept that while this condition was not originally required, a reassessment was certainly called for in the 12 months following the passage of the Rights of Nature and Mother Earth Act. It is now 18 months since the passage of that Act and there is still no hint of evidence from the Respondent parties that the Minister has initiated such a reassessment process.

In response to the relief sought, the Minister submitted that the deconstruction of the dam was too extreme and that instead the Minister would forthwith embark on the long overdue reassessment of the Paradise Dam. However, the expert evidence submitted illustrated incontrovertibly that the very survival of the lungfish species hangs on the twelfth hour cusp, breathing its last breaths. It would be wholly inadequate for this court, in the face of such urgency, to indulge the executive in yet another foray of delay and inactivity. The time for second chances has well and truly passed.

I: Conclusion

On the evidence I am satisfied that any review conducted by the Minister, if properly conducted, would and should arrive at the same conclusion as this court.

In respect of the injury suffered by the group members, I am satisfied by their evidence that individually and collectively they have suffered and continue to suffer significant injury from the construction and ongoing operation of the Paradise Dam including but not limited to:

   i. habitat destruction;
  ii. dependent ecosystem impairment;
 iii. loss of food supply;
 iv. loss of breeding grounds; and
  v. obstruction to breeding migration pathways.

Certainly, the evidence of the lungfish population is in steep decline since the construction of the Dam is incontrovertible and was accepted by both

Dr Respirator and Dr Scales. In that regard, and with the lungfish now on the endangered species register under the EPBCA Act, the ongoing operation of the Paradise Dam and its existence is entirely unsupportable.

Obviously, pecuniary damages for the individual lungfish are not adequate or applicable and thus more creative types of restitution need be entertained and engaged.

J   Orders

I make the following orders:

a.   The Paradise Dam cease operations immediately and the Respondent government parties begin preparation for the deconstruction of the Paradise Dam to be conducted in an ecologically sensitive way, and to be completed no later than 12 months from the date of this judgment.
b.   The court issue a declaration that the ongoing operation of the Paradise Dam constitutes an ongoing trespass on the group members and the class of lungfish for which they have suffered injury, loss and damage, and should be entitled to compensation.
c.   The group members be compensated for the trespass by ecologically sensitive deconstruction of the Dam and a full rehabilitation, restoration and regeneration of the Dam site such that the river ecosystem is left in the same condition as before the Dam was constructed.
d.   In terms of exemplary and punitive damages, I order that the Minister invest $5million into a 30 year breeding programme to rebuild the lungfish population to what it would be had the Paradise Dam not have been built.
e.   The Minister cover any veterinary expenses of the group members for the next 20 years.

Chapter 6

Attorney-General (Cth); Ex Rel McKinlay v The Commonwealth

Tom Round

Commentary

The High Court of Australia decided *McKinlay*[1] on 1 December 1975 – less than a fortnight before a federal election in which Gough Whitlam's Labor Government, controversially dismissed by Governor-General Sir John Kerr on 11 November, was defeated in a landslide. The Court held that the constitutional requirement (section 24) that Members of the House of Representatives be 'directly chosen by the people' did not require federal electoral districts to contain equal numbers of voters: it merely ruled out electoral colleges, of the type that choose United States Presidents or French Senators. Ironically, while a decision the other way by the Court would have complicated the impending election, the size of Labor's eventual defeat rendered the question of electoral disparities moot. Also ironically, the incoming Fraser Liberal government took up Labor's proposals to equalise electorates via statute. Proposals by the Whitlam and Hawke Labor governments to constitutionally entrench equal electorates were rejected by popular majorities at referenda in 1974 and 1988.

I have chosen to wildly rewrite the *McKinlay* judgment because the majority's reasoning embodies, for me, a paradox. It embodies the view that judges should refrain from disrupting the status quo, especially at the constitutional level,[2] because that would be undemocratic: major changes to a society's

---

1 *Attorney-General (Cth); Ex Rel McKinlay v The Commonwealth* (1975) 135 CLR 1.
2 Although ironically the single most 'disruptive' judgment delivered by the High Court of Australia since 1901 – *Mabo v Queensland (No 2)* [1992] 175 CLR 1 – involved the interpretation of the common law, not of the Constitution. (Admittedly, legislatures' ability to undo *Mabo* was limited by Constitutional factors – section 51(xxxi) restricted the federal Parliament and, given the presence of the *Racial Discrimination Act 1975* (Cth), section 109 restricted the State Parliaments).

I call *Mabo* 'disruptive' in the descriptive sense that, after 3 June 1992, Indigenous Australians found that their oral histories and traditional laws were now legally relevant; the Commonwealth required a new system of tribunals; mining companies needed to hire Native Title consultants; and universities had to radically rewrite their land law courses.

laws should be made only by an elected legislature. Yet as Justice Murphy cogently noted, what happens if the legislature itself is not democratically elected and the incumbent majority benefit from this inequality?

Which brings us to the wild law perspective.

The High Court of Australia has traditionally been deferential to the Parliament on issues affecting electoral rules – in marked contrast to supreme courts in other democracies, particularly Germany, Ireland and the United States. Whatever the subjective philosophy of the Justices, the best litmus test for predicting when the Court will and will not intervene is not so much either equal suffrage or legislative supremacy as whether judicial intervention can preserve a tolerable status quo, or easily reinstate a recent status quo, without too much disruption. Thus, new rules restricting political advertising on an unequal basis were struck down,[3] as were new rules narrowing the franchise.[4] These rules had not (or had barely) had time to come into operation. This *status quo (ante)* criterion is flexible – the unconstitutional addition of three extra seats to the House of Representatives was held severable after 13 years – but it does rule out untried new plunges into the dark, which explains why the Court repeatedly backed away from declaring electoral boundaries unconstitutional.[5] Likewise, the High Court has left Parliament a free hand in deciding how far one's wallet should affect one's electoral participation,[6] and how far voters should be required to indicate their order of preference among candidates.[7]

In *McKinlay*, the High Court did not regard voting inequality as a harm worthy of constitutional remedy – unlike, by contrast, the grave harm to the polity wreaked by the House of Representatives having 127 seats instead of 124, damage that the Court belatedly corrected after 13 years.[8] It is ironic

---

3 *Australian Capital Television v Commonwealth* [No 2] (1992) 177 CLR 106.
4 *Roach v Electoral Commissioner* (2007) 233 CLR 162; *Rowe v Electoral Commissioner* (2010) 243 CLR 1.
5 *Tonkin v Brand* [1962] WAR 2; *McGinty v Western Australia* (1996) 186 CLR 140.
6 *Fabre v Ley* (1972) 127 CLR 665; *McCloy v NSW* [2015] HCA 34; although contrast *Unions NSW v NSW* (2013) 252 CLR 530.
7 *Faderson v Bridger* (1971) 126 CLR 271; *Day v Australian Electoral Officer for South Australia* [2016] HCA 20.
8 The Constitution (s 24) sets, as a default rule, variable (in theory) by federal statute, that each State's number of Members of the House of Representatives is determined by rounding off its number of population quotas to the nearest whole number (or to five, whichever is larger). In 1964 the Menzies Liberal Government, responding to fears by its Country Party coalition partners that declining population would cost them seats, had the Parliament 'otherwise provide' that, instead, *every* remainder would be rounded up. In *AG ex rel McKellar v Commonwealth* (1977) 139 CLR 527, the High Court held this was invalid because it departed unacceptably from the constitutional ideal that the House be twice the size of the Senate.

that in Australia, the Constitution regulates in close detail the relative size of the two chambers, when they only sit and vote together on rare occasions[9] – whereas in the United States, where the balance between Houses and Senate directly affects the selection of the country's executive Government (through the weighting of States' representation in the Electoral College), there is no constitutional nexus ratio.

Some critical thinkers have argued that majority rule, although progressive in its time, has become an obstacle to further progress:

> In [industrialising] societies, majority rule almost always meant a fairer break for the poor. For the poor were the majority. Today, however, [. . . the] truly poor no longer necessarily have numbers on their side.[10]

This applies across (not only within) nation-states:

> Elites may use democratic processes to entrench their status[. . . . S]ound climate and energy planning should not treat all stakeholders in the same way. Instead, preferences and roles should be weighted to consider criteria related to equity, due process, ethics and other justice principles. This would ensure that stakeholder discussions and resulting policies serve to eradicate, rather than exacerbate, socio-economic vulnerability.[11]

The problem, though, is that to 'be weighted' is a passive verb that disguises its subject. In other words, who should do the weighting? Often cited as an advantage of 'one person, one vote, one value', as well as its inherent justice,

---

9 This 2:1 'nexus ratio' was adopted in 1900 to protect the voting weight of the Senate in joint sittings, which may be held after double dissolutions to vote on deadlocked Bills. There has been exactly one joint sitting since Federation, in 1974, and exactly three Acts on the federal statute book have been passed by the votes of Members of the House of Representatives over an opposed but smaller Senate.

    A fourth Bill, purportedly passed at the 1974 joint sitting, was later held invalid by the High Court because it had not been deadlocked long enough between the two houses. The Senate majority had expressed only ambiguous and tepid reluctance about the PMA Bill, which therefore had not 'failed to pass' as Constitution s 57 required. *Victoria v Commonwealth (PMA Case)* (1975) 134 CLR 81. As Professor Colin Howard notes, 'failure to pass is a non-event, and it is not easy to say when a non-event has happened': *Australia's Constitution: What It Means and How It Works* (Penguin, 2nd ed, 1985) 84. Had Senators expressed their opposition more forcefully, and voted explicitly to reject that Bill, they would have *helped* it pass into valid law, over their nays, at the joint sitting.

10 A Toffler, *The Third Wave* (Collins, 1980) 430.

11 J Ren, M E Goodsite and B K Sovacool, 'Climate change: climate justice more vital than democracy', 526 *Nature* (15 October 2015) 323 <www.nature.com/nature/journal/v526/n7573/full/526323a.html>.

is simplicity: it is easier to administer.[12] But we should not underestimate the amount of discretion needed simply to uphold 'one person, one vote': consider the enormous volume of jurisprudence generated under the United States' *Voting Rights Act* 1965 to determine which electoral regulations (for example, voter ID) do or do not 'result[. . .] in a denial or abridgement' of the right to vote.

Attorney-General (Cth); Ex Rel McKinlay v The Commonwealth

Judgment

Round J, dissenting

A: Introduction

I apologise for the lateness of this judgment due to factors beyond my control. However, the effluxion of time between 1975 and 2016 has enabled me to take judicial notice of a richer array of arguments and research on this pressing topic.

This suit (hereinafter *McKinlay*) is brought in the name of the Attorney-General of the Commonwealth, but at the relation of private individuals. It names, as defendants, the Commonwealth and the Australian Chief Electoral Officer.

B: Just representation

I have read the judgments of my brother Justices Barwick and Gibbs on the one hand, and Murphy on the other. I find both approaches unsatisfactory.

Justices Barwick and Gibbs take far too limited a view of what is constitutionally required to be represented, and leave too much discretion to a temporary legislative majority:

> Neither in 1901 nor subsequently has there been a universal recognition of the so-called principle that electorates should be numerically equal. [. . . I] find nothing to persuade me that the words of [Australian

12 J H Ely, *Democracy and Distrust: A Theory of Judicial Review* (Harvard University Press, 1980) 124–125.

Constitution] s 24 contain a rule that electoral divisions shall, as nearly as practicable, be equal in population or electors.[13]

On the contrary, democratic legislatures should not have, and therefore do not have, competence to deal with the franchise as they think fit. That would rightly be an absurdity. Parliamentarians hold their authority as servants commissioned by their masters – their constituents, those whom they represent, speak for, and have the capacity to bind. The courts would no more permit Parliamentarians unilaterally to alter the terms upon which they hold their office, than they would permit workingmen unilaterally to alter the terms upon which they hold their employment.

If section 24 means only that 'the House of Representatives shall be composed of members directly chosen by the *statutorily enfranchised electors* of the Commonwealth', it would mean no more than that 'everyone who is entitled by law to vote shall be entitled by law to vote'. The word 'directly' does, all agree, rule out an electoral college, but it would be a pointless tautology to exclude one single ground of limiting the suffrage – that it cannot be confined to officials who have already been directly elected, even by equal universal suffrage – while leaving 999 other grounds open to the legislature.

On the other hand, my brother Murphy's approach would tie the hands of our Parliaments far too tightly. True, he is correct when he declares that the Constitution does not give Parliament unlimited discretion over electoral laws. There exist real and substantial – if implied – and judicially enforceable limits upon the capacity of any legislature to re-make the basis of its own franchise. Parliaments may by legislation settle such details as whether electoral divisions return one member of Parliament or several, and whether electors can transfer and/or cumulate their votes among candidates. But mere statutes cannot fundamentally shift the balance of power by limiting or altering the franchise, or by arbitrarily and differentially weighting the votes of those who are enfranchised.

I part ways with Justice Murphy, however, when he quotes[14] as holy writ the United States Supreme Court's assertion that '[l]egislators represent people, not trees or acres'. As a statement of the status quo this is, alas, correct. As a normative goal to guide the interpretation of the Australian

---

13 *Attorney-General (Cth); Ex Rel McKinlay v The Commonwealth* (1975) 135 CLR 1, 45–47 (Gibbs J).

14 *Attorney-General (Cth); Ex Rel McKinlay v The Commonwealth* (1975) 135 CLR 1, 68 (Murphy J); quoting *Reynolds v Sims* 377 US 533 (1964) 562 (Warren J for the majority).

Constitution, it is inadequate. Legislators should represent trees as well as people: legislators *should* represent all living things.

Justices Barwick and Gibbs would let legislators take voting rights away from humans. Justice Murphy would block legislators from extending voting rights to non-humans. Both extremes are undesirable. The trick is to frame and interpret Constitutions in a way that leaves decision-makers leeway to do better but not to do worse.

C: Remedies

If ever a Parliament did act to substantially and seriously reduce the existing franchise, the judiciary would possess the jurisdiction and the responsibility to intervene, to declare any such Act a legal nullity, and to preserve the status quo.[15] However, every Parliament retains full authority to move in the direction of more equal suffrage – of representation for all, not only humans, but also life-forms capable of having preferences or at least interests.

This important constitutional truth has often been overlooked because, since the founding of the Commonwealth – and indeed of the several Colonies before it, which became the States – Parliaments have moved consistently in the direction of equal suffrage. It has, therefore, rarely if ever been necessary to invoke this reserve jurisdiction in a matter coming before the federal judiciary, leading some observers to the mistaken conclusion that no such jurisdiction exists.[16] A visiting Martian who sees a car being driven at very slow speeds might not realise that the vehicle has brakes.

How to represent trees, rocks and animals? Where to even begin? It is true that we cannot, at this time and perhaps ever, register the suffrages of trees, or animals, or the biosphere – or even of human beings, if they are of generations not yet born. Yet that does not mean those interests should go unrepresented.

I am not – I hope – one of Shaw's 'lunatics' who 'look forward to votes for children, or for animals, to complete the democratic structure'.[17] As well as proposals to represent nature directly, further conceptual guidance can be found by analogy in attempts to represent humans who, it is still agreed, should not be permitted to vote. The clearest-cut example is citizens under the voting age (or even not yet born), i.e., children – they are numerous

---

15 The High Court finally began to give judgments consistent with this philosophy in *Roach v Electoral Commissioner* (2007) 233 CLR 162 and *Rowe v Electoral Commissioner* (2010) 243 CLR 1.

16 J Allan, 'The three "Rs" of recent Australian judicial activism: *Roach, Rowe* and (No)'Riginalism' (2012) 36(2) *Melbourne University Law Review* 743.

17 G B Shaw, *The Intelligent Woman's Guide to Socialism and Capitalism* (Pelican, 1937).

enough to call election results into question[18]; they clearly deserve to be represented[19]; but because their lack of maturity means their ability to form and express political preferences is dubious. The debate over potential methods of representing children offers something analogous to the debate over representing flora and fauna, although some adjustments are necessary.

Five main types of mechanisms are typically adopted or proposed:

(a)   To count members of the disfranchised group when apportioning seats among regions

Almost all countries base electoral districts on total population: Australia is a rare exception, where compulsory voting makes it feasible to use enrolled voters instead. It is sometimes overlooked that using total population indirectly inflates the votes of parents (or at least of parents who are adult citizens), because their children are counted when drawing electoral boundaries: one district containing, say, 50,000 adult voters with an average of two children each will have the same population as another district containing 100,000 adult voters with an average of only one child each.[20] At the same time, a childless adult residing in the first hypothetical electorate would have proportionately greater voting power (assuming uniform turnout rates) than a parent of six or seven living in the second electorate.

So weighting districts to favour areas with more numerous biota would not solve the problem of the under-representation of those biota, as long as the franchise is exclusive to the humans who exploit or ignore those biota.

18 Unlike mentally incompetent adults and (outside the US) convicted prisoners. Australia in 2014 had 196 prisoners per 100,000 adult population (Australian Bureau of Statistics <www.abs.gov.au/ausstats/abs@.nsf/mf/4517.0>), whereas children under 15 were 19 per cent of the total population (ABS <www.abs.gov.au/ausstats/abs@.nsf/Latest products/3235.0Main%20Features102014>). By contrast, in the United States around 910 per 100,000 of the adult population are in jail: United States Bureau of Justice, *Correctional Populations in the United States* (2014) <www.bjs.gov/index.cfm? ty=pbdetail&iid=5177>. I am speaking here of children who are mentally competent for their age.

19 Unlike non-citizens, whose disfranchisement is most clearly justified if they are short-term visitors; diplomatic representatives of foreign governments; or part of an enemy occupation (see *Australian Citizenship Act 2007* (Cth) s 12(2)). However, their disfranchisement is more problematic if they are long-term residents, especially if they (a) were born in this country, (b) are refugees with a well-founded fear of persecution (which would almost certainly include denial of voting rights) in their home country, and/or (c) are otherwise stateless.

20 In real life, the presence of recently arrived (non-naturalised) migrants complicates this neat illustration. So too would long-term prison inmates and persons declared mentally impaired, except their numbers are usually minuscule, unlike migrants.

Unfortunately, those who live surrounded by the most trees are often the most eager to cut them down. The State Government of Tasmania, for example, seems to believe in salvation by logging and damming, and the Government of Queensland seems equally determined to leave not one tree standing alongside another. So – just as the interests of beef cattle would not be served by vesting plural votes in Lord Vestey – the interests of trees in rural electorates are not always helped by weighting the votes of National Party supporters.

From the Founding until 1868, the United States Constitution specified[21] that slaves were to be counted as three-fifths of a person for federal apportionment purposes. This 'Three-Fifths Clause' gave slave-owners – the 'Slave Power' – more representation in the US House of Representatives, and therefore in the federal Electoral College, than their numbers warranted, precisely because of the slaves they kept as human property.[22] Rather than the solution being to count slaves as full humans, the answer (pending abolition, of course) was not to count slaves at all when apportioning seats.

(b)   To allocate additional votes to members of the enfranchised population who can be seen as reliable proxies for that group, able and motivated to speak and vote on their behalf

More radical are proposals to specifically give individual parents extra votes to cast on behalf of their children: often called 'Demeny voting' after its most famous proponent.[23] In 2011, the Fidesz Party government of Hungary floated, but did not pursue, a similar proposal, but with a maximum of one additional vote per parent – apparently to avoid giving extra representation to Roma (also known as 'gypsies'), a minority who are unpopular among the ethnic Hungarian majority.[24]

Unfortunately, this still only goes as far as humans – and if our concern is a sustainable planet, offering incentives to encourage human reproduction may be counterproductive.

21 Article I, 2(3).
22 Garry Wills argues that when Jefferson won the Presidency over John Adams in 1800, 'at least twelve of his [Electoral College] votes were not based on the citizenry that could express its will, but on the blacks owned by Southern masters': *The 'Negro President': Jefferson and the Slave Power* (Houghton Mifflin Harcourt, 2005) 23.
23 P Demeny, 'Pro-natalist policies in low-fertility countries: patterns, performance and prospects' (1986) 12 *Population and Development Review* 335.
24 L Phillips, 'Hungarian mothers may get extra votes for their children in elections', *The Guardian* (online), 18 April 2011 <www.guardian.co.uk/world/2011/apr/17/hungary-mothers-get-extra-votes>. Pro-natalist conservatives would probably have similar qualms about encouraging *too* much procreation among Muslims, say, or unmarried teenagers.

Further, if a group are unable to communicate their wishes, it may be better to leave them disfranchised – making it clear that they remain unrepresented[25] – than to inflate the votes of some other group who claim to be speaking for them. As the blogger 'Idiot/Savant' notes, Demeny voting:

> encourages the conflation of parents' interests with those of their children (when, as we're seeing on climate change, on IP law, and all sorts of other issues, they are pretty much diametrically opposed).[26]

'Idiot/Savant' points out that giving men with wives a second vote would not be acceptable as an interim step towards female suffrage, and warns of the risk of:

> parents getting tax cuts while their children are burdened with debt, giving a false cloak of consent. [. . . B]etter to have *no-one* pretend to speak for you, than to have them speak falsely 'on your behalf' and imply your consent. [. . . T]he answer is to lower the voting age. Not everyone will count, but more will.[27]

Proxy voting is tolerable where A chooses B as A's own proxy (i.e., A has the long-term capacity to choose but is prevented by absence or ill-health from voting on particular days). It is problematic when A is unable to express a preference, or is not asked, but C decides instead that B shall be A's proxy. And why B, not D or E? Parents at least have an incentive to promote their own children's interests on some matters, but not necessarily other children's interests, and it is even less clear who can speak for animals, plants, and the planet itself. It would be inappropriate for this Court to, say, order votes polled by Green candidates to receive extra loading (120 per cent, 200 per cent?) because they represent nature.

---

25 In the sense that they cannot vote. Their interests still can and should be taken into account by those who can vote.
26 'Hungary considering Demeny voting', *No Right Turn* blog, 19 April 2011 <www. norightturn.blogspot.com.au/2011/04/hungary-considering-demeny-voting.html>. As a conservative Catholic party, Fidesz would almost certainly not regard parents as reliable proxies for their children in other contexts – for example, when women decide their offspring would have better lives if terminated *in utero* (L Cannold, *The Abortion Myth: Feminism, Morality, and the Hard Choices Women Make* (Allen & Unwin, 1998) 115) or when, like the parents of Karen Ann Quinlan or Nancy Beth Cruzan, they want their comatose son or daughter euthanized: *In Re Quinlan* 70 NJ 10 (1976) 355 A 2d 647; *Cruzan v Director, Missouri Dept of Health* 497 US 261 (1990).
27 'Against Demeny voting', *No Right Turn* blog, 6 July 2009 <www.norightturn.blogspot. com.au/2009/07/against-demeny-voting.html> (orig emph).

Any attempt to explicitly weight votes to compensate for under-representation – for example, as has been half-satirically suggested:

> A genuinely radical electoral reform would demand 1 vote for capitalists, 10 votes for male workers, 15 votes for female workers, 20 votes for working-class housewives, 50 votes for recently-arrived migrants and 100 votes for Aborigines.[28]

– runs too high a risk of exaggerating the comfort of the comfortable and exacerbating the affliction of the afflicted.[29]

Paternalism can have horrifying results: the history of the Stolen Generation in Australia shows that. But the risk of paternalism – of persons outside a group of beings deciding what is in the best interests of those beings – may be outweighed when those beings are unable to speak or otherwise communicate on their own behalf.

(c)   To have a binding, judicable, legally-enforceable, entrenched constitutional statement that the rights of that group must be taken into account, even when this involves nullifying legislation to the contrary

Ecuador (2008) and Bolivia (2011) have adopted declarations of the rights of nature.[30] Ecuador's Constitutional amendment has had some legal effect, with a court invoking Article 71 in 2011 to ban a provincial government from road-widening that would violate the rights of the Vilcabamba River.[31] On the other hand, Bolivia's *Law of Mother Earth* (a) is 'merely' a statute and (b) seems to contain loopholes that allow 'extra-officially producing unconventional [shale] gas'.[32]

28 H McQueen, 'Bourgeois Democracy: A Sham' in H Mayer and H Nelson (eds), *Australian Politics: A Third Reader* (Longman Cheshire, 1973) 177.

29 Usually, of course, plural voting and/or restrictive franchises are adopted specifically for the opposite reason – to reward those persons of perceived quality who are successful in life, usually at making money, and who are therefore presumed more diligent, more intelligent, and better able to contribute to public discussion.

30 C Kendall, 'Ecuadorians to vote on Constitution making nature a rights-bearing entity', *The Guardian* (online), 24 September 2008 <www.theguardian.com/environment/2008/sep/24/equador.conservation>; 'The rights of our planet: a vision from Bolivia' <www.worldfuturefund.org/Projects/Indicators/motherearthbolivia.html>.

31 A Schillmoller and A Pelizzon, 'Mapping the terrain of Earth Jurisprudence: landscape, thresholds and horizons' 3(1) (2013) *Environmental and Earth Law Journal* 1, 5–6.

32 'Extra-officially, because there is no law that regulates hydraulic fracturing' Observatorio Petrolero Sur report (July 2014), quoted in D Hill, 'Is Bolivia going to frack "Mother

(d) *To have an aspirational and/or interpretative declaration to that effect*

That is, a declaration that is invoked by legislators to guide their deliberations, and/or by courts to clarify ambiguous legislation and/or to determine which factors executive decision-makers must and must not consider when exercising their discretionary powers – but it cannot be invoked by courts to nullify an Act of the Legislature that explicitly provides otherwise.

Beyond simply trusting legislators to remember these principles when they are legislating,[33] a polity could establish some position or body specifically mandated to scrutinise and report on proposed and existing laws, policies and practices to highlight areas where greater consideration should be given to the interests of the disfranchised group.

'Impact statements' could be issued by a body like the Australian Human Rights Commission (under its various titles), which can intervene in (or even launch) court or tribunal hearings as well as investigate complaints.[34] Professor Thompson of La Trobe University canvasses various options:

> One measure is to appoint an ombudsman with the responsibility of bringing to the attention of government and citizens policies and decisions that could be harmful to young people and future citizens. Another proposal is to allow court actions to be taken on behalf of children and future generations by specially appointed 'guardians of the future'. Another idea is to appoint experts to a commission for overseeing government actions. The Israeli government experimented with a 'Commission for Future Generations' that was not only able to review proposed legislation but also to veto anything judged to be detrimental to the interests of future generations.[35]

In Australia, two jurisdictions (Victoria and the Australian Capital Territory) have explicit statutory Charters of Rights with this effect. Other Parliaments have not gone this far, but have taken some steps in this direction,

---

Earth"?' *The Guardian* (online), 24 February 2015 <www.theguardian.com/environ ment/andes-to-the-amazon/2015/feb/23/bolivia-frack-mother-earth>.

33 Not an utterly naïve hope. Legislative deliberations have been swayed by appeals to the *Magna Carta* (1215) in the UK Parliament and appeals to the *Declaration of Independence* (1776 and 1948 respectively in the United States Congress) and the Israeli Knesset, even though the first is only an ordinary statute and the latter two are not even legally binding documents.

34 *Australian Human Rights Commission Act 1986* (Cth) s 11 ('Functions of Commission').

35 J Thompson, 'Do future Australians deserve a vote now?' *The Conversation* (online), 24 June 2013 <www.theconversation.com/do-future-australians-deserve-a-vote-now-12202>.

for example by criteria for Parliamentary committees to scrutinise Bills,[36] presumptions that ambiguous statutes were not intended to abrogate certain rights.[37] Bolivia's Law of the Rights of Mother Earth entrusts a similar function to an ombudsman, the *Defensoría de la Madre Tierra*.

(e) To widen the laws of standing before courts and/or tribunals

Professor Stone has cogently argued the case for trees to have standing before courts and other tribunals.[38] This is not as radical as it might first appear. The Privy Council has confirmed that temples and images of Hindu deities have legal personality, including standing to sue in their own right.[39] Professor Latour's advocacy of a 'Parliament of Things' extends this thinking.[40]

However, even if laws (both procedural rules as to standing, and substantive rules about rights) are radically reconfigured to take account of non-human interests, still – barring science fictional changes to society and technology – these rights still need to be enforced by humans. Practically speaking, they will be exercised not so much by trees or wombats as by those humans whom other humans (wearing judges' wigs) deem to be more sympathetic to the interests of trees and wombats.

No legal system can ever ensure, literally, that something will be done truly automatically. What the laws can do is to stipulate – by way of conferring open (or at least widened) standing – that any person, even just one out of possibly millions, who wants to promote and further that goal, gets to do so, despite being outvoted. He or she does not require the fiat

36 E.g., Senate Standing Order 23, <www.aph.gov.au/Parliamentary_Business/Committees/ Senate/Regulations_and_Ordinances/Reports/report115/c01>, *Legislative Standards Act 1992* (Qld) s 4 (Meaning of Fundamental Legislative Principles), *Legislation Review Act 1987* (NSW) s 8A (Functions of the Legislative Review Committee with respect to Bills), and *Parliamentary Committees Act 2003* (Vict) s 17 (Scrutiny of Acts and Regulations Committee) all direct members of Parliament on the relevant committee to flag Bills that 'trespass unduly' upon, or lack 'sufficient regard' for, individual rights and freedoms.

37 For example, *Legislation Act 2003* (Cth) s 12(2) (Commencement of legislative instruments and notifiable instruments – presumption against retrospective application).

38 C D Stone, *Should Trees Have Standing? Law, Morality, and the Environment* (Oxford University Press, 3rd ed, 2010). This builds on Professor Stone's 1972 article 'Should trees have standing? Toward legal rights for natural objects,' (1972) 45 *Southern California Law Review* 450, which I would call path-breaking except that the process of breaking paths often does violence to the rights of trees.

39 *Pramatha Nath Mullick v Pradyumna Kumar Mullick* (1925) 27 BOMLR 1064; *Bumper Development Corp v Commissioner of Police of the Metropolis* [1991] 1 WLR 1362.

40 B Latour, *We Have Never Been Modern* (Harvard University Press, 1993) 142–145.

cf the Attorney-General, who presumably represents the majority of the populace.[41]

Which methods are best?

These proposals each have merits and demerits, and in some cases their full adoption is beyond the judicial function of a court. However, one clear judicial function of this court is to interpret the Constitution, the legislation and the common law of Australia. I propose henceforth that in all cases where these laws are open to different readings, we judges choose the reading that gives the maximum consideration to the interests of all life-forms on this planet. We lack the authority to fix the legislative machine, but we can assess its outputs for quality.

D: Authority for this decision

A critic might then ask by what authority the Court makes these orders. Seeking (in vain) written words in the 1901 *Constitution of the Commonwealth of Australia* that spell this out in as many words, they think this means a lack of legitimacy.

In response, I claim three sources of authority:

- First, the doctrine of necessity as a justification when survival is at stake;
- Second, a fundamental right of all living things to have their interests represented, and of sentients to have their wishes and opinions heard, when decisions are made that bind them; and
- Finally, the condition (usually in explicit words, sometimes implied, but always ultimately judicable) that every Parliament's powers are conferred for, and must in some discernible way be exercised for, the peace, order, and good government of its territory.

I: Necessity

My first and paramount justification is the doctrine of necessity. Where the future of the human race – indeed, of all life on this planet – faces a threat, survival is the supreme law.

---

41 Complications arise if, say, a thousand different citizens each put their hand up to further that goal, but in different and inconsistent ways. Thus the case for open standing is stronger when the remedy sought is clear-cut and non-rivalrous: for example, voiding an enactment *ab initio*, as opposed to quantifying damages and then dividing the pot of money among a class of plaintiffs.

It is true that traditionally, necessity had to be immediate[42] and, if it involved any violence, had to be against an active threat (not necessarily a conscious or malicious aggressor).[43]

But these side-constraints must be set aside when the future of the planet is at stake:

> one cannot enjoy freedom of speech, freedom to worship, or freedom of the press unless one first enjoys the freedom to live . . . A mistake in ruling . . . could pave the way for thermonuclear annihilation for us all. In that event, our right to life is extinguished and the right to publish becomes moot.[44]

The nineteenth-century doctrine of parliamentary sovereignty was a crude attempt at attaining this. Only humans could be heard, and it was assumed that each represented roughly equal numbers of other life-forms.

2: Bias

As my brother Murphy notes:

> If the legislature fails to ensure fairness, and there is no constitutional right enforceable in the courts, where is the remedy? There is no provision for referendum or any other remedy which can be initiated by the people.[45] [. . .] History here and elsewhere reveals that with few exceptions, legislators who hold office because of an unbalanced electoral system will not act to change the system.[46]

Too often, saying 'if your democratic rights are denied, appeal to the legislature and the ballot-box' is like saying 'if your leg is broken, walk to the hospital so they can re-set it' or 'if your phone doesn't work, call the phone company so they can fix it'. Electoral reformers may, now and then, succeed in correcting an unequal system using political pressure, but this may require

---

42 *Southwark London Borough Council v Williams and Anderson* [1971] 1 Ch 734.
43 *R v Dudley and Stephens* (1884) 14 QBD 273.
44 *US v The Progressive* 467 Fed Supp 990 (1979) 992 (Warren DJ).
45 However, even those United States that do allow voter-initiated referenda were not permitted by the Warren Court to adopt or keep unequal electorates, even though a popular majority could be construed as passively acquiescing in – or even as actively consenting to – waiver of their right to per-capita representation: *Lucas v Forty-Fourth General Assembly of Colorado* 377 US 713 (1964).
46 *Attorney-General (Cth); Ex Rel McKinlay v The Commonwealth* (1975) 135 CLR 1, 72 (Murphy J).

many years and much good luck. Meanwhile, the actions of a particular unrepresentative legislature (or series of legislatures) may be irreversible.

It is long established that even broad-sounding grants of power are subject to an implied exception that a potentially biased person or body should not be a decision-maker in matters where their self-interest is strong. The traditional remedy is to disqualify biased individuals and let others fill their place. It is not feasible to find a substitute Parliament, but instead judicial scrutiny should be heightened so that legislators are not final judges, even if they are judges, in their own cause.

3: The right to be represented

There are many competing models of 'representation'.[47] We speak of legal 'representation' even though a court's decision is not determined by vote of the parties' lawyers.[48]

The Parliament of the Commonwealth makes decisions that affect not just other humans but other species: whether, for example, to permit the mining of sand on Fraser Island, or the damming of World Heritage sites.[49] Many living things are bound by decisions purporting to be authorised by this Constitution, but have never had a vote on that Constitution on these decisions.

The Roman jurist's dictum that a law which binds all should be approved by all, was endorsed by Bracton as part of our common law heritage.[50] Since unanimous approval is often impracticable to attain, this has been relaxed in practice to requiring consultation with all and approval by a majority of them: but the goal remains.

47 H F Pitkin, *The Conception of Representation* (University of California Press, 1967).
48 Except in the rare cases where both sides agree on some point and thus remove it from the judge's jurisdiction to decide. Even then, a *pro se* litigant or party is just as entitled to refuse their agreement as is one 'represented' by an attorney. Indeed, in some cases unrepresented litigants or (especially) criminal defendants are better placed legally to argue, on appeal, that they did not really 'agree' to waive that point.
49 *Murphyores Inc v Commonwealth* (1976) 136 CLR 1; *Commonwealth v Tasmania* (1983) 158 CLR 1.
50 'That which touches all shall be allowed of all – the law that binds all, the tax that is paid by all, the policy that affects the interest of all, shall be authorised by the consent of all'. Quoted in W Stubbs, *The Constitutional History of England, in Its Origin and Development* (Cambridge University Press, 1874), 5. See also J R Pole, *Political Representation in England and the Origins of the American Republic* (St Martin's Press, 1966) 4: J W Gough, *Fundamental Law in English Constitutional History* (Oxford University Press, 1955).

It is appropriate at this point to distinguish two separate rights, one wider, the other narrower, but both manifestations of the same overarching right to be represented.

(A) THE RIGHT TO BE COUNTED

The wider right is to be taken account of in a decision-making process – to have one's interests weighed and considered by those making the decision.[51] This wider right is shared by all living things. As Bentham noted, the question is 'not, Can they reason? nor Can they talk? but, Can they *suffer*?'[52]

(B) THE RIGHT TO BE HEARD

The narrower right deriving from the right to be represented is a right to be heard – to be invited (not merely permitted) to express one's views about one's own interests, and about the best way of reconciling competing interests for the common good. This right expresses itself in different institutional forms depending on the capacity of the being concerned and the nature of the decision. Dead humans, or live plants, cannot have a vote or even a voice no matter what the forum. But dogs can have their interests taken into account by decision-makers, although they cannot of course vote

51 Different thinkers propose a variety of formulae for how equally weighed interests should be aggregated. The current debate in modern democracies can fairly be summarised as setting those who follow John Rawls's view, that the worst-off individual(s) should be the benchmark for comparison among different policies, against those who follow Jeremy Bentham's traditional utilitarian precept that the 'greatest happiness of the greatest number' means that the median should be decisive. In other words, should the 51st percentile, or the bottom one per cent, be the litmus? An older, aristocratic view associated with Nietzsche – that societies and institutions should be comparatively judged by how far they enable the most talented one per cent to attain excellence: that 'rare human types are entitled to liberties that ought to be absolutely off-limits to . . . ordinary . . . men and women': P Berkowitz, 'Other people's mothers: The utilitarian horrors of Peter Singer', *The New Republic,* 10 January 2000, 27–37, 31 – is now out of fashion. Note, however, that all of these algorithms (and others that could be devised to combine them – say, 'maximum the welfare of the bottom ten per cent' or whatever) are all compatible with weighing people equally, at least in the sense that they don't start off with any controlling prejudgment that particular identifiable individual(s) should be decisive. (Kenneth Arrow termed this the 'non-dictatorship' criterion for assessing public-choice methods: *Social Choice and Individual Values* (Wiley, 1951)).

52 J Bentham, *Introduction to the Principles of Morals and Legislation* (2nd ed, 1823), Ch 17.

nor express their interests intelligibly.[53] Children cannot vote,[54] but they are invited by statute to express their views where possible.[55]

Giving every affected entity a right to be heard is a principle that unifies the fundamental principles of representation in the legislative process and of a fair hearing *(audi alteram partem)* in judicial and quasi-judicial processes. Everyone who can speak for someone who can suffer – him or herself or another who is silent – shall be heard. Everyone who can suffer as a result of our decisions, must be represented.

It will immediately be objected, perhaps mordantly, that it is not possible to consult animals, birds or fish, let alone trees. But this is not an insuperable obstacle. Some have advocated that dead human beings should be included in the franchise.[56] Whoever would dismiss this idea should remember that the story of Antigone – often cited as a foundational text in the development of rights thinking[57] – involved the right of a dead rebel, to be properly buried.

Some entities are capable of expressing their interests in rudimentary ways (e.g., dogs, apes, and dolphins). Others cannot (e.g., fish and trees). However, not all biota have the capacity to express their interests – or, less

53 For example *Hodgens v Gunn; Ex parte Hodgens* [1990] 1 Qd R 1 and *Isbester v Knox City Council* [2015] HCA 20.

54 Although to some extent this is circular, since one criterion for defining 'adults' is those old enough to vote: those over 21 before the 1970s, those over 18 in most countries today, those over 16 in some jurisdictions. The Philippines during the Marcos era allowed 16- and 17-year-olds to vote in referenda, but their votes were advisory only, and cast in separate ballot-boxes. The numbers thereof were tallied and published – like the votes of all citizens in New Zealand's voter-initiated referenda: see *Citizens Initiated Referenda Act 1993* (NZ) s 40(2)(a) – but the legal result (rigged, admittedly, by the Marcos regime anyway) was determined solely by the ballots of those 18 and older.

55 C/f *Family Law Act 1975* (Cth) ss 68L (court may order independent representation of child's interests in divorce and custody cases) and 60CE (children not required to express their views to the court).

56 G K Chesterton argued that tradition 'is democracy extended through time. [. . . It] may be defined as an extension of the franchise. Tradition means giving votes to the most obscure of all classes, our ancestors'. *Orthodoxy* (Bodley Head, 1927), 83. Professor Tim Mulgan has brainstormed other institutional mechanisms that might give the deceased an ongoing suffrage: 'The place of the dead in liberal political philosophy' (1999) 7(1) *Journal of Political Philosophy* 52.

   Compare the *Vermillion Accord on Human Remains* (adopted 1989 by the World Archaeological Congress at South Dakota), <http://worldarch.org/code-of-ethics>; the *Queensland Criminal Code* 1899 s 236 ('Misconduct with regard to corpses'); and the 1874 (but not the 1999) Swiss *Constitution*, art 53(12) (civil authorities 'must ensure that every deceased person can have a decent burial').

57 M Cranston, *What Are Human Rights?* (Bodley Head, 1973) 9; C Douzinas, *The End of Human Rights: Critical Legal Thought at the Turn of the Century* (Bloomsbury, 2000) 23–26.

prejudicially, humans, presently the dominant species on this planet, lack the capacity or apparatus to read and interpret the interests of many species. Dogs can certainly make their immediate wishes known, and even plants can react (with what polygraphs detect as pain) when they feel or (allegedly) even apprehend cutting,[58] but many species do not show up at all on humankind's radar.

Nor are the interests of future generations necessarily obvious. Individual ungulates want to eat and to reproduce, but if they reproduce too fecundly they may leave too little food for themselves and/or their children to eat in future. The interests of an individual tree may conflict with those of all trees in a region (for example, over a controlled burn to pre-empt bushfires or to disperse seeds). There is no Lorax who can 'speak for the trees'.[59] It is a category mistake to assume that only entities that can vote should be represented (in the sense of their interests counting).

But they still have interests. A sequoia tree has an interest in not being chopped down. Fish and plants are not indifferent as to whether they feel pain, live, die, or reproduce their kind. Plants grow towards the light. Fish swim towards warmth and food.

To say that inability to speak means lack of interests that need protecting is to confuse might with right.

This of course raises many questions as to how their interests should be weighted. Should every individual animal count equally, per capita? Should four cockroaches outvote three blue whales? Should size count? Brain size only? Or body size too? What about a subclass? A particular population? Modern zoological taxonomy has moved away from Linnaeus's early view that kingdoms, classes, orders, families, genera, and species could be neatly kept at distinct levels, adopting instead the more flexible concept of a nested clade.[60] This may be more accurate, but it would complicate any attempt to enfranchise animal species.

For the present, then, we must make the best we can with the franchise for all humans, but this should not be regarded as a final stopping point. The law should continue to seek ways to ensure that (i) all living beings are represented (that is, considered) and that (ii) all rational sentient beings should be heard (that is, enfranchised). Not every living thing can be accorded

---

58 See C Backster, *Primary Perception: Biocommunication with Plants, Living Foods, and Human Cells* (White Rose Millennium Press, 2003), claiming his experiments with polygraphs show that plants not only experience but also apprehend pain.
59 Dr Seuss, *The Lorax* (Random House, 1971).
60 J Huxley, 'The three types of evolutionary process' 180 *Nature* (1957) 454.

a meaningful right to *vote*. But every living thing has a right to be *represented* – to have its interests taken into account.

What the constitution requires is not 'one human being, one vote'. Rather, it requires votes in proportion to life, to be realised progressively as this is discovered. Interests must be represented. All biota – all flora and fauna – have interests.

4: Peace, order, and good government

All powers vested in the federal Parliament by the Constitution are conferred – as is standard practice throughout British Dominions – for the 'peace, order and good government' of the Commonwealth. It is true that these words permit a wide range of diverse and even contrasting opinions as to the public good, which means they are not regularly invoked by the courts to override decisions made by majority vote.

It is a frequent misconception that, because discretion allows a wide range of choices, even opposing choices, it therefore allows anything and everything. Parents, by analogy, have lawful discretion to decide whether their children shall attend public schools, attend private schools, or be home-schooled: but if they refused to have their children educated at all, the courts would step in.

But they are not meaningless; they have limits.[61] If – to devise an example that is fantastic in every sense – a decision-maker were to use its powers to render a territory depopulated, forcing its inhabitants to leave their home-lands, it is hard to see how this could ever be considered to serve the 'peace,' the 'order,' or the 'good government' of that territory.[62]

---

61 In *Union Steamship Co v King* (1988) 166 CLR 1 at 10, the High Court acknowledged that these are usually not considered limiting words; but Justices went on to leave open whether Australian Parliaments might be bound by similar restrictions derived 'by reference to rights deeply rooted in our democratic system of government and the common law'.

62 In *R (Bancoult) v Secretary of State for Foreign and Commonwealth Affairs* [2001] QB 1067, [2001] 2 WLR 1219, the United Kingdom High Court invalidated subordinate legislation, made purportedly under the Royal Prerogative, to exile all inhabitants from the Chagos Islands (Diego Garcia). L J Hooper and J Cresswell scathingly held [2008] UKHL 61 at [142] that 'The suggestion that a Minister can, through . . . an Order in Council, exile a whole population from a British Overseas Territory, and claim that he (*sic*) is doing so for the "peace, order and good government" of the Territory, is to us repugnant'.

E: Conclusion

It is an interesting thought-experiment to wonder 'what if?'[63] – to wonder how great minds of past decades or centuries might have jumped ahead had they been given some intellectual breakthrough. John Stuart Mill, for example, believed that had the idea of a Single Transferable Vote 'by good fortune suggested itself to the enlightened and patriotic founders of the American Republic . . . democracy would have been spared . . . one of its most formidable evils'.[64] Umberto Eco's monkish protagonist mused, in the fourteenth century, that secular liberal democracy with universal suffrage might be a viable system: but then, Eco was writing fiction, and with the advantage of six centuries' hindsight.[65]

It is constitutionally mandatory that the Parliament furnish evidence of having thought about this issue – both at the retail level of assessing each piece of legislation against the criterion of earth law, and at the wholesale level of investigating political structures that can give this Court, the people and the planet some assurance that such considerations are 'baked into' the legislative process. (The more work it puts into the second, the less it will need to overcome judicial suspicion of the first).

This Court does not today offer final, prescriptive solutions. It is doubtful whether any Court will, due to the nature of its adjudicative function, be in a position to offer final, prescriptive solutions even in future centuries. But this Court does order that the Parliament direct its mind to the question.

Having districts representing only humans (howsoever weighted) will be as unsuitable for Australia in future centuries as would be a Parliament composed only of barons, or of the appointees of rotten boroughs, for Australia today.

---

63 Although see the caveats about such speculating expressed by Phillip Bobbitt in *Constitutional Stupidities, Constitutional Tragedies* (1994) 18 (a 'vapid . . . parlo[u]r game').
64 J S Mill, *Representative Government* (1861) Ch 7. As it happens, the concept of STV occurred contemporaneously but (it seems) independently to Thomas Hare in England (1857) and CCG Andrae in Denmark (1865), just as Newton and Leibniz each came up with calculus on his own.
65 '[William of Baskerville] cleared his throat, apologized to his listeners, remarking that the atmosphere was certainly very damp, and suggested that the way in which the people could express its will might be an elective general assembly'. U Eco, *The Name of the Rose* (W Weaver trans., 1980).

Chapter 7

Wild negligence

Donoghue v Stevenson

Bee Chen Goh and Tom Round

Commentary

Donoghue v Stevenson [1932] AC 562 has entered the secular canon of common law judgments, both in its outcome and in its methodology. Before it,[1] the common law required much lower standards from manufacturers: in most cases, only active incursions of violence or deception were liable as torts, and positive duties to take care could be imposed only by voluntary[2] contracts. Duties in contract were (and to some degree still are) limited by privity, allowing only the original parties to sue to enforce their bargain. Mrs Donoghue[3] would have fallen between the cracks of the old law: she had no contract with either the manufacturer or the retailer, since her friend had purchased the drink.

Previously, centuries of common law had presupposed 'buyer beware' – a simple rule that worked fine when buying fresh apples or pies from a vendor at the town fair, but less well when mass-produced products are assembled in factories miles (even whole nations) distant, and sealed in impenetrable containers required by hygiene laws.

Donoghue changed that.[4] Henceforth, manufacturers – indeed, all adults – owe duties of care to people (whom we may never have contracted with, nor

---

1 Thanks to Andy Gibson for his kind assistance on this point.
2 One could say that assuming a duty to avoid torts and statutory breaches is also voluntary, through availing yourself of the benefits of citizenship, and/or by voting to elect legislators. However, the link between an individual's consent and a potentially onerous duty is closest and clearest in contract law.
3 Mrs Mary M'Alister, an elderly widow, also sued (as Scots law allowed) under her birth surname.
4 Justice Cardozo (then on the New York State Court of Appeals, later elevated to the US Supreme Court) had reached a similar conclusion 16 years earlier in *MacPherson v Buick Motor Company* (1916) 217 NY 382. But American and British Empire jurisprudence were more isolated from each other in those days.

even met) who are vulnerable to harm from our passive negligence as well as to our active trespasses.[5]

I say 'henceforth' because, although the intricate kabuki of the common law judicial method required judges (more so in the 1930s than today) to politely pretend that this result had always been embedded in the common law, waiting to be unearthed, in practice the judgment was a substantial departure from existing expectations. This is a strike against judgments like *Donoghue* for some, but not all, jurists. When judges are praised by the consensus of hindsight it is no less often because they had the courage to replace outmoded precedents with better ones, than because they stood firm in protecting still-valid precedents against short-term spasms of change.[6] Indeed, when the precedents became 'long-hallowed' *only* because the groups they disadvantaged were denied the vote and other rights of citizenship, it seems problematic to complain that judges have refused to respect those precedents.[7]

The outcome of *Donoghue* was certainly not a matter of inescapable deduction: as one former High Court Justice noted, 'Lord Buckmaster's dissent showed that the [pre-*Donoghue*] cases were logically explicable without a general principle of negligence'.[8] Today, in hindsight, 'it is easy to overlook the facts that the law had *not* included the rule until May 1932 [. . .], that it took much argument to establish it, and that two Law Lords thought it was extremely bad law'.[9] Instead, what vindicates *Donoghue*'s neighbour principle was and has been considerations of policy and justice. Indeed, Lord Justice Atkin himself came close to forgetting to ritually recite 'Simon Says . . .' before declaring the new state of the law:

---

5 Tort law before *Donoghue* focused overwhelmingly on active incursions, absent a contractual duty to take extra care, except in rare circumstances like dealing with an inherently dangerous thing: *Rylands v Fletcher* (1868) LR 3 HL 330. Neither snails nor ginger-beer, by contrast, are dangerous in themselves.

6 Into which category fall many (*Plaintiff M70/2011 and Plaintiff M106/2011 v Minister for Immigration* (2011) 280 ALR 18 [the 'Malaysia Solution' case] – though not all (*Al-Kateb v Godwin* (2004) 219 CLR 562: *Plaintiff M68/2015 v Minister for Immigration* [2016] HCA 1) – of the High Court's and the Federal Court of Australia's decisions involving asylum-seekers.

7 For this reason, complaints that the leading Australian and United States judgments in the direction of racial equality – *Mabo v Queensland (No 2)* (1992) 175 CLR 1 and *Brown v Board of Education* 347 US 483 (1954) – drastically overturned long-settled legal expectations, are accurate but, morally, beside the point.

8 Justice M H McHugh, 'Democracy and the Law', address to the Australian Bar Association Conference (London), 5 July 1998 <www.hcourt.gov.au/assets/publications/speeches/former-justices/mchughj/mchughj_london1.htm>.

9 A MacAdam and J Pyke, *Judicial Reasoning and the Doctrine of Precedent in Australia* (Butterworths, 1998) 232 [13.16].

It is said that the law of England is that a poisoned consumer has no remedy against the negligent manufacturer. If this were [so . . . ,] I would consider the result a grave defect in the law and so contrary to principle that I should hesitate long before following any decision to that effect.

But notwithstanding its birth pangs, the *Donoghue* principle has survived eight decades; been voluntarily adopted by courts throughout the common law world; been consistently extended by judges to cover borderline cases; and been copied by legislative drafters.[10] By any measure (short only of, perhaps, ratification by a large majority in a popular referendum) Lord Atkin's view has prevailed.

However, Professor Goh has spotted the irony inherent in this much-praised judgment. We now revere Lord Atkin for being on the right side of legal history by extending our legal duties of care, and for doing so on an explicitly moral basis. Yet his Lordship's original judgment was blind to the lot of the unfortunate snail.

Imagine reading a legal judgment from a court in a slave-owning society. No need to time-travel back to Sparta or Rome: the United States South only 150 years ago, or Yemen and Saudi Arabia only 60 years ago, provide (uncomfortably) closer examples. Suppose the defendant, a vintner, killed one of his slaves in a fit of anger. Parts of the corpse ended up floating in a wine bottle that the plaintiff purchased. The deceased slave can even be identified – by a serial number branded or tattooed on her finger, perhaps.

Yet, as we read the judgments, we notice something disturbing: the judges' concern is focused solely on the disgust and nausea felt by the buyer. We are not denying the shock of finding one has ingested wine tainted with decaying human remains: we just feel there should be some acknowledgement that a fellow living creature has also suffered an even worse harm – that the loss of a slave's life should be included in our moral calculus, alongside the loss of the wine-buyer's lunch.

Compassion for all life is difficult to dispute in principle. Laws that permit humans to kill other humans in anger,[11] or to kill other living creatures

---

10 See, e.g., the *Australian Consumer Law* (Schedule 2 to the *Competition and Consumer Act 2010* (Cth)), especially ss 3 (defines 'consumer' broadly), 54 (acceptable quality) and 55 (fitness for disclosed purpose).

11 Publius Vedius Pollio (died 15 BC), a Roman aristocrat, was infamous for feeding his slaves alive to his pet eels if they displeased him. On one occasion, the Emperor Augustus was so offended that he intervened personally; but otherwise Pollio was deemed to be acting within his legal rights. R Syme, 'Who was Vedius Pollio?' (1961) 51(1) *Journal of Roman Studies* 23.

Alabama's State Constitution of 1819 did stipulate that 'Any person who shall maliciously dismember or deprive (*sic*) a slave of life, shall suffer such punishment as would

for amusement,[12] seem obviously illegitimate. But if compassion for all life means legally-enforced protection for all life, then it is difficult – perhaps impossible – to apply universally. Brutally put, some life-forms must die so that others can live. As Abraham Lincoln noted in 1864, when a 'shepherd drives the wolf from the sheep's throat,' the sheep 'thanks the shepherd as his liberator, while the wolf denounces him for the same act as the destroyer of liberty'.[13] The shepherd's decision here is zero-sum: he has not merely spoiled the wolf's sport, but condemned it to die by starvation.

Disease-spreading parasites also seem to strain the principle of compassion. If all life deserves legal protection, then here is a left-liberal pundit calling for the equivalent of genocide:

> imagine that we could wipe out just a handful of mosquito breeds – the most unpleasant ones. What if we could press a button to destroy an invasive, deadly species like . . . the awful Anopheles mosquitoes, which transmit malaria and seem to have evolved as human parasites? If we got rid of these disgusting critters, wouldn't everyone be better off? No one bit their nails when we cleared the world of polio and rinderpest. Should mosquitoes get special treatment just because they're insects?[14]

I am no expert on Buddhist philosophy and jurisprudence. Like most tertiary-educated westerners, I'm vaguely aware that cows in India are not eaten for meat by the Hindu majority,[15] and that Jains wear special shoes

---

be inflicted in case [of . . .] a free white person [. . .]; except in case of insurrection of such slave' (Art VI ('Slaves') s 3). But prosecutions were rare and penalties were light. In Missouri, for example, not one of the (few) prosecutions of slave-owners for crimes against their slaves resulted in a conviction: H C Frazier, *Slavery and Crime in Missouri, 1773–1865* (McFarland, 2009) 127.

12 See the rewritten judgment of Justice Rogers in *Levy v Victoria* (Chapter 20).

13 'Address at Sanitary Fair', Baltimore, Maryland, 18 April 1864.

14 D Engber, 'Let's kill all the mosquitoes: now is the time to wipe the disease-carrying critters off the face of the Earth,' *Slate* (online), 29 January 2016 <www.slate.com/articles/health_and_science/science/2016/01/zika_carrying_mosquitoes_are_a_global_scourge_and_must_be_stopped.html>. By contrast, Jeffrey Powell (Professor, Yale University) proposes a more pacific (if paternalistic) solution: 'Rather than just killing mosquitoes, a more effective and lasting strategy would be to genetically change them so they can no longer transmit a disease-causing microbe'. 'To fight Zika, let's genetically modify mosquitoes: the old-fashioned way to fight insect-borne diseases', *The Conversation* (online), 29 April 2016 <www.theconversation.com/to-fight-zika-lets-genetically-modify-mosquitoes-the-old-fashioned-way-57789>.

15 Although United States anthropologist Marvin Harris famously argued that there are sound utilitarian, environmental reasons why farmers in India do not kill their cattle for food: 'more calories go up in useless heat and smoke during a single day of traffic jams in the United States than is wasted by all the cows of India during an entire year [. . .] If you

that minimise their risk of crushing ants. But Thailand, Sri Lanka and other predominantly Buddhist nations in Southeast Asia do not demonstrate any qualms about eradicating mosquitoes or smallpox.

Perhaps, then, compassion for life does not translate into an absolute ban on ending the lives of other creatures. It could coexist with a right to kill when (and only when) necessary to protect life (for food, or for removal of predators or parasites) and a duty to minimise the pain involved – a sort of Just War doctrine applied across species.[16] More minimally, it could require that the presumptive right of every creature to live should be taken into consideration, even if it ends up being overridden, in the way that administrative decision-makers are required to take account of objections even if they end up overruling them.

As well as constraining actions and decision-making processes, an ethic of compassion for life could indicate the proper attitudes to hold. Christian philosopher Wendell Berry has argued that:

> To live we must daily break the body and shed the blood of Creation. When we do it knowingly, lovingly, skilfully, reverently, it is a sacrament. When we do it ignorantly, greedily, clumsily, destructively it is a desecration.[17]

Native American hunters, enacting rituals expressing both gratitude and regret for the animals they killed,[18] are a better model to emulate than Clarence Darrow, who quipped: 'I have never killed anyone, but I have read some obituary notices with great satisfaction'.[19]

Professor Goh's rewritten judgment seems to be more in this spirit. I do not read her as demanding legal sanctions against the human responsible

---

want to see a real sacred cow, go out and look at the family car'. *Cows, Pigs, Wars and Witches: The Riddles of Culture* (Random House, 1974) 31.

16 Thus, Just War doctrine allows soldiers to wound or kill enemy combatants who refuse to surrender; but not to torture or humiliate them if they do surrender, nor to deliberately wound or kill enemy civilians. For a representative statement, see *Catechism of the Catholic Church* [2302–2317]. For similar views shared by the late philosopher Norman Geras, a non-Catholic socialist, see 'Our morals: the ethics of revolution' (1989) *Socialist Register* 185.

17 W Berry, *The Gift of Good Land: Further Essays Cultural and Agricultural* (Counterpoint, 1981) 281.

18 Compare Sinclair Lewis's reflections in the American Civil War: 'Slavery had been a cancer, and in [1860, there] was known no remedy save bloody cutting. There had been no X-rays of wisdom and tolerance. Yet to sentimentalise this cutting, to justify and rejoice in it, was an altogether evil thing'. *It Can't Happen Here: A Novel* (Doubleday, 1936) Ch 13.

19 C Durrow, *The Story of My Life* (Da Capo Press, 1932) Ch 10.

for carelessly drowning the snail, but rather as reminding us to remember the value of the snail's life.[20]

How far one's duty to feel compassion translates into a mandatory duty not to kill is open to debate. It is not hard to feel suspicion that decision-makers who are supposed to protect lives as far as possible have not really made every effort or explored every avenue when they and their own are not at risk of death. This view finds expression, for instance, in the widely-held (but inaccurate) suspicion that countries would never go to war if politicians' own sons and daughters were the ones fighting as soldiers, especially as conscripts.[21]

In light of this, Professor Goh's discussion of reincarnation is highly relevant. It seems to me that Western law would almost certainly give greater weight to a snail's rights if belief in reincarnation were widespread – if the judge could look upon the snail and think 'That might have been me, or one of my loved ones'.

20 Viewed one way, demands for (heavier) legal punishments can seem vengeful, petty or even bloodthirsty, confirming Nietzsche's warning to 'Beware those in whom the desire to punish is strong'. *Thus Spoke Zarathustra* (1883: rep Viking, 1954) 100. Thus journalist Phoebe Maltz Bovy notes that 'privilege' is 'not usually the most helpful term' for discussing lenient rape sentences because it 'suggests that it would be a better world if *everyone* were subject to police brutality, or if *everyone* with a drug problem were treated as criminals rather than patients': 'Brock Turner's Privilege: the law should apply to all, everywhere, equally,' *The New Republic* (online), 9 June 2016 <https://newrepublic. com/article/134102/brock-turners-privilege>.

But such demands can be viewed more positively – at least when the alternative is either no punishment at all or a laughable 'slap on the wrist'. Thus the Black Lives Matter (BLM) campaign seeks the prosecution of police officers and others who kill African Americans, at least in part to deter others in future from resorting too quickly to deadly force. As Irish journalist Nuala O'Faolain noted, condemning a four-year sentence for a man who impregnated a 14-year-old girl: 'The length of a sentence makes no difference to the harm already done. But sentences are all we have. They are the sanctions which express our estimate of this or that crime'. 'Crime and Punishment? The X Case Sentence' (1992), report in *Are You Somebody?* (Sceptre, 1996) 392.

21 Compare the widely-covered song *Fortunate Son* by Creedence Clearwater Revival (1969) ('I ain't no Senator's son') and the film *Fahrenheit 9/11* by Michael Moore (Lionsgate, 2004). In fact, politicians and opinion leaders like Herbert Asquith, Arthur Balfour and Rudyard Kipling did not embrace pacifism after losing sons in the First World War. More recently, John McCain, Jim Webb, Sarah Palin and seven members of Congress were parents of soldiers serving in Iraq: K Kiely, 'Lawmakers have loved ones in combat zone', *USA Today* (online), 23 January 2007 <http://usatoday30. usatoday.com/news/washington/2007–01–22-iraq-congress_x.htm>. Josef Stalin even refused to ransom his own son when Yakov was taken POW by the Germans in World War II.

In this sense, Professor Goh's view is part of a family of theories that hold that justice is best secured by requiring decision-makers to share the risks that their decisions impose on others.

These theories do not consider a mere abstract commitment to justice to be sufficient; nor do they rely on sympathy or empathy as an emotion. Rather, they work on the basis that decision-makers must actually have, or risk having, their own feet held to the fire. As Justice Robert Jackson of the United States Supreme Court said, 'there is no more effective guarantee against arbitrary and unreasonable government than to require that the . . . law which officials would impose upon a minority must be imposed generally'.[22]

The problem facing Professor Goh's commendable move to instil greater compassion into our legal system is that belief in reincarnation as an empirical concept is not widespread in the West. No notable Western philosopher in two thousand years, other than Arthur Schopenhauer, has seriously espoused it.[23] This may well represent 'a refusal to take seriously the wisdom behind the myth of reincarnation, a belief of . . . two-thirds of humanity including . . . the ancient Greeks'.[24] But nonetheless it is hard-wired into the law of those Western nations that dominate the global legal scene. Even India and Japan, two of the world's most populous nations,[25] predominantly Hindu and Buddhist respectively, do not base their civil or criminal laws on a belief in reincarnation. *A fortiori*, Western nations are (today) not inclined to legislate concepts that cannot be scientifically verified in the laboratory.[26] Some areas of wild law – climate disruption and biodiversity – benefit from this faith in replicable empirical experiments, but the rights of snails would not.

Certainly many Western philosophers – most famously Peter Singer – defend the rights of animals.[27] But Singer and his school do not believe in reincarnation, nor base their philosophical structure on it. On the contrary, Singer is zealously utilitarian and hostile to religious or metaphysical constraints that might impede a society from maximising the greatest happiness

---

22 *Railway Express Agency v New York* 336 US 106 (1949) 112.

23 P Abelson, 'Schopenhauer and Buddhism' (1993) 43(2) *Philosophy East and West* 259; B Magee, 'A Note on Buddhism' in *The Philosophy of Schopenhauer* (Oxford University Press, 1983).

24 R R Singh, *Death, Contemplation and Schopenhauer* (Ashgate, 2012) 64.

25 And both less officially hostile to religious 'superstitions' than the world's most populous nation, the People's Republic of China.

26 This does not make them untrue: but it does make them harder to legislate in the face of opposition by large numbers of skeptics.

27 *Animal Liberation: A New Ethics for Our Treatment of Animals* (HarperCollins, 1975).

of the greatest number of its members.[28] He includes animals among those members, but only to the extent that they approximate humans in sentience.[29] Apes make the grade; snails do not.

It is an interesting thought-experiment, though, to speculate how credible proof of reincarnation – if ever established to the satisfaction of courts and legislatures – would shake up legal systems like Australia's. Succession law would need to be revised, both in the private sphere of wills and estates, and in the public sphere of the Crown: 'the [monarch] is dead, long live the monarch' would be applied literally, and rather than verifying the true paternity of each newborn prince or (since 2015) princess, might instead involve a search for the new embodiment of the deceased king or queen, similar to the identification of each new Dalai Lama in Tibet.

Abortion and stem-cell research could well cease to be controversial (as indeed they are uncontroversial in Japan[30] – and even in South Korea, which has a large Christian minority but is culturally Buddhist).[31] On the other hand, so too might the death penalty, which is also widely accepted in Japan. Like the Time Lords in *Doctor Who* – a humanlike race who are capable of 'regenerating' in a new body to escape fatal injuries – we might come to regard death as a 'man flu'.[32]

28 'We have no need to postulate gods who hand down commandments to us, because we understand ethics as a natural phenomenon'. P Singer, *Ethics* (Oxford University Press, 1994) 5. Singer's use of 'gods' as a synecdoche for all religious or metaphysical beliefs betrays a bias in favour of Western (Judaeo-Christian/Islamic) definitions of religion: broadly construed, his naturalism excludes a belief in karma – or any other supernatural consequences that persist after you die – as well. 'Religions are united not by belief in God but by belief in life after death'. S Batchelor, *Buddhism Without Beliefs: A Contemporary Guide to Awakening* (Bloomsbury, 1997) 34.
29 'The capacity for suffering and enjoying things is a prerequisite for having interests at all [. . .] in any meaningful way. It would be nonsense to say that it was not in the interests of a stone to be kicked along the road by a schoolboy. A stone does not have interests, because it cannot suffer'. P Singer, *Practical Ethics* (Cambridge University Press, 1979) 7.
30 Japan combines absence of legal restrictions with the widespread *mizuko-kuyo* ceremony for aborted and miscarried foetuses: J Wilson, *Mourning the Unborn Dead: A Buddhist Ritual Comes to America* (Oxford University Press, 2009); W LaFleur, *Liquid Life: Abortion and Buddhism in Japan* (Princeton University Press, 1993).
31 'Unlike its Asian neighbours, Korea has a huge and powerful Christian community [. . .] Yet this has not translated into a moral movement against stem-cell research. [. . .] Korean Christians – who have spent that century under occupation, at war, and then rebuilding a destroyed and colonized nation – have been busy with more practical moral questions of human rights, justice, and economic development'. D Plotz, 'The Seoul of clones: solving a biotech mystery – why South Korea leads the world in stem-cell research' *Slate* (online), 19 October 2005 <www.slate.com/id/2128361>.
32 Clara Oswald: 'You killed that man!' The Doctor: 'It was him or you! [. . . T]he difference is, when *you* die, you stay dead. [. . .] Death is Time Lord for man-flu'. *Doctor Who*, series 9, episode 12, 'Hell Bent' (broadcast 5 December 2015).

Western jurists who embrace a Buddhist ethic are rare: and the most famous example may have been idiosyncratic rather than representative. Christmas Humphreys, Britain's most prominent convert, who drafted the widely-accepted 'Twelve Principles of Buddhism' (1942), saw no contradiction between his beliefs and his day job as a prosecutor, even though this included procuring convictions in capital cases – most famously (and quite dubiously) of Derek Bentley, Timothy Evans and Ruth Ellis. At the same time, Humphreys only accepted his 1968 appointment as a judge after, and because, Britain had suspended the death penalty.[33]

It seems that his strict approach to role ethics helped Humphreys reconcile both *dharma* and *karma*. He believed that 'as a prosecutor, it was his task merely to establish guilt. Sentencing was, of course, a matter for the judge. To Humphreys, it was karma that had made him a prosecutor, just as it was karma that had lead criminals to commit crimes'.[34] As Humphreys later told Ruth Ellis's son, 'I have my part to play. Defending counsel has his. The judge has his. The jury have theirs . . . [M]ercy never came into it'.[35] Asked how he, as a Buddhist and a judge, could impose even a non-capital sentence, Humphreys replied: 'I am the man in the dock'.[36]

Donoghue v Stevenson

Judgment

Before Goh J

This case has been brought by the Plaintiff, Australian Earth Laws Alliance Ltd (AELA), on behalf of a dead snail found inside a ginger-beer bottle, against the Defendant, Svenson & Co, for negligence causing it fatal injury.

Alas, the dead snail, of course, cannot institute this suit in its own right. Death is a terrible thing. This Court takes cognizance that AELA is able and willing to file this suit on behalf of the dead snail as a test case on contemporary principles of law based on an ecocentric conception of the right of a sentient being to life.

The facts of this case, briefly, are as follows. One Mrs Donoghue went with a friend to a café whereupon she consumed some ginger-beer mixed with ice-cream. The ginger-beer came in an opaque bottle and so no one

---

33 C Humphreys, *Both Sides of the Circle: The Autobiography of Christmas Humphreys* (Allen & Unwin, 1978) 234.

34 D P Horigan, 'Christmas Humphreys: a Buddhist judge in twentieth-century London' (1996) 24 *Korean Journal of Comparative Law* 15.

35 M Weller, *Ruth Ellis: My Sister's Secret Life* (Robinson, 2005) 202.

36 H Tennyson, 'Interview with Christmas Humphreys' (1977) 52 *The Middle Way* (Buddhist Society of England) 26.

could see its contents. No one would, of course, anticipate that it contained a dead snail. When Mrs Donoghue continued to pour to empty the ginger-beer over her ice-cream, fragments of the decomposed snail appeared.

The Plaintiff's claim on behalf of the dead snail is founded on the tort of negligence, and the Plaintiff urges this Court to impose tortious liability on the Defendant based on Buddhist principles.

It is alleged by the Plaintiff that the Defendant owed the snail a duty of care not to cause it injury, let alone fatal injury, and that this snail, like all sentient beings, was otherwise entitled to a right to life.

The Defendant claims it has had no knowledge of how the snail died, nor how it managed to get inside the ginger-beer bottle. The Defendant further denied responsibility on the ground of negligence, as the law currently stands, and alleges that the Plaintiff's suit is vexatious and frivolous.

At the outset, for the sake of clarity, may I state that this Court will not be deterred by novel arguments in establishing new frontiers in tortious liability, especially with regard to the evidence posited before me.

What I have been asked to consider here is this: can the principles of the law on negligence be (1) re-interpreted or (2) extended to include non-human but sentient beings to give rise to a new or expanded foundation based on ecocentric sensibilities?

Given global movements on the law relating to the environment, this legal question posed above is quite timely. We are, up to now, too accustomed to an anthropocentric worldview of the environment. And we are none the better for it. Perhaps, I may add, we human beings are the authors of our own misdeeds against the natural environment.

As part of this contemporary global environmental consciousness, we need to take heed of wild law philosophy embedded within Earth jurisprudence which compels us to direct our attention to ecocentric sensibilities. In this context, human beings are perceived of as being part of the larger earth community, as opposed to anthropocentrism focusing on human beings as the central or exclusive functionary. Indeed, wild law enlightens us to the legal conception that creatures of the earth are treated as subjects of the law, rather than as objects of the law. And as subjects they possess rights, and demand our respect. In this vein, Buddhist thought is akin to wild law philosophy.

Therefore, it is time now, before it is too late, to move towards an ecocentric conception founded on Buddhist principles.

Now, the law of negligence has come a long way since 1932. Lord Atkin's famous dictum espousing the modern law of negligence is well-known enough, with 'the lawyer's question, Who is my neighbour?'

Good rhetoric in aid of effective advocacy is hard to come by these days. In this case, Counsel for the Plaintiff has ably argued that the snail, before death struck, had a life. It was a living thing. It was moving about its own business, no matter how slow its pace. It was definitely a sentient being.

The question of law urged before this Court is this: can a snail, being a non-human sentient being, be within the Atkinian conception of 'persons' (clearly meaning 'human persons' in the original Atkinian conception) in order to trigger the tort of negligence, if successfully proven? Or, to phrase the question differently: would it not be possible to extend or expand the idea of 'persons' to mean or to include also 'non-human sentient beings'?

In my opinion, the snail is within the Atkinian contemplated 'persons', according to Buddhist ways of interpretation. My reasoning follows.

In the circumstance before me, I would interpret the word 'persons' to mean 'beings' or, more appropriately, 'sentient beings' to be caught within the Atkinian conception of the neighbour principle. In other words, it is tantamount to substituting 'sentient beings' for 'persons' in that famous Atkinian espousal.

I base my judgment fundamentally on the Plaintiff's novel arguments for an ecocentric conception of the law of negligence advanced on compassionate Buddhist principles and precepts.

Indeed, compassion lies at the very heart of Buddhism.

More significantly, for the purpose of this judgment, I would like to entertain the Buddhist conception of the right of every sentient being to life, and with it, associated rights. Moreover, the Buddhist path is the path of non-violence – *ahimsa*.

A brief outline of Buddhism will suffice for the present moment.

Buddhist teachings are generally acclaimed to be atheistic. Buddhism is more known for its philosophy, a way of life, than as an institutionalized form of religion. His Holiness the 14th Dalai Lama himself prefers the word 'spirituality' to 'religion' when he discusses Buddhism.

Buddhism has been around for over 2,500 years. Its founder was Prince Siddhartha Gautama, an Indian prince who gave up his royal entitlements to a life in search for, and pursuit of, the ultimate truth. One may say, in hindsight, he gave up physical comforts for metaphysical reality.

Fortunately, that supreme moment of truth came when the Prince meditated under the Bodhi Tree in Bodh Gaya and attained ultimate enlightenment. The Buddha simply refers to the 'Awakened One': one who is realized; one who has achieved enlightenment.

Buddhist tenets are founded on four fundamental truths, known as the Four Seals. These Four Seals may be summarized as (1) Impermanence (2) Suffering (3) Inter-dependence and (4) Enlightenment.

Closely linked with the Four Seals are the Four Noble Truths: (1) The Noble Truth of Suffering (2) The Noble Truth of the Causes of Suffering (3) The Noble Truth of the Cessation of Suffering and (4) The Noble Truth of the Path to Cessation of Suffering.

In light of the above precepts, among other dharma teachings, Buddhists endeavor to adhere to 'The Noble Eightfold Path': (1) Right View, (2) Right

Intention, (3) Right Speech, (4) Right Action, (5) Right Livelihood, (6) Right Effort, (7) Right Mindfulness, and (8) Right Concentration.

For the purpose of this judgment, I will focus only on Right Action. Right Action carries six aspects: again, for our purpose here, two are relevant. These two are: (1) abstinence from killing all sentient beings; (2) such abstinence implies a commitment to compassion, and respect for the life of all sentient beings.

Indeed, Buddhism has, at its concomitant core value, karma-based ethics which emphasize respect for life. And, central to this value, is not to do harm, or cause to inflict any harm. This is based on the eternal principle that every cause carries its effect manifest lifetime over lifetime.

On karma, and in the words of Professor Ron Epstein:[37]

> Karma is the causal network of intentional actions, both mental and physical, that is the foundation of Buddhist ethical understanding. The foremost principle of Buddhist karma-based ethics is ahimsa, the principles of non-harming and of respect for life. This does not only refer to respect for human beings, but also for every manifestation of life on the planet, especially sentient life.

I am convinced that the snail is considered a being with sentient life within the Buddhist conception of sentient beings.

And I am further prepared to consider the Plaintiff's arguments here based on the Buddhist philosophy of a sentient being's right to life in extending the Atkinian modern law of negligence.

So, now, let me turn to the Buddhist precept of the right to life of every sentient being.

In Buddhism, it is important that we recognize every sentient being is endowed with the gift of life. Some lives may be short, some long. That depends on the type of being one has reincarnated into in this life. The proper thing is to respect every living being, every sentient being on earth.

Such respect necessitates that one does not cause harm to come to another, as alluded to above. This is especially so when we consider that the harm inflicted is physical harm, which not only results in injury, but death, as in this present case. This is totally against the teachings of the Buddha.

The Buddha teaches that all sentient beings possess Buddha nature. All are capable of enlightenment. The right to life is enjoyed by one and all sentient

---

37 R Epstein, 'Environment issues – a Buddhist perspective' (2005) 35 *Vajra Bodhi Sea* 23.

beings. It is the right of every sentient being to reach enlightenment, to attain *Nirvana* and to be rid of *samsara*. Ultimately, it is the goal of every sentient being to be liberated from being trapped in this cycle of birth and death and rebirth. It is in such a cycle that suffering resides.

The Buddha espouses that, *inter alia*, all life is interconnected. The taking of any life is, therefore, strictly prohibited in Buddhism. As life is interconnected, one depends on another or the other for survival. This, basically, is the Buddhist concept of dependent origination. Simply put, the interdependence of all things great and small on earth.

Not only is life of every sentient being to be respected, it is important to note that compassion is required. This is because compassionate respect is a refrain, and an injunction against inflicting harm.

In this connection, I further note the Principle of Compassionate Ecology, tirelessly advocated by His Holiness the 14th Dalai Lama, which is sought to be an ingrained Buddhist precept: for the Buddha teaches that all life is sacred, sacrosanct, and worthy of protection. Respect and compassion are, therefore, integral to the ideals of the sanctity of life.

I now turn to the emergence of the Principle of Universal Responsibility, another tenet espoused by His Holiness the 14th Dalai Lama, in relation to the call by His Holiness in gathering global efforts for the preservation of our physical environment.

This Principle of Universal Responsibility has gained a worldview as propounded and popularised by His Holiness the 14th Dalai Lama. His Holiness teaches that, as human beings, we owe the Earth family and its entire sentient beings a universal responsibility to take care of and nurture them. Our Earth is the only home we have. It is like our mother. It is our source, our beginning and our ending. We are both victor and victim in our attitude, and our actions, towards the environment.

As His Holiness the 14th Dalai Lama said, in an address delivered in New Delhi on 2 February 1992: 'So if we have a genuine sense of universal responsibility, as the central motivation and principle, then from that direction our relations with the environment will be well balanced'.

His Holiness further articulated, in relation to environmental conservation and protection: 'The key point is to have [a] genuine sense of universal responsibility, based on love and compassion, and clear awareness'.

In a similar vein, and in verse and poetry, His Holiness has also poetically shared this important Buddhist philosophy, as extracted below:

> In the remoteness of the Himalayas
> In the days of yore, the land of Tibet
> Observed a ban on hunting, on fishing
> And, during designated periods, even construction
> These traditions are noble

For they preserve and cherish
The lives of humble; helpless, defenceless creatures

Playing with the lives of other beings
Without sensitivity or hesitation
As in the act of hunting or fishing for sport
Is an act of heedless, needless violence
A violation of the solemn rights
Of all living beings

Being attentive to the nature
Of interdependence of all creatures
Both animate and inanimate
One should never slacken in one's efforts
To preserve and conserve nature's energy[38]

In light of the above, I accept the Plaintiff's submission based on this emerging Principle of Universal Responsibility, founded on the Buddhist idea of a compassionate ecology.

Earlier, I alluded to both the ideas of karma and reincarnation in the Buddhist worldview. Karma, as we know it, entails cause and effect. In the Buddhist view, karmic debts may be paid or repaid in one's given lifetime now, or it may be held over different or many lifetimes, depending on individual circumstances.

Further, in the Buddhist view, karmic consequences necessarily entail being reborn, or reincarnated. Reincarnation may not mean one returns to a human birth. On the contrary, depending on one's good deeds and intentions or otherwise in the current lifetime, one may be reincarnated as a beast or an insect in the next life. The Buddha's own preview of his countless reincarnations in numerous forms just before he achieved Enlightenment is testimony to this point.

Accordingly, lack of respect for the life of another, exhibited in the taking of life of another, inflicting harm on another, mindlessly or wilfully, would attract its consequences.

In the context of the current case, who is to say that the dead snail was not, in fact, a natural 'person' in a previous life, as a Buddhist would have us believe? And if so, that would bring this snail squarely within the Atkinian conception of tortious liability without further argument. Well, for now, enough said of reincarnation.

---

38 Poem released on the opening of the International Conference on Ecological Responsibility: *A Dialogue with Buddhism*, New Delhi, 2 October 1993.

In any event, when the above Buddhist principles are taken into account, and merge with the modern law of negligence, we see a natural synthesis.

The Atkinian-Buddhist approach to the ecocentric conception of the tort of negligence may be re-stated thus: 'You must take reasonable care to avoid acts or omissions which you can reasonably foresee would be likely to injure your neighbour. Who *or what*, then, in law, is my neighbour? The answer seems to be – *sentient beings* so closely and directly affected by my act that I ought reasonably to have them in contemplation as being so affected when I am directing my mind to the acts or omissions which are called in question' (italics are mine).

Returning to the facts of the instant case, I have no doubt that the death of the snail in question was attributed to a mindless act or omission committed by a servant or agent of the Defendant.

The snail meant no harm when it wandered its way into the Defendant's premises. Neither did it expect that harm would come to it.

The Defendant was in the business of manufacturing ginger-beer for consumption. Hygiene and cleanliness would be of an utmost concern for the health of the end consumers. Great care needed to have been exercised to ensure that not only the preparation of the ginger-beer met with health standards, but that no foreign matter be carelessly allowed to enter the ginger-beer bottles.

Alas, on that fateful day in question, the Defendant not only negligently caused the death of the snail in the present case, but the Defendant also negligently caused the dead snail to be bottled inside an opaque container, the contents of which no one knew until the drink was consumed.

Earlier, I began my judgment by saying the dead snail could not have instituted this lawsuit. Neither could it have counted on any next-of-kin to do so. In an anthropocentric worldview of our physical environment, its death must lie where it fell.

If humanity persists in that restrictive view, we will not only be mindless, but equally heartless.

This Court is persuaded by the novel arguments advanced by the Plaintiff in the instant case advocating an ecocentric conception of environmental law based on Buddhist perspectives.

It has provided the Court with an opportunity to examine and analyse an ancient spiritual wisdom, Buddhism, in the context of a re-formulation of the modern law of negligence.

The Buddhist perspectives relied on here are compatible with our understanding of the framework of the modern workings of the tort of negligence. The gloss to this tort is the idea that all sentient beings, not just human beings, are considered potential victims of wrongdoing.

In summary, the expanded tort of negligence from an anthropocentric to an ecocentric conception based on Buddhist perspectives demands that we respect life no matter what, that we respect all God's creatures great and small, no matter where. And further, such respect embeds compassionate care and a restraint against inflicting harm.

In my final analysis, the dead snail was, in law, the Defendant's 'neighbour'. The Defendant, therefore, owed it a duty of care. The Defendant breached that duty, and now must answer for that lapse.

I, therefore, hold the Defendant accountable in the tort of negligence against the snail for causing it fatal injury.

Based on the foregoing, I find for the Plaintiff, and the Defendant will bear the costs for both parties.

Chapter 8

Shaw v McCreary

Edward Mussawir

Commentary

The case that I chose to write a judgment about for this project, *Shaw v McCreary*,[1] is one that has captivated me for a number of reasons. I must admit firstly that this case, on first glance, might not immediately appear to lend itself to the contemporary aims of the movement of 'Earth jurisprudence' in Australia. The case is more than 100 years old, Canadian, and relates to a narrow area of common law concerning civil liability for animals. It concerns the question of responsibility for injury inflicted by an escaping bear as between a husband and wife: the husband being the keeper of the bear and the wife being the owner of the property on which it was kept. All the relevant legal 'persons' in this case are therefore human, and the environmental implications are somewhat peripheral. Yet the meaning of the bear – as I would like to show in this case – plays a crucial role, and it does so in a way that makes us think slightly differently about the notion of 'anthropocentrism' in law.

The reason that I think it is an important case to revisit in the context of the contemporary narratives that seem to be mobilised against 'legal anthropocentrism' is, I think, that it allows us to sharpen our focus on the implications of these narratives and in particular on what within the art and technique of law might still be relevant to them. We are able to obtain this focus more clearly in my opinion, not necessarily in cases that are contested on the ground of major political, social, ethical or environmental challenges and struggles, but in those cases which we might tend to otherwise see as inhabiting 'ordinary' situations if not for the subtlety of meaning that, in litigation, reserves them for the topic of jurisprudence.

I am aware that, perhaps because of this, this piece may not be especially satisfying for those who seek from the 'wild law' movement a certain programme translatable to struggles carried out on the legal front. Part of

---

1 *Shaw v McCreary* [1890] 19 Ontario Reports 39.

the appeal to me of the current project is its acknowledgement, modestly enough, that the aims of whatever movement there is in 'Earth jurisprudence' may be met just as well by holding to the playful iteration of legal judgment than by bringing such judgment toward a transcendent polemical or philosophical plane: one that is content with advocacy, with calls for reform, or with appeal to a more or less abstract idea of justice. I take the invitation to write judgment then, rather than as one to hold a defined critical stance on juridical logic and this form of legal writing from the outside, as instead one to inhabit and work within it. In this way, one can show a certain care for the law that as critics, we can easily forget to take responsibility for, but also to potentially 'unwork' or unravel part of it from within.

Shaw v McCreary, I think, is a wonderful case to undertake this task, because one finds oneself opened onto a field of delicate juridical problems, seemingly cleared of their ideological implications, and where one nonetheless finds an intimate encounter with an animal, a bear – offering us the challenge of describing its legal outline outside anthropocentrism but between the rights and duties of spouses, the respective duties of a 'bear-keeper' and a 'bare-owner' of property. In the judgment that follows, my aim is to try and give the animal a more central and explicit place within the technical legal reasoning involved, than what seems to have been possible in the original and appellate decisions. To do this, rather than directly rehearse the philosophical positions from which it is possible today to critique the anthropocentrism of the doctrine of common law in a general sense, I have tried to chart a more modest path through the doctrinal logic involved. My hope was that in doing this it might be possible, within the language and logic of law itself, to open a space in which the juridical outline of an animal might take clearer shape. To the extent that modern legal reasoning has tended to close down that space, if not to erase the animal completely from its forms of knowledge, I believe that the following judgment responds to some of the concerns of a jurisprudence centred in 'wild law'.

Shaw et al v McCreary et al

Husband and wife – Animals – Liability of wife of owner of animal ferae naturae for escape from her separate property – Negligence

This case concerns a decision at common law to hold a woman, Mary McCreary, responsible for injury inflicted by a wild animal: a bear.

A: Cause and jurisdiction

The matter arises from a decision of the Chancery Division of the Supreme Court of Ontario in 1890 to extend liability for the injurious action of a bear to the owner of the property from which the bear had escaped, the owner of the property being the wife of the keeper of the bear.

The purpose for reconsidering this case today is an academic one: to assess the reasoning in this case as it was determined in 1890 and, as we shall see, to assess the area of common law relating to civil liability for animals, not primarily from the perspective of the parties themselves for whom there is no longer any dispute – but rather, if it is possible, from the perspective of the *animal* of such form of liability. The concern is, in other words, to address a form of judicial bias that may not have been evident at the time of hearing: a bias inherent in common law reasoning more generally and which may be called 'legal anthropocentrism'.

The basis for reconsidering the possible anthropocentrism in this case stems from a contemporary movement that has been called in academic writing 'wild law' or 'Earth jurisprudence'. The stated central concern of this movement is to offer a challenge to 'the dominant human-centred focus' and the 'hegemony of anthropocentrism' in the common law.[2] For present purposes, this movement offers the occasion for a certain kind of jurisprudential reflection: one that invites an attention to the practices of common law judgment from an ecological or 'Earth-centred' perspective. More specifically, it confirms that the place of the wild animal in common law is something that deserves a renewed jurisprudential attention.

B: Questions for determination

The primary question for determination is this: was there an 'anthropocentrism' in the decision of the Supreme Court of Ontario when it held that the owner of the property on which a bear was kept by her spouse could be liable for injury inflicted by the bear escaping onto the street and injuring the plaintiff?

To address this anthropocentrism, clearly the right of the animal itself is fundamental and also problematic. It is worth mentioning here however, that this question need not be taken definitively. The problem of whether an animal should be treated as a 'person' or 'subject of rights' in a proceeding is one that should depend on the specific juridical circumstances and issues.[3]

---

2 N Rogers and M Maloney, 'The Australian Wild Law Judgments Project' (2014) 39(3) *Alternative Law Journal* 172.
3 We can afford to put to one side then the various forms of advocacy and rhetoric through which animals either have been in the past or contemporarily brought under the auspices of subjectivist legal processes and protection. The medieval quasi-judicial phenomena famously chronicled by E P Evans in *The Criminal Prosecution and Capital Punishment of Animals* (Lawbook Exchange, first published 1906, 1998 ed.) sit alongside the forms of legal advocacy that today argue for animals as subjects of liberal rights, such as those

And in the current case, the designation of subjectivity to the animal does not necessarily help us to conceptualise this right. Even when our concern is to address and to challenge the anthropocentrism of law, the current case is not one which requires us to personify an animal in order to help resolve the legal issues.

The primary problem is rather to address the centrality or marginalisation in which the animal finds itself in the judicial reasoning in question: a reasoning that attempts solely to consider whether the defendant Mary McCreary is liable for the harm occasioned by the bear.

C: Facts

The relevant facts of the case are set out in the original reported judgment in the Ontario Reports as follows:

> This was an action brought by John Shaw, an infant, by Matthew Shaw, his father, as his next friend, and the said Matthew Shaw against John McCreary and Mary McCreary for damages caused by a bear owned by John McCreary and kept on the premises of Mary McCreary, getting out on the public street and attacking and injuring the plaintiff John Shaw . . . The evidence shewed that the defendants were husband and wife, and that the husband had brought the bear to the premises where she and her husband resided, they being owned by the wife as her separate estate; that the bear being so kept there, without objection on the part of the wife, had escaped to the street and had attacked and thrown down and severely bitten the plaintiff John Shaw. The action was tried at the Toronto Winter Assizes on January 16th 1889.[4]

D: Direction of the trial judge

The Chief Justice in the original instance charged the jury in the following way:

> Under the circumstances I think the defendant McCreary [i.e., John] is responsible. The action is brought against him and his wife on the

arguing for the extension of the writ of *habeas corpus* to animals such as chimpanzees, elephants, dolphins etc., which are kept in captivity. It would be necessary however to confine each of these phenomena to their distinct rhetorical idioms in order to clear, rather than crowd, the jurisprudential landscape in which one considers the meaning of animals and therefore their relation, if any, to what the Western tradition calls 'right'.

4 *Shaw v McCreary* [1890] 19 Ontario Reports 39, at 39.

ground that the wife owned the property, and Mr Fraser [counsel for the plaintiffs] pressed me very strongly with the argument that the owner of the property is responsible for anything that takes place on that property, at least for allowing a ferocious animal to be on it. That may be so in ordinary cases, but in my opinion, considering that the owner of the property in this case and John McCreary were husband and wife, I do not think the wife is obliged to disobey the positive injunctions or wishes of her husband. That leaves it, in my opinion, that the responsibility rests on him.[5]

The Chief Justice accordingly dismissed the action against the wife. From that decision the plaintiffs appealed to the Divisional Court on the ground that they were entitled to recover not just as against the husband John McCreary but also against the wife Mary McCreary.

E: Decision of the Divisional Court

The decision of the trial judge was overturned on appeal in the Chancery Division (the Divisional Court) by Boyd C and Justice Ferguson. The two judges concurred in their reasoning.

Boyd C noted that the common law relating to liability for wild animals was established in two relevant English cases: *Besozzi v Harris*[6] and *Wyatt v Rosherville Gardens Co.*[7] where in each instance the keeper of a bear had been held to be strictly liable regardless of whether the defendants thought that the bear who injured the plaintiff had a docile nature. According to the principle in *Besozzi*, everyone is presumed to know that a wild animal such as a bear has a savage nature and that anyone keeping such an animal is strictly bound to keep it so that it does not cause injury. This is in contrast to animals considered tame or domestic for which knowledge on the part of the keeper of its vicious propensity is required to establish liability.

The case of *M'Kone v Wood*[8] was applied, confirming the principle that a person who is not the owner of the animal may also be liable for it if he or she 'harboured' the animal on his or her premises or allowed it to 'be or resort' there, such action rendering this person liable 'as if he were really the owner of the beast'.[9]

---

5 *Shaw v McCreary* [1890] 19 Ontario Reports 39, at 39–40.
6 *Besozzi v Harris* (1858) 1 F & F 92.
7 *Wyatt v Rosherville Gardens Co* (1886) 2 Times L R 282.
8 *M'Kone v Wood* (1831) 5 C & P 1.
9 *Shaw v McCreary* [1890] 19 Ontario Reports 39, 44.

Neither of the judges found that there was any evidence that Mary McCreary had been acting under coercion from her husband to allow the bear on the property.

The effect of the Ontario Statute R S O Chapter 132 sections 3 and 14 was to give a married woman, as Boyd C describes it, 'all the rights of a *feme sole* in respect of her separate property as against all the world, including her husband'.[10] This meant that there was no reason in either of the judges' opinions to exclude the wife from liability in the way the Chief Justice had done. Justice Ferguson explained that the statute meant that the wife was legally capable of enjoying the property 'as if she were sole and unmarried'.[11]

Boyd C also considered that allowing a wild animal to be kept on one's property could at the very least be analogised to keeping any other substance which, if it escaped from that property, could cause injury to neighbours such as under the principle in *Rylands v Fletcher*.[12]

The court ordered that the matter be sent down for further trial with respect to the wife unless consent was given allowing the verdict to include both the defendants with all the costs of the action.

F: The subject of civil liability for animals: the person of bare-owner and the person of keeper of a bear

This brings me to consider whether any particular 'anthropocentric' bias can be found in the reasoning of the Divisional Court. To answer that question, let me consider the judgments firstly through an attention to the subject of civil liability for wild animals specifically through the personality of the 'owner' and the 'keeper'.

The legal doctrine that constructs the person of the 'keeper' of a wild animal, for the purposes of civil liability, is not the same as that which constructs the 'owner' or the 'possessor' of property. In a majority of cases this distinction may not be necessary. The keeper of the animal, the owner of the animal and the owner of the property on which the animal is kept are most often one and the same person. This identity is even tighter in the case of wild animals such as bears, since private ownership in a wild animal is said to arise and persist only to the very extent that this wild animal is held captive.[13] Thus, ownership in a wild animal is of a qualified nature and is lost as soon as it escapes and regains its natural liberty, by contrast to a domestic animal which may remain ours even when it has left our property

10 *Shaw v McCreary* [1890] 19 Ontario Reports 39, 43.
11 *Shaw v McCreary* [1890] 19 Ontario Reports 39, 45.
12 *Rylands v Fletcher* (1868) UKHL 1.
13 *Case of Swans* (1592) 7 Co Rep 15 b.

and gone onto another's. For domestic animals then, there is no difficulty in conceptualising the owner of an animal and the owner of the property on which it is held as two distinct persons.[14] Whereas for wild animals the difficulty is greater since they tend to be imagined as belonging to the earth or soil over which property-ownership is already a form of appropriation. As such the limits of the legal personae in owning or keeping wild animals are not as firmly delineated.

This distribution of ownership and possession however, may or may not indicate to us something about who is or who can be considered a 'keeper' for the purposes of civil liability for wild animals. The consequence of the distinction seems to be that, for a wild animal, so long as all the legal signposts point to the identity of the same 'keeper', it appears as though the question of the ownership stays out of the picture. However, when it is necessary to think the liability of the owner of property as distinct from the liability of the keeper, as in the present case, one seems to have to resort to a device that the legal doctrine doesn't immediately supply. Unless it is being 'kept' as such, the owner of property bears no responsibility for a wild animal. But if the animal is being kept by another person without disagreement from the owner of the property on which it is kept, what responsibility lies with this 'bare-owner'?

The respective approaches of the Chief Justice and that of the Divisional Court seem to follow separate avenues around this problem. The Chief Justice seems to take advantage, for example, of an additional legal circumstance of the case: the fact that the two parties happen to be husband and wife. He thus employs a vestige of the fiction of *coverture* in order to ground the basis that the husband and wife cannot be treated as separate persons in the proceedings. For him, conveniently, there can only be one relevant person responsible for the animal, that being the keeper of it: the husband. Once that keeper can be identified, any liability arising out of the personality of the owner of the property is to be rejected.

The solution preferred by the Divisional Court on the other hand is to divide the keeping of the bear between two owners: the owner of the animal and the owner of the property. By doing this, the judges are able to find the wife liable in her *sole capacity* as owner of the property from which the animal escaped, by virtue of having full rights of ownership as against her husband according to the statute that granted such rights to married women and having consented to the animal being kept there. By linking the liability of the wife to the more general liability for owners who happen to allow

---

14 Indeed, one can validly hypothesise that it is livestock (domestic animals) that first functioned in Western law as the image under which all moveable property was later able to be conceived. Hence the derivation of *pecunia* (money, property) from *pecus* (livestock).

dangerous substances to be held on their property according to the principle in *Rylands*, this approach has the advantage of bypassing the problem of having to instead discover a form of joint 'keeping' on behalf of the husband and wife as the product of an assortment of circumstantial elements (the respective frequency of visiting, feeding it etc., or taking on such tasks in the absence of the other) while still maintaining the stipulation, undisputed on the facts, that the husband was the sole 'owner' of the bear, to the qualified extent in which a wild animal may be 'owned'. It also avoids the juridical awkwardness of making liability turn, whether one way or another, on the familial or matrimonial relation of the two 'owners': the fact of Mary McCreary being the wife of the owner of the bear for the Divisional Court has apparently no significance for her liability as owner of the property.

G: The matrimonial significance of the bear in the directions of the trial judge

It is necessary to highlight at this point some aspects to the Chief Justice's approach at first instance that were not dealt with specifically by the Divisional Court but which are worth acknowledging for the purpose of assessing whether, if at all, there was an anthropocentrism to the decision handed down in the Divisional Court.

The judges of the Divisional Court found that the trial judge (the Chief Justice) had erred upon two relevant grounds: 1) that he had failed to acknowledge the relevant absence of evidence that the wife had been coerced into keeping the bear on the property by her husband; and 2) that he disregarded the fact that, by application of the relevant statute, the wife's ownership of the property was not restricted in any way by the fact of being married. The suggestion that the Chief Justice had overlooked these factors, however, is not necessarily borne out definitively by the record.

Let us consider the directions of the trial judge from a perspective that places the bear closer to the centre of the juridical logic in question. It is true that, abstracted from the nature of the animal in question, the Chief Justice's logic appears based solely on *coverture* or a subordination of the legal volition of the wife as against her husband. Yet, other than this logic, nothing indicates that the judge was unaware of the statutory provisions that would affirm the wife's capacity to hold the property as a sole owner. Instead, the Chief Justice's logic is explained more consistently by acknowledging the modification of reasoning introduced by the bear itself. From this perspective for instance, the judge, rather than taking ownership as a sign of a civil responsibility for the animal – assigning in effect a 'patrimonial' duty to the property-owning wife – is capable instead of reverting to an older juridical register: imagining the ownership and the form of the wife's responsibility under a distinctly 'matrimonial' logic. Under this older juridical symbolic, the bear is able to constitute not just any 'substance' which is objectively dangerous to bring to the property independently of any marital relation, but indeed constitutes for the trial judge itself a *solemn spousal gift* that, even in the face of potential demands from outsiders, could not legally have been refused. In this way, we cannot account for the logic of the Chief Justice purely in an unwarranted reversion to the doctrine of *coverture*.

The basis for this reasoning becomes apparent when one looks to the meaning of the bear itself in cultures that tend to follow a matrimonial rather than patrimonial pattern. Studies of the custom of peoples such as the Nivhk and Ainu attest to a form of juridical meaning of the animal linked to matrimony: young bears were raised in the home by the women, often suckling human milk and kept in cages and then ritually killed upon reaching a certain age.[15] The bear used in the ceremony for the Amur people had to be raised by the clan of 'wife-givers' (as hosts) and killed by a member of the clan of 'wife-takers'.[16] To be injured by the bear in the ceremony was taken, contrary to any compensable harm, as having been bestowed as a mark of honour. One may even turn for evidence of this older symbolic, as Johann Jakob Bachofen does, in ancient Western literature. In Ovid's *Metamorphosis*, Polyphemus courts Galatea with the offer of a set of bear-cubs: 'I came upon twin cubs of a shaggy bear that you can play with: so alike you can hardly separate them. I came upon them and I said: I shall keep these for my mistress'.[17] And conjecture also abides the enigmatic remains of the *arkteia*, the Attic bear-rite in which young Athenian girls participated in 'playing the bear' before reaching the age of marriage: a rite that was already probably a mysterious relic by the time it appeared in Aristophanes' *Lysistrata* in 411 BC.[18]

From this point of view, we can see the Chief Justice refusing to base liability for the bear on the positive legal fiction constructed by the statute – that Mary McCreary holds the property 'as if' she were unmarried – instead preferring to recognise the archaic matrimonial state of affairs under which this animal could alone be the sign of what is lawfully held or received. The bear is held on the wife's property in other words, not like any other

---

15 See L Black, 'The Nivhk (Gilyak) of Sakhalin and the Lower Amur' (1973) 10(1) *Arctic Anthropology* 1; M M Balzer, 'Sacred Genders in Siberia: Shamans, Bear Festivals, and Androgyny' in S P Ramet (ed.), *Gender Reversals and Gender Cultures: Anthropological and Historical Perspectives* (Routledge, 1996) 164; K Kindaichi and M Yoshida, 'The Concepts behind the Ainu Bear Festival (Kumamatsuri)' (1949) 5(4) *Southwestern Journal of Anthropology* 345.

16 Black, above n 15.

17 See Ovid, *Metamorphoses* 13, 834 quoted in J J Bachofen, *Der Bær in den Religionen des Alterthums* (Ch Meyri, 1863) 8–9.

18 See Aristophanes, *Lysistrata*, lines 641–647. For Bachofen's analysis of the phenomenon see Bachofen, above n 17, 24–28.

property, but as a particular matrimonial rite. It is as though the woman cannot be required to chase a bear like she would potentially chase a husband from her land.

H: The shape of the bear under the patrimonial logic of the Divisional Court

With these factors in mind, one may begin to isolate something of the 'anthropocentric' contour to the logic of the Divisional Court: a logic in which the meaning of the animal, in other words, tends to become abstracted from its juridical setting.

The decision of the Divisional Court overturns the momentary 'aberration' in the matrimonial logic of the trial judge in order to restore a properly 'patrimonial' relation under which ownership of the land carries a control and a duty of management tantamount to 'keeping' the animal: the ultimate authority to consent to have it harboured there or to have it removed. From this perspective, the wife is considered abstractly to hold *all the rights of a stranger . . . as against her husband*' (at 43 [emphasis added]) according to Boyd C. The judges were in this way also able to accept the submission of the appellant that '[t]he fact of their living together can make no difference' (at 41) to her liability. Husband and wife are here, for all intents and purposes, strangers to one another and the legal logic is constructively complete: an owner has the ultimate right to refuse or to accept a wild animal, like any object, onto her property. It is interesting to note, as an aside, that the author of a critique of this case which was published in the *Canada Law Journal* shortly after the judgment noted that this logic was not taken far enough by the judges of the Divisional Court when, having successfully cast the wife as a 'bare owner' of the property, they still decided to make this abstract owner of the property responsible for the actions of the true owner or keeper of the bear. In the opinion of that anonymous commentator, a property owner should not be held liable for an animal that is essentially 'kept' by another on the owner's land, whether or not that owner consents to it being there. The author thus attempts to divorce the bear from a property that the Divisional Court apparently was still imagining as a kind of collective *household* directly managed by the owner of the property (as head of the household), and tries to displace thereby any remaining remnant of a matrimonial significance of the animal, where it may be taken as the sign of a happy marital home, persisting in the logic of the court.[19] A 'bare-owner' – an owner

19 See Anonymous, 'Liability for Injuries by Mischievous Animals' (1890) 26 *Canada Law Journal* 421.

without any other qualification – under this logic can derive nothing other than a benefit from the virtue of his or her ownership of property.

A subsequent English case of *North v Wood*[20] seems to partly confirm this commentator's scepticism of the patrimonial approach, since in that case, where equivalent roles were played by a father (who was the owner of the property) and a daughter (who was the owner of a vicious dog kept on the property), the father was held *not* to be liable since the daughter was of sufficient age (17 years) to be considered its sole keeper.

One may also be tempted at this point to raise the additional objections that scholars, especially within the wild law movement, have made to the anthropocentric nature of the whole patrimonial institution of ownership itself and particularly toward the idea of the right of ownership of 'man' over nature, including wild animals. This movement would see value in thinking of these entities not as objects of human ownership and exploitation at all, but as members of a common natural community designated by 'Earth'. This approach, which to an extent informs this judgment, clearly attempts to grapple with an important challenge to the concept of legal subjectivity. It has advantages also in potentially understanding the injury in question in a broader way than the way in which the parties themselves have conceived or defined it, and it may in some cases authorise legal action to be taken to remedy that injury. However, when the law is called only to resolve a private civil dispute, and to apportion liability for a personal harm, it is not automatically useful to extend the idea of harm beyond the bounds in which it has been constructed for litigants within the forms of civil law. In the present case, I believe, the wild law perspective must find a slightly different route than one that attempts to institute additional 'rival' subjects to those litigants. Whether one subscribes to the idea or not, it doesn't necessarily aid an understanding of the legal issues – specifically the question of the liability of Mary McCreary to the plaintiff – to imagine the suspension of the institution of ownership over nature and to treat the wild animal in question as incapable of being owned. To concern oneself with the question of anthropocentrism in the common law, one should still come at this issue precisely within the same forms that this law provides. To unseat something of the centrality of the 'human' here may still be possible, but it does not mean sidelining the claims which are formally offered for determination by private parties to a civil proceedings.

---

20 *North v Wood* [1914] 1 KB 629.

I: The animal in the law of scienter and negligence

I would suggest that a more reliable approach to the problem may be to try to isolate the casuistry in which the bear serves a necessary jurisprudential function and thus to trace an outline of it that may be able, more effectively, to re-enter the rubric and jurisdiction of modern law. Allow me to mention two examples that may be instructive in this regard.

First, it is worth noting that the common law relating to civil liability for animals in various jurisdictions since the original decisions in this case, has undergone some significant modifications. The importance of these modifications for a jurisprudence that attempts to think the status and meaning of animals is very often understated. The tendency in the modern reform of this area of law has been to abandon the form of the old common law actions for liability for animals in favour of a more general duty such as the duty of care owed, for instance, in the modern law of negligence. In the old form of action such as for our purposes the action of *scienter*, the animal itself was an essential juridical component: the unlawful act in question was strictly formulated as the *keeping of a dangerous animal*. The terms used in this form of action was 'knowingly keeping' an animal in the habit of biting etc. What is striking is the way in which the modern law experiences the presence of this formulary animal as a uniquely confusing and idiosyncratic figure. Glanville Williams criticized the irrational patchwork in the array of common law actions for liability for animals.[21] And a number of jurisdictions such as New South Wales, South Australia and New Zealand have even taken the step of abolishing it altogether by statute, leaving it covered by the torts of negligence or nuisance.

There is one thing that may be helpful to observe in this legal development. In the difference that separates the action of *scienter* (in which the animal is a definitive legal element) from that of negligence (in which it constitutes no more than one factual circumstance in some breach of a duty of care), there lies the precise juristic outline of the animal: an animal that has an existence in law and *only in law*. It is precisely this outline that becomes something necessary when it is a question of identifying the latent anthropocentrism behind a common law decision. One cannot criticise a court for marginalising the interests of an empirical animal that, at the end of the day, it has no adequate means of knowing. But one *can*, on the other hand, point to the purely juristic animal that common law seems to be methodically

21 G Williams, *Liability for Animals: An Account of the Development and Present Law of Tortious Liability for Animals, Distress Damage Feasant and the Duty to Fence, in Great Britain, Northern Ireland and the Common-law Dominions* (Cambridge University Press, 1939).

erasing from its forms of action and, by doing so, denying itself the means of knowing how to treat an animal as a matter of right.

J: Traces of the bear in the Roman law of pauperies

Reconstructing the meaning of this animal is not an easy task. I would like to introduce a second example here which, I suggest, may serve to assist us. That example comes from the Roman law relating to what is known as the action for *pauperies*.

In *Digest* 9.1.1, the classical Roman jurist Ulpian from a fragment on the *Edict* describes the action of *pauperies*. This action, dating back supposedly, according to this fragment, to the *Twelve Tables* (c 450 BC), was a form of action in relation to damage caused by an animal, the chief remedy for which was noxal surrender: handing over the animal itself. The bear makes a curious appearance under the discussion of this action. Ulpian explains that the nature of the *pauperies* action concerns 'damage done without any legal wrong on the part of the doer'[22] and that it therefore lies in cases where the 'harm constitutes the delict itself':[23] in cases other than where one can point to some personal fault or negligence such as when, while a dog is being taken out on a lead, it is improperly restrained or taken to an improper place.[24] The jurist adds to this that 'if a bear breaks loose and so causes harm, its former owner cannot be sued because he ceased to be owner as soon as the wild animal escaped. Accordingly, if I kill the bear, the corpse is mine'.[25]

The compilers of the Digest evidently took this curious inclusion of the bear here as the example of what was, for them, a blanket exclusion of wild animals from the *pauperies* action. The point is thus introduced in the text of the Digest in this way: 'But [this action] will not lie in the case of beasts which are wild by nature: therefore if a bear breaks loose . . .' and so on. Bernard Jackson notes that this is a likely interpolation by the compilers.[26] It was because of this alteration he explains, which would otherwise leave *ferae* (wild animals) peculiarly out of the picture of liability, that it was necessary for the Byzantine compilers to insert a general catch-all in a separate area of law in the Digest: after a discrete list of animals included in the Edict of the Aedile. The Edict of the Aedile provided for reparations for harm done when anyone had kept near a highway a 'dog, any wild boar, wolf,

---

22 *Digest* 9.1.3.
23 *Digest* 9.1.1.
24 *Digest* 9.1.5.
25 *Digest* 9.1.10.
26 B S Jackson, 'Liability for Animals in Roman Law' (1978) 37(1) *Cambridge Law Journal* 122, 136 (n 94).

bear, panther, lion'.[27] The compilers thus craftily patched in after this provision the cento: 'and generally any dangerous animal, whether at large or so bound or chained'.[28] It is in this way that, because the Roman jurisprudence survives for us only by having been reduced to a code, one can only guess at the subtlety to the original significance of Ulpian's bear even if we can witness something of the effect of its subtraction: the shadow that is left of it on the page of positive law.

Jackson conjectures that it may have been posed by Ulpian for purely academic reasons,[29] perhaps to explain that the nature of the *pauperies* action (involving the remedy of noxal surrender) was one for the *corpus* of the animal itself brought properly against whoever the owner of the animal was *at the time the action was instituted*, rather than against the owner or keeper of the animal at the time the damage occurred.[30] Thus, the *pauperies* action, being one that *traces* the animal in the same way that modern law traces 'proceeds', can only go so far. It *can* presumably trace an animal such as a bear which may have been received by this or that subsequent owner through a recognised mode of transfer: by purchase, gift, bequest, inheritance etc., to go as far as catching for example the good-faith purchaser of an animal who may have had no knowledge of the injury it had caused. However, within the procedural limits of the pursuit that it sets up, the action cannot, for a bear that breaks loose, either revert to the previous owner or outstrip the hunter who would next take it: i.e., the one who would take it *pro suo*, as his own, by first possession, by natural right.[31] The point is that this natural right is not thought by the Roman jurist as a universal, fundamental norm, imposing itself on the narrow institutional world of the civil law. If the bear stands in for a wildness or nature in the juridical thought, it is precisely in order to help resolve a crucial technical point in relation to the action for *pauperies*.

---

27 *Digest* 21.1.39.
28 Jackson observes that the skill of the employment of this cento lies in the fact that the words might very well have meant in their original context: 'or generally any other harm caused by the animal' Jackson, above n 26, 136 (n 98).
29 Ibid.
30 *Digest* 9.1.12.
31 Ulpian thus evokes the casuistry concerning acquisition of things through capture and possession of wild animals: whether for instance to establish possession of a wild animal it is enough to have it within sight, to have injured it while continuing the chase, or that someone who steps in and takes it can be considered to have stolen it, or to have actually physically captured it. See *Digest* 41.1.5. Yan Thomas notes that the juridical category *pro suo* in Roman law covered exclusively the capture of wild animals and the accretion of alluvial land. Y Thomas, '*Imago Naturae*. Note sur l'institutionnalité de la nature à Rome' in *Théologie et droit dans la science politique de d'état moderne. Actes de la table ronde organisée par l'École française de Rome avec le concours du CNRS* (École Française de Rome, 1991) 201, 216.

It is reincorporated by the jurist, in other words, as a narrow institutional category in service of a purely procedural casuistry: offering to fill the detail of the shape and limit of liability constructed by a legal action that pursues not the owner as tortfeasor, but the animal itself through and across the necessarily non-tortious terrain of its various owners.

K: Conclusions

Let's try finally to project something of this outline of the juristic animal onto the issues in question in the present case.

It is difficult to isolate the bear in its juridical independence and to reconstruct the jurisprudence in which it may still operate as a decisive figure in this case. However, the examples we have canvassed allow us to see to what extent the decision of the Chief Justice and the decision of the judges of the Divisional Court – in their attempts to think the doctrinal solution – reveal an attempt at the same time to hide the tracks of the bear. It seems, as I will indicate, as though it is only through an anthropocentric legal interpretation that wishes to further erase the presence and meaning of this animal from its jurisprudence, that the bear's actions could have come to have been made to loosely correlate with a wrong on the part of the wife as sole and bare-owner of the property.

The preference of the Divisional Court to frame the liability of Mary McCreary for the bear kept on her premises according to the principle in *Rylands*, i.e., under a more abstract or amorphous analogy than under the *scienter* principle, is instructive.[32] For the judges of the Divisional Court, the principle of law in the case is thought to be more clearly figured by dissolving the bear into water: leaving it with no greater juridical significance than any other dangerous substance, let alone any other dangerous animal. It becomes equivalent, ultimately, to any circumstantial 'risk' for which an owner has a duty to take relevant precautions. I have noted the anthropocentrism to this juridical erasure of the animal.

At the same time, there is something that this decision has in common with that of the Chief Justice. It is not an inability to imagine the abstract rights and duties of property owners, keepers and spouses, but an inability to imagine the precise juridical effect of the animal on those rights and duties. It is as if the 'bare-ownership' of the wife does not seem to afford the

---

32 The revision that the principle in *Rylands* has undergone, at least in Australian jurisprudence, has taken this abstraction only further. This jurisprudence has sought to alleviate the uncertainty of the various qualifications and exceptions in the principle by bringing it under the more abstract universalism of a principle of foreseeability in negligence. See e.g., *Burnie Port Authority v General Jones Pty Ltd* (1994) 179 CLR 520.

'keeping', harbouring, protection etc., of a husband and a bear in the same way that a husband (by analogy to his harbouring of the bear) is supposed to have both property and a wife. One should, I think, be wary of adopting a reasoning here that is content to resort to a naturalistic fantasy in order to think the relation of civil liability between a woman and a bear: the one being used as metaphor for the other. It is in a similar way that one may be wary of a reasoning that tends to want to find in an 'Earth-mother' the primordial source of an all-encompassing right, only thereby to accuse woman of being a mere imitator.[33] The Chief Justice reasons that a wife could not lawfully have refused the bear, while the Divisional Court, as mentioned, cannot think the question of liability without erasing it. The husband and his bear find themselves incapable of being imagined either as ornaments or as passing visitors to the woman's property.

I would venture to overturn the decision of the Divisional Court on this basis and hold that Mary McCreary is not liable for the damage caused to the plaintiff. This, of course, is not because she somehow lacked the authority to refuse or remove the bear from her property. Rather this bear restores a distinct contour to a legal principle that was, in effect, aiming to see civil liability 'follow' the animal itself, beyond and across the terrain marked by such purely human rights and responsibilities as ownership. Instead of highlighting the opaqueness to an outmoded *scienter* action, it offers us a refinement of it, a distillation of its central mode of operation. The action based on the rule in *Rylands* fails since a bear is not analogous to a large volume of water which, whether it can naturally escape from the land or not and cause damage, can never have a mind for escaping or for causing havoc. But if we can cast it in its independent legal guise, the bear may help us imagine the distinct way in which the continuing civil actions for liability for animals work, that is: analogously to the activity of a hunter who doesn't have a 'wrongdoer' so much as an animal in its sights. The action is thus blind to a mere property-owner in her sole vocation as owner of the land on which the animal is kept.

33 See especially the critique of the decontextualised use of a quote from Plato that 'it is not the earth than imitates woman, but woman who imitates the earth' in N Loraux, *Born of the Earth: Myth and Politics in Athens* (Selina Stewart trans., Cornell University Press, 2000) Chapters 7, 8 and 9.

Part II

Mining, climate change and communities

Chapter 9

Coal mines and wild law
A judgment for the climate

Felicity Deane and Katie Woolaston

Commentary

If you were to consider the term 'wild law' in relation to Australia's environmental laws, you would certainly be drawn to Australia's primary environmental legislation, the Federal *Environmental Protection and Biodiversity Conservation Act 1999* ('the EPBC Act'). There is a very specific need for a 'wild law' interpretation of the EPBC Act, given its primary purpose is the protection of biodiversity within a sustainable development framework. Specifically, the actions and developments considered within the EPBC Act raise questions about governance and policy goals, and wild law provides a way to answer these questions.

To begin, the wild law philosophy promotes the belief that many problems within human society stem from the anthropocentric view of human beings at the centre of existence, with a corresponding lack of emphasis on the importance of a flourishing environment.[1] The theory underpinning wild law supports the idea that humans are only one part of a wider Earth community, rather than being the centre of it.[2] Wild laws are laws that capture the importance of preserving the natural environment for the benefit of all Earth's ecosystems and natural entities.[3] They recognise the rights of all beings to coexist and fulfill their respective roles within the natural world. These laws may take many forms, as there is no prescription for their structure. Indeed, diversity of law and governance is one of the principles underlying the development of wild law.[4] In relation to how this philosophy influences legal reasoning, we can suggest that any decision-maker must prioritise a flourishing environment and make decisions accordingly.

1 P Burdon, 'Earth Jurisprudence' (2009) 106 *Chain Reaction*, 41.
2 Ibid.
3 C Cullinan, *Wild Law: A Manifesto for Earth Justice* (Chelsea Green Publishing Co, 2nd ed, 2011) 10.
4 Ibid., 76.

Within the EPBC Act is a structure that dictates the decision-making process for the approval of actions and developments that negatively affect biodiversity. Unfortunately, the wild law premise, that the environment be prioritised, is rarely seen in the decision-making processes undertaken in accordance with the Act. The judgment below is a revision of one such case wherein the interests of the natural world were overlooked for the benefit of economic development. The case is the *Wildlife Preservation Society of Queensland Proserpine/Whitsunday Branch Inc v The Minister for Environment and Heritage*.[5]

The case concerned an environmental challenge to two planned coal mines in Queensland's Bowen Basin – the Bowen Coal 'Isaac Plains Project' and the QCoal 'Sonoma Project'. Together, the proposed open-cut mines would extract over 18 million tonnes of coal per annum, over a lifespan of nine years. The challenge was led by the Proserpine/Whitsunday Branch of the Wildlife Preservation Society of Queensland, and supported by the Environmental Defenders Office Queensland ('the applicants'). They submitted that the proposed mines be assessed under the EPBC Act. The applicants proposed that this was necessary as the resultant greenhouse gas emissions – produced by the mining operations and from the coal that would be mined as a result of these projects – would have adverse impacts on the Great Barrier Reef.

Justice Dowsett heard the matter and found that the proposed mines could proceed because there was no obligation to consider the cumulative impacts of greenhouse gases within the decision-making process of the EPBC Act. Instead, it was found that the only relevant consideration was the direct emissions produced as a result of the mines. Justice Dowsett found that the direct impacts on the reef from those emissions were not only likely to be insignificant, but were also 'speculative'.[6]

A rewrite of this judgment from a wild law perspective was desirable, due to the environmental impacts of the original decision. As noted by Chris McGrath:

> Despite the massive emissions of greenhouse gases involved, both the Isaac Plains Coal Mine and the Sonoma Coal Project were determined not to be controlled actions and no conditions were imposed upon them to reduce or off-set their greenhouse gas emissions. The decision in this case shows that the emissions from the use of the coal from the mines are effectively not regulated under the EPBC Act, which indicates an

5 *Wildlife Preservation Society of Queensland Proserpine/Whitsunday Branch Inc v Minister for the Environment and Heritage & Ors* (2006) 232 ALR 510.
6 Ibid., 520.

important gap in the ability of that Act to genuinely protect the matters of national environmental significance it recognizes as warranting protection.[7]

Wild law, as the name suggests, would not require strict adherence to legal precedence, but rather promotes flexibility in decision-making, with a view to ensuring the natural world can exist, thrive and regenerate. Whilst some amendment to the current legal framework is desirable from a wild law point of view, we have attempted to demonstrate that wild law principles can be incorporated into the current system. As such, in rewriting this judgment we have not attempted to amend any of the existing legislation but instead, have shown how the current framework could be interpreted more 'wildly'. We have done this by allowing section 75 a liberal interpretation, and have drawn on the principles of Ecologically Sustainable Development ('ESD').[8] It is clear that in drafting the EPBC Act, the legislators intended for the principles of ESD to have meaning and to be given consideration in the application of the Act. However, unfortunately this has not been the case. We submit that ministerial approval or judicial review rarely makes note of the principles of ESD, in direct contrast to the need for their consideration under wild law. The lack of attention given to the principles of ESD has seen a number of controversial decisions that impact negatively on matters of national environmental significance.[9] This decision was arguably one of the stepping stones to larger cases that have seen arguments associated with a need to mitigate climate change rejected.[10] Although climate change is recognized as a key threatening process under the EPBC Act, the decision of Justice Dowsett referenced here has meant that it is not itself a trigger to initiate a controlled action, unless there is a direct identifiable impact on

7 C McGrath, *Federal Court Case Challenges Greenhouse Gas Emissions from Coal Mines* (18 June 2006) Environmental Law Australia <http://envlaw.com.au/wp-content/uploads/whitsunday19.pdf>.

8 The EPBC Act defines ESD through a number of principles, including; a) decision-making processes that integrate long-term and short-term economic, environmental, social and equitable considerations; b) lack of full scientific certainty should not be used as a reason for postponing measures to prevent environmental degradation where there is a threat of serious and irreversible harm; c) the principle of inter-generational equity; d) conservation of biological diversity and ecological integrity should be a fundamental consideration in decision-making. See *Environmental Protection and Biodiversity Conservation Act 1999* (Cth) s 3A.

9 See e.g., *Adani Mining Pty Ltd v Land Services of Coast and Country Inc* [2015] QLC 48; *Xstrata Coal Qld Pty Ltd v Friends of the Earth, Brisbane Co-Op Ltd* [2012] QLC 13; *Lawyers for Forests Inc v Minister for the Environment, Heritage and the Arts* [2009] FCA 330.

10 Ibid.

a protected matter. Similarly, the decision removed any possibility of using one part of a cumulative process as a means of justifying a controlled action, even where that one part is a significant contributor to the overall process. This is despite evidence suggesting that the most damaging environmental effects are likely to come from a combination of multiple actions over time, which has been known for decades.[11]

In support of this, it has long been recognized by the courts that greenhouse gas emissions and the resulting climate change impacts pertaining to major development proposals fall squarely within the framework of ESD. Further, allowing developments with significant resultant greenhouse gas emissions is not consistent with these principles. The issues that are faced by applicants in these types of matters are two-fold. First, there is an absence of a legislative basis to prioritise climate change impacts over other concerns (usually economic), and second, there is generally a lack of scientific evidence demonstrating a causal relationship between emissions from a single project and a 'significant impact' on a protected environment. A 'wild' interpretation of the legislation, and specifically the principles of ESD, could assist in overcoming these issues. In particular, the precautionary principle requires that the lack of scientific evidence is not a reason for preventing environmental damage. In Australia the precautionary principle is said to include two threads.[12] That is, the precautionary principle requires that preventative action is taken in the face of uncertainty, and further, this principle necessitates that there is a reversal of the burden of proof in environmental matters. Both these elements place sustainability and environmental protection above economic development.

Although the emphasis in this judgment has been on the precautionary principle, it is imperative that we attempt to discern ways that the other principles of ESD can be interpreted wildly and used more consistently in the legal framework. For example, the principle of inter-generational equity would require a more thorough analysis of the long-term effects of an action, not only from an environmental point of view, but from an economic and social one as well. With respect to coal mining, the long-term environmental effects are becoming increasingly clear. But what is not discussed as often is

---

11 Council on Environmental Quality, *Considering Cumulative Effects under the 'National Environmental Policy Act'*, (January 1997), US Department of Energy, Office of National Environmental Policy <http://energy.gov/sites/prod/files/nepapub/nepa_documents/RedDont/G-CEQ-ConsidCumulEffects.pdf>.

12 See Justice K Bell, 'The Precautionary Principle: what is it and how do courts use it to protect the environment?' (Paper presented at Environmental Defenders Office, Seminar Series 2010, 13 July 2010) 3.

the diminishing long-term economic benefit.[13] A wild interpretation of the EPBC Act would ensure that such other issues were given more appropriate consideration.

Since the decision in the *Wildlife Preservation Society of Qld* case there has not been a dramatic shift to embrace the wild law approach, nor even the consideration of greenhouse gas emissions in approvals for actions with a potential impact on climate change. There have been some positive movements, such as the requirement for governments to take all mining downstream emissions into account; this, however, has not been enough to prevent approvals of mines on a widespread basis.[14] For example, in the case of the *Anvil Hill Project Watch Association Inc v Minister for the Environment and Water Resources*,[15] an application was dismissed on the basis that the application could not be distinguished from that of the *Wildlife Preservation Society of Queensland* case. This is clearly a disappointing result of the abovementioned decision.

A decision in the New Zealand High Court,[16] and subsequent appeal in the Supreme Court could, alongside the *Wildlife Preservation Society of Qld* case, be labelled an environmentally damaging decision. This decision did not take into account dissenting judgments abroad, but rather maintained the view that climate change related impacts should not be taken into account in mining approvals.[17] Of course, this case was decided in relation to the New Zealand legislation and therefore the circumstances surrounding this and the *Wildlife Preservation Society Qld* case differ significantly. However, it remains a point of interest as a legislative challenge associated with activities resulting in significant greenhouse gas emissions. In particular, the majority view of the Supreme Court in this case defined cumulative effects in such a way as to exclude those effects of other activities, instead deciding that such effects could only be those attributable to the action in question.[18]

A wild law revision of this judgment enables the creation of a legal tangent from that which currently exists. Instead of the gradual projection towards

---

13 See e.g., A Lucas, 'Stranded assets, externalities and carbon risk in the Australian coal industry: the case for contraction in a carbon-constrained world' (2016) 11 *Energy Research and Social Science*, 53.

14 *Gray v Minister for Planning* (2006) 152 LGERA 258 ('Anvil Hill Case').

15 *Anvil Hill Project Watch Association Inc v Minister for the Environment and Water Resources* (2007) 243 ALR 784.

16 *Royal Forest and Bird Protection Society of New Zealand Inc v Buller Coal Ltd* [2012] NZHC 2156.

17 See for example: *London Borough of Hillingdon v Secretary of State for Transport* [2010] EWHC 686 (Admin); *Pembina Institute for Appropriate Development v Canada (Attorney-General)* (2008) FC 302) cited in S Schofield, 'Greenhouse Gas Emissions from Coal' (2013) 10 *New Zealand Law Journal* 377.

18 Schofield, above n 17.

awareness and acceptance of the environmental problems associated with actions such as those of the aforementioned mining operations, a wild law perspective forcibly requires confrontation of any degradation. This confrontation would not allow the continued oversight of this global problem, masked with claims of economic development. A wild law interpretation of the EPBC Act would require that any actions would necessarily have to ensure the environment was not damaged, and if taken to the fullest extent of its understanding, would require that any action promote a flourishing environment, reversing any degradation or adverse impacts. It is an approach that may one day be both desirable and necessary.

Although the purpose of the Act is environmental sustainability in development, there remains a lack of consideration for the greater environment in its interpretation, and instead the emphasis is placed on economic development. A wild law approach would commence with a thorough consideration of the principles of sustainable development contained within the EPBC Act, and would end with the conclusion that the environmental impact far outweighs the resultant economic development. Indeed, any economic development would necessarily require that the environmental costs are not only taken into account, but prioritised over wealth generation. It is through this wild lens that we reconsider *Wildlife Preservation Society of Queensland Proserpine/ Whitsunday Branch Inc v Minister for Environment and Heritage.*

Finally, we should note that in rewriting this judgment we have focused on the elements of the case that specifically concern the matter of climate change mitigation. Whilst the applicant made other arguments concerning certain endangered animals and other threatened ecological communities, we have not addressed them in the judgment. We note that these issues are equally relevant as environmental concerns from a wild law perspective, however, the scope of this judgment revision does not allow us to address all these issues in detail. We have also relied on the law as it was at the time of the original decision. Since this decision there have been amendments to the EPBC Act, such as section 527E. Section 527E(1)(a) of the Act states that an event or circumstance is an 'impact' of an action if it is a direct consequence of the action. An indirect consequence of an action is also an 'impact' if (subject to s 527E(2), which is not relevant here) the action is 'a substantial cause of that event or circumstance'; s 527E(1)(b). To summarise, an event or circumstance will be an indirect impact when any secondary action is reasonably foreseeable as a consequence of the primary action and that the resulting event or circumstance is a reasonably foreseeable consequence of the secondary action. Arguably this addition to the EPBC Act would have required more consideration of the secondary actions associated with the coal mining, that is the combustion of the coal and the resultant greenhouse gas emissions, and is a step toward a wild law approach (albeit a small one).

WILDLIFE PRESERVATION SOCIETY OF QUEENSLAND
PROSERPINE/WHITSUNDAY BRANCH INC v MINISTER FOR
THE ENVIRONMENT AND HERITAGE

FEDERAL COURT OF AUSTRALIA
DEANE J and WOOLASTON J
  5 June 2006, Brisbane
[2006] FCA 736

Judgment

4: Statutory background

The Commonwealth EPBC Act was enacted to implement the provisions of the *Convention on Biological Diversity 1992* and other international environmental agreements into Australian law. The objects of the EPBC Act are set out in s 3(1) as follows:

a) to provide for the protection of the environment, especially those aspects of the environment that are matters of national environmental significance; and
b) to promote ecologically sustainable development through the conservation and ecologically sustainable use of natural resources; and
c) to promote the conservation of biodiversity; and
    (ca) to provide for the protection and conservation of heritage; and
d) to promote a co-operative approach to the protection and management of the environment involving governments, the community, land-holders and indigenous peoples; and
e) to assist in the co-operative implementation of Australia's international environmental responsibilities; and
f) to recognise the role of indigenous people in the conservation and ecologically sustainable use of Australia's biodiversity; and
g) to promote the use of indigenous peoples' knowledge of biodiversity with the involvement of, and in co-operation with, the owners of the knowledge.

With respect to subsection (b), promoting ecologically sustainable development, section 3A states:

The following principles are principles of ecologically sustainable development:
    decision-making processes should effectively integrate both long-term and short-term economic, environmental, social and equitable considerations;

a) if there are threats of serious or irreversible environmental damage, lack of full scientific certainty should not be used as a reason for postponing measures to prevent environmental degradation;

b) the principle of inter-generational equity – that the present generation should ensure that the health, diversity and productivity of the environment is maintained or enhanced for the benefit of future generations;

c) the conservation of biological diversity and ecological integrity should be a fundamental consideration in decision-making;

d) improved valuation, pricing and incentive mechanisms should be promoted.

We will return to the importance of these objectives later in this judgment.

This matter specifically concerns approval of an action (the mines) that may or may not be 'controlled actions'. If the mines are deemed to be controlled actions, the relevant impacts of the action must necessarily be assessed under Part 8 of the Act.

Section 67 contains the definition of a 'controlled action', as well as the definition of a 'controlling provision':

> An action that a person proposes to take is a controlled action if the taking of the action by the person without approval under Part 9 for the purposes of a provision of Part 3 would be (or would, but for section 25AA or 28AB, be) prohibited by the provision. The provision is a controlling provision for the action.

Section 68 outlines a referral mechanism for persons or entities that are proposing to undertake an action which they think might be a 'controlled action'. If they think their action might be a controlled action for the purposes of the Act, they must submit the action to the Minister for his determination on whether or not the action is a 'controlled action'. The Minister then must make a determination based on section 75, the most relevant parts of which are outlined below:

1) The Minister must decide:

    a) whether the action that is the subject of a proposal referred to the Minister is a controlled action; and

    b) which provisions of Part 3 (if any) are controlling provisions for the action.

    . . .

    Considerations in decision

2) If, when the Minister makes a decision under subsection (1), it is relevant for the Minister to consider the impacts of an action:

a) the Minister must consider all adverse impacts (if any) the action:

   (i) has or will have; or
   (ii) is likely to have;
      on the matter protected by each provision of Part 3; and

b) must not consider any beneficial impacts the action:

   (i) has or will have; or
   (ii) is likely to have;
      on the matter protected by each provision of Part 3.

The EPBC Act requires that in addition to determining whether the action is controlled, the Minister must also determine what the 'controlling provision' is under Part 3. So for example, in this case, the applicant is arguing that the action is controlled, and the controlling provision is section 12. Section 12 limits actions that are able to be taken that affect a World Heritage Area, and it is relevant here as we must consider the impacts on the Great Barrier Reef (a World Heritage Area). However, regardless of the applicant's argument, the Minister must consider all 'controlling provisions' to ensure that the action does not fall within any of their scope. In this regard the Minister must ensure that the action does not have a significant impact on a World or National Heritage Area, Wetlands of National Importance, Listed Threatened Species and Communities, Listed Migratory Species, the Great Barrier Reef Marine Park, and so on, to all the matters listed in Part 3. As a result, the EPBC Act requires circular section reasoning in the determination of controlled actions and ministerial approval.

For the purposes of considering this application, we recognise that the Minister was first required to decide whether the mines were 'controlled actions' using the criteria in section 75(2), and second, to ensure that none of the controlling provisions under Part 3 applied. Crucial to the Minister's reasoning is the meaning of 'adverse impacts' within the context of the EPBC Act, and specifically, section 75.

The context of 'adverse impacts' is not clarified within the EPBC Act itself. This means we must more broadly consider the principles of statutory interpretation to clarify the meaning of it. Further, the objects of the Act and consequently the principles of ESD become of the utmost importance. As noted above, these concepts are considered in more detail below.

B: The application

The application in the present case is brought in accordance with the *Administrative Decisions (Judicial Review) Act 1977* (Cth). The applicant seeks review of two decisions by the Minister's delegate pursuant to section 75 of the EPBC Act. In each case, the delegate decided that the referred proposal was not a controlled action, and so did not need to go through the approval process outlined in Part 9. The first referral was in relation to a proposal by the second respondent ('Bowen Coal') to develop a new coal mine near Moranbah ('the Isaac Plains Project'). The second referral concerned a proposal by the third respondent ('QCoal') to develop a new coal mine near Collinsville ('the Sonoma Project').

The applicant had two grounds for review. First, Mr Flanigan, the respondent's delegate, did not take account of the:

> adverse impacts the Isaac Plains Coal Project and the Sonoma Coal Project are likely to have on the matters protected by Part 3 of the EPBC Act due to the mining, transport and use of the coal from the mines emitting a large amount of greenhouse gases contributing to global warming.[19]

The applicant's second ground for review was that in making the decision Mr Flanigan erred in law in that he treated the expression '"all adverse impacts the action is likely to have on the matter protected by each provision of Part 3", in section 75(2) of the EPBC Act, as not including the adverse impacts the [proposals] are likely to have on the matters protected by Part 3 of the EPBC Act'.[20] It was submitted that he should have considered the greenhouse gases from the mining, transport and use of the coal from the mines, and its contribution to global warming as an adverse impact.

In challenging the Minister's decision in relation to the mines, the applicant submitted that the respondent's delegate failed to take account of the potential impacts from the mined coal's greenhouse gas emissions on global warming and the consequential impacts on matters of national environmental significance. These matters of national environmental significance include World Heritage areas, specifically, the Great Barrier Reef World Heritage Area and the Wet Tropics Heritage Area. These impacts were a result of the mining, transport and burning of the coal. Specifically, the applicant urged consideration of the fact that over the proposed 9-year life of the mine, 18 million tonnes of coal would be extracted and exported. The applicant asserted that the ultimate purpose was to burn such coal in power generation and that the 'production of greenhouse gases is almost certain to occur as a result of the action and can reasonably be imputed as within the contemplation of the proponent of the action'.[21] With respect to the impact

---

19 *Wildlife Preservation Society of Queensland Proserpine/Whitsunday Branch Inc v Minister for the Environment and Heritage & Ors* (2006) 232 ALR 510, 518.
20 Ibid., 519.
21 Ibid., 513.

on the World Heritage Areas, the applicant asserted that Australia was vulnerable to the impact of climate change and stated that:

> [p]otentially significant impacts include those of our agricultural productivity, coastal communities, threats to human health, and the imposition of further survival pressures on a range of native plants and animals.[22]

Further, the applicants submitted that Australia had obligations to uphold as signatory to the *Convention on Biological Diversity* and the *World Heritage Convention*. In summary, the applicant claimed that the development should have been declared a 'controlled action', and therefore subject to stricter controls and requirements than what was ultimately found necessary.

In answer to these arguments, the first respondent relied on a number of documents to demonstrate consideration of the issues raised by the applicant. An officer of the Department considered the proposal and the applicant's submission. He prepared a minute in which he advised the delegate to determine that the proposal was not a controlled action. Of the applicant's submission, the officer observed:

> The mine site is located in the catchment of the Isaac River which flows south into the Fitzroy River, and hence in the sea at Rockhampton (a distance of about 300 km). The proposed action incorporates measures to minimize impacts on the quality of water being discharged into the Isaac River. The Shoalwater and Corio Bays Ramsar site lies about 50 to 100 km north of the mouth of the Fitzroy River. The ecological character of the Ramsar site will not be affected given the minimal nature of discharges and distances/dilution factors involved.[23]

The officer further stated:

> The comments from the Wildlife Preservation Society of Queensland suggest that the impacts of climate change, as a consequence of the burning of coal produced from the mine, on the world heritage values of the Great Barrier Reef World Heritage Area should be considered. The nature of induced climate change from the referred coal mining operation, and impacts on world heritage values, are speculative.[24]

22 Ibid.
23 Ibid., 514.
24 Ibid.

He then recommended that the delegate decide that the proposed action was not a controlled action. The delegate, Mr Flanigan, accepted that recommendation, writing on the minute the following words:

> I regard the likelihood of significant impacts on NES arising from the marginal addition of greenhouse gases to be extremely small, in addition to speculative.[25]

The applicant was presumably advised of the decision and requested reasons pursuant to section 13 of the ADJR Act.

As part of the reasons, under the heading 'Findings on material questions of fact and reasons for my decision', Mr Flanigan said:

> I formed the view that significant impacts on the heritage values of the Great Barrier Reef World Heritage Area or on the ecological character of the Shoalwater and Corio Bays Ramsar site are not likely given the nature and location of the proposed action. In this respect, I found that the mine area is within the catchment of the Isaac River, which flows south into the sea at Rockhampton (a distance of about 300 km). I considered that the nature of any indirect impacts on world heritage values or the ecological character of a Ramsar site associated with the referred action are speculative.[26]

Following these determinations Mr Flanigan referred to various specific matters, not including greenhouse gas emissions or climate change. The applicant suggested that the absence of any such express reference suggests that those matters were not considered. That said, in cross-examination, Mr Flanigan indicated that he considered the matters of greenhouse gas emissions and climate change as '*indirect impacts*' of the action and that he subsequently examined them within the scope of that category of concerns. Evidence of this is apparent from the note Mr Flanigan made on the Departmental minute. Such matters were referred to in the minute as '(s)econdary or consequent impacts'.[27]

Mr Flanigan has sworn an affidavit in these proceedings. Within this affidavit he explained the process by which he considered the greenhouse gas emissions and climate change issues. He described himself as having 'a sound general knowledge and understanding of the issues associated with

25 Ibid., 515.
26 Ibid.
27 Ibid., 514.

greenhouse gas emission and climate change'.[28] He set out in some detail his process of reasoning in this case, which he considered substantial. He said in cross-examination, 'It takes longer to write it down and get it in writing than to think it'.[29]

C: The decision

In deciding this case there are two arguments we are being asked to contemplate. First, whether the Minister took proper account of the 'adverse impacts the Isaac Plains Coal Project and the Sonoma Coal Project are likely to have on the matters protected by Pt 3 of the EPBC Act due to mining, transport and use of the coal from the mines emitting a large amount of greenhouse gases contributing to global warming'. Second, whether the Minister made an error in law when he treated the expression '"all adverse impacts the action is likely to have on the matter protected by each provision of Pt 3", in section 75(2) of the EPBC Act, as not including the adverse impacts the [proposals] are likely to have on the matters protected by Pt 3 of the EPBC Act due to the mining, transport and use of the coal from the mines emitting a large amount of greenhouse gases contributing to global warming'.[30] In deliberating on these arguments we suggest that both grounds for review are asking the same question, that is, did the Minister appropriately consider the relevant controlling provisions under Part 3 to ensure that the action was not a controlled action?

Above we made note of the relevant provisions of the EPBC Act. For the purposes of this judgment the relevant parts of section 75 of the EPBC Act require that the Minister consider the impacts of an action in determination of whether something is a controlled action. The impacts can be either certain or likely, and those impacts must be upon matters protected in Part 3 of the Act.

In the present circumstances the Minister was being asked to consider whether the increase in greenhouse gas emissions from the burning of the coal, which we may consider a secondary action, resulting from the mining of the coal, a direct action, could be taken into account to determine whether there would be an adverse impact on matters of national environmental significance, and in particular, the Great Barrier Reef World Heritage Area.

28 Ibid., 516.
29 Ibid.
30 Ibid., 518–19.

The term 'impact' has been judicially considered in *Minister for the Environment and Heritage v Queensland Conservation Council* ('the *Nathan Dam* case').[31] In the context of a section 75 decision, the Full Court stated[32] that the term means 'the influence of effect of an action', and can readily include the indirect consequences. They found that it was not confined to the effects of the action on the matter protected, but extends to effects which are sufficiently close to the action, such as that they have consequences on the protected matter.

In understanding the scope of the word 'impact' it does not matter that burning of coal in this instance is outside the control of the proponent of the action. As noted by the Full Court in the *Nathan Dam* case,[33] '"all adverse impacts" includes each consequence which can reasonably be imputed as within contemplation of the proponent of the action, whether those consequences are within control of the proponent or not'. This judicial analysis gives us a good understanding of the breadth of meaning of 'impact' within the scope of the EPBC Act.[34]

In this regard, there is significant accepted evidence to suggest that greenhouse gas emissions will affect, and are affecting, the Great Barrier Reef in a negative manner. Although we accept this as fact, supporting evidence can be found in the most recent report published by the Intergovernmental Panel on Climate Change. The report notes that:

> Coral reefs in the Australian region are subject to greenhouse-related stresses, including increasingly frequent bleaching episodes, changes in sea level, and probable decreases in calcification rates as a result of changes in ocean chemistry. . . . Coupled with predicted rises in sea level and storminess, bleaching-induced coral death also could weaken the effectiveness of the reefs in protecting the Queensland coast and adversely affect the biodiversity of the reef complex.[35]

It is important to note that the science suggests the greenhouse gas emissions from the coal from the mines in question will not, by themselves,

---

31 *Minister for the Environment and Heritage v Queensland Conservation Council* (2004) 139 FCR 24.
32 *Minister for the Environment and Heritage v Queensland Conservation Council* (2004) 139 FCR 24, 53, 57.
33 *Minister for the Environment and Heritage v Queensland Conservation Council* (2004) 139 FCR 24, 57.
34 The term 'impact' was not defined in the Act until the introduction of section 527E on 19 February 2007.
35 Intergovernmental Panel on Climate Change, *Climate Change 200: Impacts, Adaptation and Vulnerability. Contribution of Working Group II to the Third Assessment Report of the Intergovernmental Panel on Climate Change* (Cambridge University Press, 2001).

cause the above noted impacts. Rather, it is the atmospheric accumulation over time that will lead to detrimental impacts on such ecosystems as the Great Barrier Reef. In this regard, when the Minister's decision is examined, we consider that the primary question to be determined by him was whether one part of a cumulative effect is significant enough to be taken into consideration when making a determination under section 75. In doing so, it is necessary that consideration be given to the principles of ESD outlined in section 3A. On this issue, the applicant argues that the Minister has failed to adopt a 'common sense' approach to causation of greenhouse gas emissions impacts in light of the subject, scope and objects of the Act. Following on from this observation, the question we must respond to here is two-fold. First, did the Minister give any consideration to the impacts of greenhouse gas emissions on the protected matter, and second, did he give the *appropriate* consideration to these impacts on the protected matter in arriving at this decision?

It appears that we can dismiss the first issue to this question reasonably quickly. As noted above, Mr Flanigan indicated that he considered the matters of greenhouse gas emissions and climate change as '*indirect impacts*' of the action, and that he subsequently examined them within the scope of that category of concerns. If we accept this evidence from Mr Flanigan, which we do, then it appears that our answer to the first question is in the affirmative. The impacts of greenhouse gas emissions on the protected matter were indeed considered, even if they were not contained in the Statement of Reasons released by the Minister. This says nothing of Mr Flanigan's expertise in this area and we note that Mr Flanigan's admission of possessing 'a sound general knowledge and understanding of the issues associated with greenhouse gas emission and climate change' does not necessarily put him in a position of greater knowledge than that of the general public. However, there is little doubt that the matter in question was considered notwithstanding this lack of knowledge, and therefore we will move on from this point to address the second issue of the application.

The second issue is far less easily dismissed and therefore we begin with a discussion of the requirements for making a decision in accordance with section 75 of the EPBC Act. We suggest that to answer the question we must first know what is 'appropriate consideration' in the context of this decision. Evidence of what is appropriate consideration brings us to the matter of whether the impacts are likely to be significant.

In the *Nathan Dam* case, the judge at first instance considered the meaning of 'significant' for the purposes of section 75 of the EPBC Act. The judge in this case adopted a wide interpretation and suggested that only those impacts 'which lie in the realm of speculation' can be excluded from any decision made in accordance with the Act. On appeal, this broad interpretation was not accepted, and the appeal judges did not provide explanation of the meaning of 'significant' in the context of this decision. As such, little guidance is provided from this particular decision. However, this case does indicate that the objects of the EPBC Act are relevant in consideration of section 75.

Of all the principles of ESD discussed in section 3A of the EPBC Act, the precautionary principle has most relevance in the current context. The precautionary principle itself is derived from the *Rio Declaration on Environment and Development*. Principle 15 states the principle in these terms:

> In order to protect the environment, the precautionary approach shall be widely applied by States according to their capabilities. Where there are threats of serious or irreversible damage, lack of full scientific certainty shall not be used as a reason for postponing cost-effective measures to prevent environmental degradation.[36]

It is well accepted that the precautionary principle is more than just an aspiration within the EPBC Act, and is a relevant consideration in environmental decision-making. Justice Stein found as much in *Leatch v National Parks and Wildlife Service and Shoalhaven City Council*[37] in relation to the *National Parks and Wildlife Act 1974 (*NSW):

> While there is no express provision requiring consideration of the 'precautionary principle', consideration of the state of knowledge or uncertainty regarding a species, the potential for serious or irreversible harm . . . and the adoption of a cautious approach in protection of endangered fauna is clearly consistent with the subject matter, scope and purpose of the Act.[38]

In *BGB Properties Pty Ltd v Lake Macquarie City Council*,[39] McClellan CJ stated that consideration of the precautionary principle (along with the other principles of ESD) 'does not preclude a decision to approve an application in any cases where the overall benefits of the project outweigh the likely environmental harm'. We differ with the learned judge's reasoning in this instance on the basis that likely environmental harm should far outweigh other benefits, unless the costs of fully restoring the environment, and therefore enabling it to continue to flourish, are included as a cost to any potential developer.

36 *Rio Declaration on Environment and Development*, UN Doc (1992) A/CONF 151/26 (vol I); 31 ILM 874 art 15.
37 *Leatch v National Parks and Wildlife Service and Shoalhaven City Council* (1993) 81 LGERA 270.
38 Ibid 384.
39 *BGB Properties Pty Ltd v Lake Macquarie City Council* (2004) 138 LGERA 237, 262.

In *Telstra Corporation Ltd v Hornsby Shire Council*,[40] Preston CJ was required to consider whether Telstra had applied the precautionary principle in the design and installation of telecommunications towers, and whether the Council had a right to refuse the installation based on that same principle. After analysing the principle Preston CJ stated:

> The function of the precautionary principle is, therefore, to *require* the decision-maker to assume that there is, or will be, a serious or irreversible threat of environmental damage and to take this into account, notwithstanding that there is a degree of scientific uncertainty about whether the threat really exists (emphasis added).

We consider this statement applies equally to the breadth of the application of the precautionary principle within the EPBC Act. In this regard, a common sense application of the principle suggests that lack of full scientific certainty surrounding the impact of one action that has a known cumulative effect on a matter of national environmental significance should not be used as a reason for allowing that action. With particular reference to Part 3 of the Act, and the definition of 'significant impact', the precautionary principle ought be used to give the term 'significant' a wide meaning, where there is a lack of scientific evidence on the significance of that action. Along with expanding the meaning of 'significant', the precautionary principle has specific application in the logic of the decision of this case.

The precautionary principle is generally applied with a goal of the prevention of serious or irreversible harm to the environment. Therefore, any threat to the environment requires careful consideration, whether the threat be direct, indirect, or one part of a cumulative process. Relevant here is that the principle requires the decision-maker to be cautious. The applicants have conceded that the precise nature of the impact of the greenhouse gas emissions on global warming, and therefore on matters of national environmental significance is 'difficult to determine'. Although this admission was on the part of the applicants, as we agree with its premise we suggest that the precautionary principle is highly relevant to any decision-making process surrounding actions that increase greenhouse gas emissions significantly.

We note that the respondent was not provided with any estimate on the potential increase in greenhouse gas emissions from the mining, transport and use of the coal from the proposed actions, at the time the decision was made. However, the lack of information provided to the respondent in this instance does not exclude the greenhouse gas emissions from consideration in the determination of whether the action's impact is 'significant'. Indeed,

---

40 *Telstra Corporation Ltd v Hornsby Shire Council* (2006) 67 NSWLR 256, 273.

in such instances where insufficient information is provided, the precautionary principle is particularly relevant. In a situation where the Minister was not provided with any scientific evidence concerning the nature and extent of the emissions, but where there is no doubt that the action will lead to an increased concentration in the atmosphere, and as it is commonly understood that greenhouse gas emissions are adversely impacting the Great Barrier Reef, precaution must necessarily be taken. We agree that the Minister and his delegate have not provided any evidence to suggest they used caution in considering the impact of the mines, and the resulting greenhouse gas emissions (both direct and indirect), on the Great Barrier Reef. Instead, they have dismissed the impact as indirect and therefore irrelevant, a dismissal that we suggest is erroneous under the EPBC Act.

If we consider the laws of the Earth in relation to the EPBC Act and the precautionary principle, we can take this conclusion several steps further. To wildly interpret the EPBC Act is to say that the regulations associated with the Act must necessarily prevent damage to the environment in any way that may have a permanent impact. As such, the term 'significant' could be replaced with the word 'any' insofar as the impact has some form of permanence. Again, we can extend this reasoning to the precautionary principle. A wild law interpretation of the precautionary principle would require caution at a much higher level than that which is currently accepted in the application of this international requirement. In this regard, we suggest that any wild law interpretation of this principle would demand the 'balancing of environment and economic competing interests test' be replaced with a requirement to consider whether *any* action would have *any* impact on the environment that may be irreversible. In the event that an action may have a possibility of irreversible damage, any application for its approval must necessarily be refused.

As a result of the above observations, and in applying the precautionary principle we conclude that the proposed action is a 'controlled action' and that the Minister erred in law in not declaring it as such. The greenhouse gas emissions that will be released into the atmosphere as a result of the activities associated with this action have the potential to increase atmospheric concentrations, and consequently the proposal must undergo greater scrutiny prior to any approval. This aligns with the objects and purpose of the EPBC Act.

Chapter 10

Quantifying the environmental impact of coal mines

Lessons from the Wandoan case, *Xstrata Coal Queensland Pty Ltd v Friends of the Earth Brisbane Co-Op*

Julia Dehm

Commentary

Xstrata's proposed Wandoan mine was highly contentious as it would have been double the size of Queenland's (then) largest coal mine and would have been one of the largest mines in the Southern hemisphere. Further, the mine's infrastructure investments would have enabled opening up the vast Surat Basin in Queensland. The proposed mining would have required the completion of the 'Southern Missing Link' railroad to connect the mine to the port at Gladstone. Such infrastructure investment would have made other, smaller coal projects in the region viable. For these reasons, the opponents, Friends of the Earth – Brisbane Co-Op, considered opposing the Wandoan mine as critical and took on this high-profile, crowd-funded legal challenge. The Environmental Defenders Office Queensland, who acted in this case in the Land Court of Queensland, described the litigation as a 'David and Goliath' battle.[1] Several landowners affected by the mine were also part of the litigation. They were partially successful, in that parts of their lands containing dwellings were excised from the mining lease area.

The key issue discussed in this commentary and which is the focus of the rewritten judgment is the Land Court's failure to engage adequately with arguments that the mine should not be allowed to proceed due to its climate change impacts. Friends of the Earth Brisbane Co-Op made arguments that the mine should be refused pursuant to sections 269(4)(j)–(l) of the *Mineral Resources Act* 1989 (Qld) (MRA), due to the adverse environmental impact cause by these operations, the prejudice to the public right and interest, and the inconsistency with the principles of ecologically sustainable development

---

1 Media Release, 'Challenge to Wandoan Coal Mine launched' (Environmental Defenders Office Queensland, 25 February 2011) <www.edoqld.org.au/news/challenge-to-wandoan-coal-mine-launched/>.

as set out in the *National Strategy for Ecologically Sustainable Development*. Further, they argued that the Court must perform its functions so as to best achieve the objectives of ecologically sustainable development (section 5 *Environmental Protection Act* 1994 (Qld) (EPA)) and thus should recommend refusal of the unsustainable mining activity.

Legally, the battle was won by Xstrata, as the Land Court decision (discussed below) allowed the mine to proceed.[2] However, different 'laws', namely the economics of supply and demand in the end prevented the project from going ahead.[3] In November 2012 Xstrata conceded that the future of the mine could be in doubt due to poor market conditions.[4] This admission came after Glencore International took over Xstrata for AU$30 billion, creating the world's largest commodity-trader.[5] In March 2013 there were further indications that the mine's future hung in the balance, with the Chief Executive of Glencore-Xstrata telling investors the company was taking a risk averse approach, putting off 'greenfield projects', and focusing on 'brownfield developments' and existing mines instead.[6] In September 2013 Glencore-Xstrata announced its plans to shelve the project because of the declining cost of thermal coal in international markets, then down to AU$80 per tonne.[7] The company assessed the project to not be 'economically sound'

2 The decision in this case has, however, also, as Jacqueline Peel and Hari Osofsky observe, generated an 'antiregulatory legislative response'. Soon after the decision, the *Environmental Protection (Greentape Reduction) and Other Legislative Amendment Bill 2012* was introduced to the Queensland Parliament. The Act, which came into force in March 2013, means that decision-makers are no longer required to take into account the *National Strategy for Ecologically Sustainable Development* principle that 'the global dimension of environmental impacts of actions . . . should be recognized and considered'; see J Feel and H M Osofsky, *Climate Change Litigation: Regulatory Pathways to Cleaner Energy* (Cambridge University Press, 2015) 303.

3 Following the decision Friends of the Earth – Brisbane Co-Op Limited had lodged an appeal (8 May 2012), however, this appeal ended because the mine was discontinued before a legal decision could be reached.

4 O Jacques, 'Xstrata admits Wandoan Coal may crumble in tough market' *The Chronicle* (online), 22 November 2012 <www.thechronicle.com.au/news/massive-wandoan-coal-mine-doubt-toowoomba/1632299/>.

5 See A Duffy, 'Glencore-Xstrata create world's largest commodity trader' *Australian Mining* (online), 22 November 2012 <www.australianmining.com.au/news/glencore-xstrata-create-world-s-largest-commoditie>.

6 P Manning, 'Xstrata chooses brown over green' *The Sydney Morning Herald* (online), 7 March 2013 <www.smh.com.au/business/xstrata-chooses-brown-over-green-20130306-2flnl.html>.

7 A Heber, 'GlencoreXstrata parks Wandoan coal project, cuts capex' *Australian Mining* (online), 11 September 2013 <www.australianmining.com.au/news/glencorexstrata-parks-wandoan-coal-project-cuts-ca>, see also P Ker, 'Wandoan coal project scrapped' *The Age* (online), 11 September 2013 <www.theage.com.au/business/wandoan-coal-project-scuppered-20130910–2ti7x.html>.

given lower than anticipated returns, high costs and further concerns about cost-blowouts in such a major project.[8] The project has thus been put 'on hold', although not formally scrapped or withdrawn from the planning system.[9] This was celebrated by environmentalists as a key victory, especially as without Xstrata's Wandoan 'cornerstone' project, the 'South Missing Railway Link' was unlikely to be built, thereby making a further 70 million tonnes per annum of small mine proposals less economically viable.[10]

A: The decision

The original judgment by President MacDonald considered the requirement in section 269(4)(j) of the MRA that the Court should consider whether 'there will be any adverse environmental impact caused by those operations and, if so, the extent thereof' extremely narrowly.[11] It interpreted the term 'those operations' as operations 'carried on under the authority of the proposed mining lease'[12] enumerated in section 234 MRA as:

(a) to mine the mineral or minerals specified in the lease and for all purposes necessary to effectually carry on that mining;

(b) such purposes, other than mining, as are specified in the mining lease and that are associated with, arising from or promoting the activity of mining.[13]

The Court further looked to section 6A of the MRA to define the verb 'mine':

(1) *Mine* means to carry on an operation with a view to, or for the purpose of

(a) winning mineral from a place where it occurs; or

(b) extracting mineral from its natural state; or

(c) disposing of mineral in connection with, or waste substances resulting from, the winning or extraction.[14]

---

8 Ibid.
9 See J Hepburn, 'Wandoan coal project scrapped' *Chain Reaction* No 119, November 2013 <www.foe.org.au/wandoan-coal-project-scrapped>.
10 Ibid.
11 *Xstrata Coal Queensland Pty Ltd v Friends of the Earth – Brisbane Co-Op Ltd and Department of Environment and Resource Management* [2012] QLC 013 ('*Xstrata*').
12 Ibid., 524.
13 Ibid., 525.
14 Ibid., 526.

Based on this extremely literalist approach to statutory interpretation, President MacDonald then stated:

> It is apparent from these statutory provisions that 'the operations to be carried on under the authority of the proposed mining lease' are confined to the physical activities associated with winning and extracting the coal from the place where it occurs or from its natural state. It is these activities which will be carried on under the authority of the proposed mining leases. The 'operations' referred to in s 269(4)(i) and (j) do not, on a proper reading of the legislation, extend to the transportation of the coal to ports and to the burning of the coal in power stations overseas.[15]

As such, the President excluded from the Court's consideration the Scope three emissions from burning coal and focused only on the 'impact' of the Scope one and two emissions from mining. The narrow literalism of this argument produces the absurd result that the environmental impacts of the burning of mined coal cannot be considered.

The President additionally held that the terms 'impact' and 'environmental' in section 269(4)(j) MRA should be interpreted narrowly. The Court accepted the definition of 'impact' as defined in the Macquarie Dictionary put forward by Friends of the Earth namely as 'influence or effect [exerted by a new idea, concept, ideology, etc.]'.[16] The Court however, refused to apply the reasoning in the *Nathan Dams* case[17] where 'impacts' were deemed to include activities not within the control of the project proponent, provided they were within the contemplation of the proponent. This distinction was justified on the basis of the difference between the Queensland *MRA* and the Commonwealth *Environmental Protection and Biodiversity Conservation Act* 1999 (EPBC Act), and the different focus between 'operations' in the former and 'action' in the latter.[18]

The Court also distinguished between the test of causation under the EPBC Act and the lack of a test of causation under the MRA to find that 'FoE have been unable to point to any specific environmental impact caused by these operations'.[19] In order to support that claim, the Court quoted the expert testimony of Professor Ian Lowe, who wrote in this report:

---

15 Ibid., 528.
16 Ibid., 533.
17 *Minister for the Environment and Heritage v Queensland Conservation Council Inc* (2004) 139 FCR 24.
18 *Xstrata*, 548.
19 Ibid., 549.

It is not possible to link these emissions to any particular impact on a specific part of the environment in Queensland, Australia or globally, other than to contribute to greenhouse gases in the atmosphere and thereby contribute to global warming and climate change. The impacts of greenhouse gas emissions from this mine should, therefore, be understood as contributing to the cumulative impacts of global warming and climate change.[20]

The fact that Professor Lowe was unable to 'attribute any environmental consequence to any specific project and agreed that it would be speculative and unscientific to do so'[21] was used to justify the lack of causation. The President contrasted this evidence with the evidence given by another expert witness, Dr Malthausen, who had 'purported to calculate the effects of the project's GHG emissions by calculating the temperature increase attributable to burning the amount of coal from the project and estimated that 23,000 people would be inundated by the consequent rise in sea levels'.[22] The President assessed that '[i]n the light of Professor Lowe's evidence, Dr Meinshausen's attempted quantification of the specific impacts of the project was unconvincing' and therefore chose to not accept his evidence.[23]

However, the evidence presented by Lowe and Meinshausen need not be understood as in conflict with one another. Whilst Lowe's pointed to the impossibility of linking specific emissions to a *particular* impact on a *specific* part of Queensland's environment, he emphasised that the mine should be understood as contributing to the cumulative impacts of climate change. Meinshausen's evidence was an attempt to (conservatively) quantify the effect of the cumulative impacts of the additional greenhouse gases from this mine. His quantification of the cumulative impact of the mine is not an assessment of the *specific* impacts of the emissions from that mine, but rather an analysis of the impact the quantity of emissions generated by the mine could have on the climate system. From a scientific point of view, what matters to the climate system is the emission, and not *where* or *when* or *over what time frame* it occurs. From this, Meinshausen was able to estimate that the quantity of emission the mine would produce would lead to an increase in projected atmospheric concentration by 0.11ppm by mid-century, and thus offer a 'best estimate (that the) additional warming due to the boom of the mine would be about 0.006°C by 2020'.[24] From this he could

---

20 Ibid., 550.
21 Ibid., 551.
22 Ibid., 552.
23 Ibid., 552.
24 Exhibit 103, para 32, cited in FoE Closing Submissions, ibid., 149.

estimate that the emissions from the mine could generate a small sea level rise of 0.23 centimetres in equilibrium, with the resultant effect of flooding an additional 23,000 people's homes. These estimates do not link in a line of direct causation *specific* emissions and *particular* effects, but rather project and model the impacts of increased in aggregate cumulative emissions.

The demand to demonstrate a direct chain of causation between emissions from a mine and specific harms denies the 'complex nature of the climate change problem as one that arises because of the *cumulative* effect over time and space of numerous emissions of greenhouse gases from a range of sources'.[25] In the Wandoan case the Court failed to recognise how traditional legal ways of connecting action to harm are incapable of properly addressing 'over-determined' harms such as climate change. Rather than supporting novel deployments of science of demonstrated relations between cause and effect, the Court demanded as a legal requirement something that could not be scientifically demonstrated. As such, the Court shied away from the urgently necessary task of developing legal concepts so that they can address the 'wicked' nature of climate change as an 'over-determined' problem.

B: Scope three emissions

The Wandoan case is one of many in which environmental activists pushed for a broader assessment of the 'impact' of coal mining to also include the contribution that burning extracted coal has on climate change and associated environmental harms. Similar litigation efforts continue; at the time of writing hearings were about to commence in *Australian Conservation Foundation (ACF) v Minister for the Environment*,[26] where ACF are arguing that the Federal Minister for the Environment failed to properly consider the impacts of the mine on climate change and on the Great Barrier Reef, when he re-approved the Adani Carmichael Coal Mine Project in Queensland Galilee Basin. The Federal Environment Minister has reportedly argued in response that the coal from the mine – Australia's largest – would have no 'substantial' impact on climate change, and consequently also not on the Great Barrier Reef.[27]

There is a substantive history of litigation in Australia concerning Scope three emissions and their climate 'impacts', under both the Federal EPBA

---

25 See J Peel, 'Issues in Climate Change Litigation' (2011)(1) *Carbon and Climate Law Review* 15, 17.

26 QUD1017/2015. See a summary on the Environmental Defenders Office Queensland website <www.edoqld.org.au/carmichael-coal-mine-federal-court-challenge/>.

27 M Slezak, 'Greg Hunt: no definite link between coal from Adani Mine and climate change' *The Guardian* (online), 6 May 2016 <www.theguardian.com/australia-news/2016/may/06/greg-hunt-argues-theres-no-definite-link-between-coal-and-climate-change>.

Act and state regulatory schemes.[28] The 2004 *Australian Conservation Foundation v Minister for Planning*[29] decision by the Victorian Civil and Administrative Tribunal was a 'vitally important (decision for) climate change jurisprudence in Australia'[30] because it redressed the failure of International Power Hazelwood to include the impacts of greenhouse gas emissions from burning coal in its environmental impact statement for a proposal to develop an additional coalfield, and thereby 'paved the way for greenhouse gas emissions produced through future burning of the coal to constitute relevant reconsiderations'.[31]

However, although jurisprudence has established that Scope three emissions need to be included in environmental assessments,[32] the real barrier has been in having the 'impact' of such Scope three emissions recognised as significant enough to prevent destructive projects. In the 2006 case *Wildlife Preservation Society of Queensland Proserpine/Whitsunday Branch Inc v Minister for Environment and Heritage*[33] an environmental group argued that two proposed mines should be considered a 'controlled action' under the EPBC Act because the increased greenhouse gas emissions produced from the mine would cause climate change and thereby impact on the Great Barrier Reef. In this case, the Scope three emissions had been included in an environmental assessment; however, the Minister found the alleged impacts to be too speculative and not 'likely' and the Court concurred. Justice Dorsett wrote that he was 'far from satisfied that the burning of coal at some unidentified place in the world, the production of greenhouse gases from such combustion, its contribution towards global warming and the

28 I note however, that President MacDonald drew a clear distinction between the use of the term 'impact' in the EPBC Act and the Queensland legislation, *Xstrata*, 569.

29 [2004] VCAT 2029 (29 October 2004).

30 T Bach and J Brown, 'Recent developments in Australian climate change litigation: forward momentum from down under' (2008)(Winter) *Sustainable Development Law and Policy* 39, 41.

31 Ibid.

32 *Gray v Minister for Planning* (2006) 152 LGERA 258 was a challenge by environmental activist Peter Gray of the New South Wales Director of Planning's decision to accept for public display the environmental assessment by Centennial Coal Company Limited regarding their proposed Anvil Hill Mine on the basis that the assessment failed to include the Scope three emissions of the mine. The decision on 27 November 2006 by Judge Nicola Pain of the New South Wales Land and Environment Court replied on ecologically sustainable development principles and found that the assessment should have included Scope three emissions. (For a discussion, see A Rose, '*Gray v Minister for Planning*: The rising tide of climate change litigation in Australia' (2007) 29 *Sydney Law Review* 725).

33 *Wildlife Preservation Society of Queensland Proserpine/Whitsunday Branch Inc v Minister for Environment and Heritage* (2006) 232 ALR 510.

impact of global warming upon a protected matter, can be so described [as an impact of the proposed coal mines]'.[34]

In Queensland, a previous case had challenged the expansion of an Xstrata mine in Central Queensland, arguing that the resulting emissions would contribute to climate change and that Xstrata should be required to avoid, reduce or offset those emissions. In the initial judgment in the Queensland Land and Resources Tribunal, *Re Xstrata Coal Queensland Pty Ltd*,[35] President Koppenol recommended the mine be approved without any conditions and relied in part on material denying climate change.[36] He wrote, 'I am not satisfied that . . . a demonstrated causal link between this mine's GHG emissions and any discernable harm – let alone "any environmental degradation" – caused by global warming and climate change has been shown'.[37] The case was successfully appealed to the Supreme Court of Queensland,[38] however, it could not be remitted for rehearing as the Queensland government passed legislation that validated the approval of the mine.[39]

The evidence presented in the Wandoan case of analysing the impact of the Scope three emissions from the mine in the context of a global 'carbon budget' was an attempt to address these barriers.

C: Carbon budget

The most interesting element of the Wandoan case was the expert evidence put forward about the mine's contribution to the global 'carbon budget' that did not feature in the Court's reasoning. The concept of a global 'carbon budget' was still quite novel at the time of the Wandoan decision, but has subsequently been further developed and popularised. Key to the popularisation of the concept was a *Rolling Stone* article by environmentalist Bill McKibbon, published just a few months after this decision was handed down.[40]

The concept of a global 'carbon budget' was also utilised in the Intergovernmental Panel on Climate Change (IPCC) in its Fifth Assessment Report (AR5). It concludes that cumulative anthropocentric greenhouse gas

34 Ibid., 524.
35 *Re Xstrata Coal Queensland Pty Ltd* [2007] QLRT 33 (15 February 2007).
36 See the critiques of the Stern Review and the Fourth Assessment Report of the Intergovernmental Panel on Climate Change *Summary for Policy Makers* 16–20.
37 *Re Xstrata Coal Queensland Pty Ltd* [2007] QLRT 33 (15 February 2007) 21.
38 *Queensland Conservation Council Inc v Xstrata Coal Queensland Pty Ltd* [2007] QCA 338 (12 October 2007).
39 *Mining and Other Legislation Amendment Act 2007* (QLD), section 43B inserted a new section 579A into the *Environmental Protection Act* 1994, which validates the amended environmental authority MIM800098402.
40 B McKibbon, 'Global warming's terrifying new maths' *RollingStone* (online), 19 July 2012, <www.rollingstone.com/politics/news/global-warmings-terrifying-new-math-20120719>

emissions since the industrial revolution need to be limited to one trillion tonnes (1000gtCO$_2$e). Given that by 2011 approximately 515GtCO$_2$e of this 'budget' had already been emitted, it found, similarly to the materials canvassed in this judgment that only 485GtCO$_2$e could be emitted if dangerous climate change was to be avoided. The IPCC AR5 also endorsed the emphasis that Meinshausen placed on cumulative emissions rather then timetable based emission reductions.

The concept of a global 'carbon budget' has subsequently been deployed by climate movements to argue why the vast majority of proven fossil fuel reserves need to stay underground. This rewritten judgment draws heavily on the expert evidence presented, but not legally utilised in the Wandoan case, to suggest ways in which the law can draw upon and utilise such scientific information or support different lines or legal reasoning. Such legal innovations are urgently necessary if we are to still minimise the already devastating impacts of climate change.

LAND COURT OF QUEENSLAND

Xstrata Coal Queensland Pty Ltd and others v Friends of the Earth – Brisbane Co-Op Ltd and others and Department of Environment and Resource Management [2012] QLC 013

Judgment

Delivered on: 27 March 2012

A: Background

This case concerned several objections to mining leases and associated environmental authority granted to Xstrata Coal Queensland Pty Ltd, ICRA Wandoan Pty Ltd and Sumisho Coal Australia Pty Ltd for an open cut coal mine 350 kilometres north-west of Brisbane, in the Surat Basin near the township of Wandoan. This mine would produce 1.2 billion tonnes of thermal coal over its 30 years lifetime, approximately 30 million tonnes per year, intended for export. The mining leases and associated authority were granted pursuant to *Mineral Resources Act 1989* (QLD) (MRA) and the *Environmental Protection Act 1994* (QLD) (EPA).

B: The objection

Friends of the Earth Brisbane – Co-Op Ltd (FoE) have lodged an objection based on the contribution that the mine will make to global climate change and related ocean acidification. Their objection is based on the considerations in section 269(j)–(l) of the MRA and sections 5 and 223 of the EPA. They have argued that the Scope three emissions need to be considered in an assessment of impact under the MRA and EPA, and that the public right and interest is prejudiced due to the contribution the mine will make to climate change. Therefore, they argue, the mine is not consistent with principles of ecologically sustainable development as set out in the *National Strategy for Ecologically Sustainable Development*.

The other objectors are landowners with neighbouring properties to the proposed mine site and their objections are based upon noise, dust, groundwater and other issues.

Given this, the Court recommends that the mining lease should be withdrawn due to the impact extracting and burning coal will have on the global climate. Therefore, it is not necessary for this Court to consider the other more narrow points of objection raised.

C: Climate change

1: The EIS

The environmental impact statement (EIS) prepared by the proponents describes the greenhouse gas emissions of the proposed mine over its lifetime. It sets out the Scope one (direct greenhouse gas emissions from sources that are owned or controlled by a company), Scope two (indirect greenhouse gas emissions from the generation of purchased electricity consumed by the company) and Scope three (all other indirect greenhouse gas emissions resulting from the company's activities, but occurring from sources not owned or controlled by the company) emissions. The Scope three emissions of this project refer to those from the transportation of coal to the ports, its shipping and the end-use of the coal in electricity production. The majority of these emissions would take place in Asia and South America where it is proposed the coal extracted would be exported to. Under the international greenhouse gases accounting framework established by the Kyoto Protocol to the United Nations Convention on Climate Change (UNFCCC) these greenhouse gas emissions are therefore not attributed to Australia, but to the importing jurisdictions.

The scope one and two emissions released over the proposed mine's lifetime have been estimated as 17.7 $MtCO_2e$.[41] The Scope three emissions related to the mine have been estimated as 1.32 $GtCO_2e$ over its lifetime, with 99 per cent of these emissions due to the combustion of coal. The

---

41 Xstrata Coal, *Wandoan Coal Project: Environmental Impact Statement* (December 2008), section 14.4.

combined Scope one, two and three emissions from the mine is estimated to represent 0.17 per cent of annual global greenhouse gas emissions.

2: The meaning of 'impact'

This Court is required to consider the 'adverse environmental impact caused by these operations'. FoE submitted that such an assessment should consider Scope one, two and three emissions. The legislation defines 'operations' as 'the operations to be carried out under the authority of the proposed mining lease' (section 269(4)(i)), namely 'to mine the mineral or minerals specified' (pursuant to section 234). The applicant submitted that a consideration of 'impact' should only address the environmental effects of the mining operation as defined in section 6A as 'winning mineral from a place where it occurs', 'extracting mineral from its natural state' or 'disposing of mineral'. This proposed approach to statutory interpretation goes counter to the purposes and objectives of the Act. In this context therefore, any consideration of the 'impact of these operations' must necessarily take into account the effects of activities that are inherently an ancillary result of mining operations, namely the burning of the extracted coal. Further, any common sense consideration of the impacts of 'extracting material from its natural state' must necessarily assess the uses that will be made of the material extracted, and the fact that *but for* the process of mining, the greenhouse gas emissions from burning this coal would not enter the atmosphere.

There is little statutory guidance in how to interpret the term 'impact' and I accept that the ordinary usage of the term as defined by the *Macquarie Dictionary* as 'influence or effect [exerted by a new idea, concept, ideology, etc.]' should be adopted. Several cases have interpreted the terms 'impact' in relation to the *Environmental Protection (Biodiversity Conservation) Act 1999* (Cth) (EPBC Act) and it is appropriate to look to this jurisprudence for guidance. The Full Court of the Federal Court has held that '"impact" in its ordinary meaning can readily include the "indirect" consequences of an action and may include the results of acts done by persons other than the principal actor'.[42] The Court indicated that 'all adverse impacts' 'includes each consequence which can reasonably be imputed as within the contemplation of the proponent of the action, whether those consequences are within the control of the proponent or not'. Therefore any assessment of the 'adverse environmental impact caused by these operations' must consider indirect downstream effects including all Scope three emissions, which are an inevitable consequence of the extraction and mining of coal.

---

42 *Minister for the Environment and Heritage v Queensland Conservation Council Inc* (2004) 139 FCR 24, 53–57.

The proponents have also argued that these Scope three emissions will result from the burning of coal internationally, outside of Australia and that these impacts are therefore outside the jurisdiction of this Court. This Court is required to assess the 'adverse environmental impact' of these operations. The MRA and the EPA both define 'environment' as 'ecosystems and their constituent parts, including peoples and communities' (section 8(a)). When considering, therefore, the impact of these operations on climate change, this Court is required to adopt a global conception of the 'environment', given the global nature of the climate system. Regardless of which country or jurisdiction the coal mined from this proposed mine would be burnt in, it will have the same impact on the global climate system. In assessing the 'adverse environmental impacts' of this proposed mine, therefore, this Court is required to consider all impacts arising from the extraction of this coal from the ground, regardless of where the coal is burnt and its greenhouse gas emissions released into the atmosphere.

3:   Climate science and the question of adverse impact

The Intergovernmental Panel on Climate Change (IPCC) in its *Fourth Assessment Report* (2007) found that global mean temperatures increased by 0.74°C by between 1906–2005 and that this increase is very likely (>90 per cent) due to anthropogenic emissions of greenhouse gases from the combustion of fossil fuels, agriculture, and land-use changes. The proponents did not contest this scientific evidence of anthropocentric climate change.

I was very impressed by the scientific material submitted by the objectors on climate change and the expert witnesses they put forward. Emeritus Professor Ian Lowe told the Court that present atmospheric levels of greenhouse gases are 390ppm and increasing by approximately two ppm annually. He stressed that greenhouse gas emissions are 'additive',[43] and that the impacts of greenhouse gas emissions from this proposed mine should be assessed in terms of how they contribute to the cumulative impacts of global warming and climate change.[44] He argued that but for the process of mining, the $1.3GtCO_2e$ that would be released from the burning of this coal would otherwise be trapped underground in geological structures, completely isolated from the atmosphere and unable to contribute to global warming.

Professor Lowe also gave evidence that it is wrong to argue, as the proponents do, that 'the mining of this coal will not make any difference to global

---

43 I Lowe, 'A brief summary of the science of global warming and climate change', report prepared for an objections hearing in the Land Court of Queensland regarding the proposed Wandoan Coal Mine, 3 August 2011 ('Lowe evidence') [32].
44 Ibid.

warming because if this mine does not proceed the coal will just come from another mine somewhere in the World'. In response, Professor Lowe argued that 'coal is a finite resource, so the mining and use of the coal from this mine will release to the atmosphere fossil carbon that would otherwise be trapped in the ground'.[45] He further gave evidence that:

> Such reasoning also ignores the growing recognition that reasonable and practicable measures should be required to avoid, reduce or offset the greenhouse gas emissions from all human activities, including the proposed mine. Global warming and climate change are massive problems for society that, ultimately, need to be addressed through action at the level of individual projects such as this proposed mine.[46]

I accept the scientific evidence presented by Professor Lowe and his argument that any consideration of the 'impact' of the greenhouse gas emissions from the proposed mine needs to consider the cumulative impact of greenhouse gas accumulation in the atmosphere.

The arguments raised by the proponents speak to the complex legal and policy challenges that climate change represents. Climate change has been described as a 'diabolical policy challenge'[47] and as a 'super-wicked problem'[48] for policy makers because of the multiple and diverse sources of greenhouse gas emissions. It is therefore an 'over-determined' problem that presents acute collective-action challenges. The fact that greenhouse gas emissions are produced by millions of activities all around the world cannot be deployed as an argument to evade responsibility for emissions that are within our control to prevent.

International climate law has articulated that the stabilisation of greenhouse gases at a 'safe' level is a 'common concern' for all.[49] This principle suggests an international obligation to take responsibility for mitigation wherever possible. This responsibility is not diminished because other jurisdictions fail to take comparable action. I reiterate Professor Lowe's comment that 'we cannot influence what other jurisdictions do, but we can act responsibly ourselves'.[50] Moreover, one would hope that by showing leadership on such questions we may encourage other jurisdictions to do likewise.

---

45 Ibid., [33].
46 Ibid.
47 R Garnaut, *Garnaut Climate Change Review* (Cambridge University Press, 2008) xvii.
48 R Lazarus, 'Super wicked problems and climate change: restraining the present to liberate the future' (2009) 94 *Cornell Law Review* 1153.
49 *United Nations Framework Convention on Climate Change*, opened for signature 5 September 1992, 771 UNTS 107 (entered into force 21 March 1994) Preamble.
50 Transcript, 30 August 2011, 6–50.

In order to consider whether this proposed mine would have an 'adverse impact' on the global environment, this Court does not need to establish a clear link between the emissions from this specific mine and specific climate change impacts. Such demands for direct causal attribution do not align with the science on climate change. It is enough for this Court to be satisfied that the emissions from this mine will increase the cumulative greenhouse gas emissions in the atmosphere, that these accumulations will have negative climate impacts and that these impacts will manifest in specific ways, including in Queensland.

The key questions for this Court is therefore not one of causation but of the degree of impact the emissions from this proposed mine will have. That is, this Court is required to assess the degree to which the greenhouse gas emissions as a result of this proposed mine would contribute to global climate change and to evaluate whether this contribution is acceptable or impermissible. Such an evaluation present tremendous challenges.

In evaluating the degree of impact the proposed mine would have, this Court is greatly guided by the expert evidence provided by the objector's witnesses. Professor Lowe asked the Court to place the potential CO_2e emissions from this mine in context. He presented some startling figures to this Court, namely that the annual average emissions from the proposed mine (41.7 million tonnes) equates to:

- about 7.4 per cent of annual Australian emissions;
- approximately 0.085 per cent of current annual global emissions;
- more than than the annual emissions of New Zealand (approximately 32.6 million tonnes in 2007); or
- more than (at peak production) the annual emissions of Ireland (approximately 44.3 million tonnes in 2007).

Professor Lowe also presented evidence to this Court that the lifetime emissions from this proposed mine ($1.311GtCO_2$) is equivalent to:

- two years and four months of Australian emissions;
- one year and seven months of Germany's annual emissions (at 2009 levels);
- two years and eight months of UK emissions;
- three years and five months of French emissions;
- over 39 years of New Zealand emissions; or
- about two point seven per cent of the current annual global emissions.[51]

51 Exhibit 102, 33, 36.

4: The concept of a 'carbon budget'

In assessing whether this mine's contribution to global greenhouse gas emissions should be legally considered impermissible, this Court was helpfully guided by the evidence of Dr Malte Meinshausen.[52] His evidence helpfully introduced the concept of a 'carbon budget' to this Court. In his expert evidence he presented to this Court the information about the global 'carbon budget' or the maximum amount of greenhouse gas emissions that can be emitted between now and 2050 without causing 'dangerous climate change'. He defined 'dangerous climate change' as warming greater than 2°C above pre-industrial levels.[53] The objective of limiting warming to 2°C was articulated in the controversial 2009 Copenhagen Accord[54] and confirmed in the 2010 Cancun Agreements by the Conference of the Parties (COP) to the UNFCCC.[55]

Dr Meinshausen presented evidence to this Court quantifying the remaining global 'carbon budget' for between 2000 and 2050 that cannot be exceeded, in order to have a two per cent, a 25 per cent and and 50 per cent chance respectively, of limiting global warming to 2°C. Further, he explained that $357GtCO_2e$ of this 'carbon budget' has already been used up in the

52 Dr Meinshausen is based at the School of Earth Sciences, University of Melbourne, Australia and the Potsdam Institute for Climate Impact Research, Potsdam, Germany. Dr Malte Meinshausen, 'Contribution of the Wandoan Coal Mine to climate change and ocean acidification', report prepared for an objections hearing in the Land Court of Queensland regarding the proposed Wandoan Coal Mine, 3 August 2011 ('Meinshausen evidence').

53 Although this Court accepts Dr Meinshausen's evidence and the claim that 2°C of warming has come to be internationally seen as the threshold of 'dangerous' climate change, I note that many activists, policy-makers and scientists consider limiting warming to 2°C as already too dangerous, suggesting that nothing over 350ppm can constitute a 'safe' level of atmospheric greenhouse gas emissions. See for example J Hansen, M Sato, P Kharecha, D Beerling, R Berner, V Masson-Delmotte, M Pagani, M Raymo, D L Royer and J C Zachos, 'Target atmospheric CO_2: where should humanity aim?' (2008) 2 *The Open Atmospheric Science Journal* 217–231. Both the Copenhagen Accord and the Cancun Agreements also refer to a commitment to reconsider this target and to consider limiting warming to 1.5°C. These clauses are the product of important lobbying work done by small-island States in particular, for which climate change poses an existential threat. Given that current atmospheric concentrations are 390ppm, this suggests that the atmosphere's capacity to absorb greenhouse gas emissions has *already* been exceeded.

54 The Copenhagen Accord was a last minute agreement reached by a small group of countries behind closed doors. It was opposed by several countries, and as consensus on the Accord could not be reached it was simply 'noted' by the COP in Decision 2/CP.15 'Copenhagen Accord' FCCC/CP/2009/11/Add 1 (30 March 2010).

55 Decision 1/CP.16 'The Cancun Agreements: outcome of the work of the Ad Hoc Working Group on long-term cooperative action under the Convention' FCCC/CP/2010/7/Add 1 (15 March 2011).

Carbon budget 2000–2050 (GtCO$_2$)	Carbon budget 2011–2050 (GtCO$_2$)	Probability of exceeding 2°C (%)	Probability of remaining beneath 2°C (%)
886	529	20	80
1000	643	25	75
1437	1080	50	50

Figure 10.1 Carbon budget to 2050 to achieve less than 2°C warming

Source: Malte Meinshausen, 'Contribution of the Wandoan Coal Mine to climate change and ocean acidification' (Report prepared for an objections hearing in the Land Court of Queensland regarding the proposed Wandoan Coal Mine), available at http://envlaw.com.au/wp-content/uploads/wandoan12.pdf, 10.

period between 2000 and 2011. Therefore, he showed that in order to have a 20 per cent chance of limiting warming to 2°C, no more than 1080GtCO$_2$ can be emitted (see table above). There is no reason to contest this evidence.

5: The precautionary principle

Climate science is based on probabilities, and various scenarios are predicted with varying degrees of likelihood. In assessing different scenarios this Court is required to apply the precautionary principle. The 'precautionary principle' forms a key part of international environmental law[56] and has been incorporated into Australian law.[57] This Court, therefore, will consider the impact of this mine in relation to a global carbon budget that provides an 80 per cent chance of staying within 2°C.

This Court therefore needs to assess whether the emission of 1.32 GtCO$_2$e from this mine would make an impermissible impact on the global environment, in a context where the total global 'carbon budget' is 529 GtCO$_2$e for the 2011–2050 period. Dr Meinshausen's evidence documented that the

56 See in particular, Rio Declaration on Environment and Development UN Doc A/CONF 151/26 (vol I); 31 ILM 874 (1992).
57 In 1992 the precautionary principle was listed as one of the four principles intended to inform environmental policy and programme as part of the Intergovernmental Agreement on the Environment (IGAE) between the Commonwealth, States and Australian Local Government Association. The principle of precaution was also included in the National Strategy for Ecologically Sustainable Development released in 1992. This states 'where there are threats of serious or irreversible environmental damage, lack of full scientific certainty should not be used as a reason for postponing measures to prevent environmental degradation'.

Carbon budget 2011–2050 (GtCO₂)	Probability of exceeding 2°C (%)	Probability of remaining beneath 2°C (%)	Percentage contribution of mine to remaining budget based on 1.311 GtCO₂ of emissions (%)	Fraction mine contributes to budget based on 1.311 GtCO₂ of emissions
529	20	80	0.25	1/403rd part
643	25	75	0.20	1/490th part
1080	50	50	0.12	1/824th part

Figure 10.2 The contribution of the Wandoan Coal Mine to the carbon budget up to 2050 to achieve less than 2°C warming (Meinshausen evidence)

Source: Malte Meinshausen, 'Contribution of the Wandoan Coal Mine to climate change and ocean acidification' (Report prepared for an objections hearing in the Land Court of Queensland regarding the proposed Wandoan Coal Mine), available at http://envlaw.com.au/wp-content/uploads/wandoan12.pdf, 18.

percentage contribution of this mine to the total global 'carbon budget' is 0.25 per cent, or a 1/403rd part (see table above).

6: 'Unburnable carbon'

During my evaluation, I have found other aspects of Dr Meinshausen's evidence helpful in approaching this question. Dr Meinshausen also presented evidence to the Court that if all the proven recoverable reserves of coal were dug up and burnt, this would be equivalent to adding an extra 2780GtCO₂e of greenhouse gasses to the atmosphere. Another way of looking at this information is that to stay within the 'carbon budget' *at most* one fifth of the proven reserves of coal can be extracted and burnt. *This means that planetary survival requires that four-fifths of the known reserves of coal must remain under the ground.* If we were to burn all the available coal resources (compared to reserves) it would lead to much higher temperature rises and an unrecognisable future.[58]

58 This analysis is confirmed by the recent *Unburnable Carbon* report prepared by the Carbon Tracker Initiative. The total carbon potential of all the fossil fuel reserves in the world comes to 2795 GtCO₂, 65 per cent from coal, 22 per cent from oil and 13 per cent from gas. The fossil fuel reserves that are owned by the top 100 listed coal and top 100 listed oil and gas companies comes to 745 GtCO₂. Extracting all these reserves would already overshoot the global carbon budget. See Carbon Tracker Initiative, *Unburnable Carbon – are the world's financial markets carrying a carbon bubble?* <www.carbontracker.org/wp-content/uploads/2014/09/Unburnable-Carbon-Full-rev2-1.pdf>.

I note that the proponents do not contest this evidence about the unsustainability of continued fossil fuel extraction. Xstrata in their evidence acknowledges that 'current and predicted levels of greenhouse gas emissions associated with the burning of fossil fuels ... under a business-as-usual scenario are unsustainable'.[59]

I note that aspects of Dr Meinshausen's evidence suggested that the cumulative effect of emissions from all the *already existing* coal mines, gas and oil fields might already overshoot the global 'carbon budget'. This evidence therefore suggests that *any new fossil fuel development is highly irresponsible.*

There are also strong principles of global equity why Australia should take a proactive stance in preventing further fossil fuel emissions. The principle of 'common but differentiated responsibility and respective capabilities' articulated in the *United Nations Framework Convention on Climate Change* states that countries in the global North, which have the greatest historical responsibility for climate change and the greater capacity to take action, should take the lead in mitigation.

D: Conclusions

This Court finds that the approval of the Wandoan Coal Mine represents a significant step towards exceeding the critical thresholds of atmospheric CO_2 levels. Given that having a reasonable change to limiting temperature rise to 2°C requires that 80 per cent of proven fossil fuel reserves remain underground, the extraction of further coal from this mine is highly irresponsible.

The objectors, FoE, submitted that '[i]t is difficult to imagine an issue associated with a recommendation that raises more squarely the public right and interest'.[60] The future livability of our planet is at stake in these questions as well as the lives and livelihoods of millions. As such, this Court finds that several good reasons have been demonstrated as to why this mine proposal, and other similar mine proposals, should not be allowed to go ahead.

---

59 Exhibit 47, page 67 of the extract of the Xstrata plc Sustainability Report 2010 attached to the affidavit of Ms McCarthy cited in *Friends of the Earth Brisbane Co-Op Ltd Closing Submissions*, [198].
60 *Friends of the Earth Brisbane Co-Op Ltd Closing Submissions*, FoE closing submissions, [43].

Chapter 11

Coast and Country Association of Queensland Inc v Minister for Environment and Heritage Protection[1]

Kate Galloway

Commentary

A: Introduction

Mining law and its close relative environment and planning law tend to resonate in the public imagination as intimately connected with land. Yet as creatures of statute, these fields of law are largely procedural. Their operation falls within the realm of executive discretion, to be exercised in accordance with a complex statutory framework. Integral to the framework is provision for objections to mining activity, requiring adjudication between the miner, the decision-maker, and the objectors.

Arising from the statutory framework, litigation concerning the carrying out of mining activities tends to be framed as judicial review of the executive decision. This draws the parties and the judges into the statutory framework, and the legal exercise is one of statutory interpretation rather than anything approaching an intuitive examination of the land, of nature, or of human experience. Ecology, often intrinsic to objections, is extrinsic to the decision-making process and its revision by the Court.

The role of the Court as interpreter of parliamentary intention is thus ostensibly bound to adhere to the boundaries it perceives as enacted by the legislation. Within this approach to legal reasoning there is ostensibly no room for recognition of nature as a subject of the law, or for an injunction on activities that damage ecology – such as climate change – unless these have originally been provided for within the declared intent of the statute itself.

---

1 *Coast and Country Association of Queensland Inc v Minister for Environment and Heritage Protection* [2015] QSC 260.

In the Queensland decision of *Xstrata v Friends of the Earth*[2] for example, the miner agreed on, and the Court accepted the effect of, climate change. But the decision could not take account of the effects of climate change wrought by the mining activities because this was a factor too remote to fall within the prioritization of environmental harms afforded by Parliament. The status of climate change as a (non)factor was affirmed in *Hancock Coal v Kelly*[3] where the Court effectively prioritized the economic benefits of the mine proposal over environmental considerations. The Court's decision in this respect was thus based not on environmental considerations, but rather involved a question of statutory interpretation as to the relative weight afforded to the relevant factors.

This largely technical approach is typical of environmental law.[4] Using such an approach, the same conclusion has been reached more recently in Queensland where the Court noted that there would be no net effect of the mining on climate change as the coal will simply be sourced elsewhere, and that where emissions did exist they would be negligible. In each case, the Court found that the public interest test under the *Mineral Resources Act 1989* (Qld) ('MRA') would not be satisfied.[5] Similar conclusions have been reached elsewhere.[6]

B: Legal context

Hancock Coal involves objections to coal mining activity in the Galilee Basin in Queensland. The miner – backed by an initial government approval – argued that its Alpha Mine involved 'significant infrastructure' that would open up further mining activity within the Galilee Basin. The objections centred on groundwater, climate change, economics, ecology, and surface water.[7]

In hearing the objections, the Land Court declared its status as a 'creature of statute'[8] tasked with implementing the role prescribed by the legislature, namely determining objections under the MRA and the *Environmental*

2 *Xstrata Coal Queensland Pty Ltd v Friends of the Earth – Brisbane Co-Op Ltd (No 2)* [2012] QLC 67 ('*Xstrata v Friends of the Earth*').
3 *Hancock Coal Pty Ltd v Kelly (No. 4)* [2014] QLC 12 ('*Hancock Coal*').
4 See generally, K Tranter, 'Return to Green Foundations: Liberation and Survival' (1999) 8(2) *Griffith Law Review* 280.
5 *Adani Mining Pty Ltd v Land Services of Coast and Country Inc* [2015] QLC 48 pp. 447, 456.
6 See e.g., *Anvil Hill Project Watch Association Inc v Minister for the Environment and Water Resources* (2007) 243 ALR 784.
7 *Hancock Coal* [2014] QLC 12 [131].
8 *Hancock Coal* [2014] QLC 12 [8].

Protection Act 1994 (Qld) ('EPA'). The Court found against the mining permit, or in the alternative, in favour of the mining permit provided that the relevant statutory processes under the *Water Act 2000* (Qld) made provision for groundwater.

While explicitly involving environmental considerations, the decision engages in the legal process of statutory interpretation. In doing so, and in accordance with accepted principles of statutory interpretation, it declares the paramountcy of parliamentary intention.[9]

What the Court does not disclose is the implicit privileging of anthropocentric considerations, principally within the market economic paradigm. The Court's job of interpretation of the statute masks the complicity of its role in the subjugation of the natural world as a discrete bundle of resources for the satisfaction of unchecked human economic growth.

C: Contribution of a wild law perspective

The recommendations made in *Hancock Coal* were the subject of a judicial review in the Queensland Supreme Court.[10] Relevant to this chapter, the basis of the application was that the Member should not have allowed the mine at all, taking into account scope three greenhouse gas emissions,[11] and a 'net benefit' test taking all relevant criteria into account. On appeal, the Court upheld the original decision. For scope three emissions, it was a relevant consideration that the net effect of the emissions would remain the same, whether the coal were mined in Queensland or not. It was correct for the Member to have taken account of what was likely to occur overseas in terms of alternative sources of coal producing the same (or worse) emissions.[12] As for the 'net benefits test', the Court found that there was no such language used in the statute. It was not a concept, therefore, which the Court must take into account. Instead, the Court was constrained only by the factors listed in section 269(4) of the MRA.[13]

This rewriting of the judicial review explores the application of the law of statutory interpretation through a wild law lens, invoking the elements

9 See e.g., *Hancock Coal* [2014] QLC 12 [144].
10 *Coast and Country Association of Queensland Inc v Smith; Coast and Country Association of Queensland Inc v Minister for Environment and Heritage Protection* [2015] QSC 260 (4 September 2015) ('*Coast and Country*').
11 'Scope three emissions' of greenhouse gases are 'an optional reporting category that allows for the treatment of all other indirect emissions . . . a consequence of the activities of the company . . . [occurring] from sources not owned or controlled by the company'. *Hancock Coal* [2014] QLC 12 [204].
12 *Coast & Country* [2015] QSC 260 (4 September 2015) [46].
13 Ibid., 21.

of Earth jurisprudence.[14] It challenges these two bases for assessing the Minister's decision on the grounds that they fail to represent an ecological view of the processes at play in the mining activities that are the subject of the dispute. The key difference between this judgment and the original appeal is the shift from an implicit anthropocentric (and economic) bias in the decision, towards a reading of the statutory framework that upholds ecological integrity in a holistic way.

Contrary to the law's tendency to compartmentalise the environment into discrete objects, Earth jurisprudence situates the earth at the centre of the moral community.[15] No mere object of human dominion, the earth is itself a moral subject, part of an interconnected, or networked, community that is described in scientific terms by the discipline of ecology.

For the positivist, Earth jurisprudence might appear to contradict legal doctrine. However, there is a long history of its principles embedded within the law. In the English tradition, the *Forest Charter* (1217) represented the intersection of the people, the environment, and the law. The 'forest' was more than the limited concept of a stand of trees. The Charter covered 'not only woodland but also heath, marsh and villages in or nearby woodland'.[16] In other words, it was concerned with what might now be called the environment in its broadest sense.

The legal effect of the Charter together with *Magna Carta* was to constrain the exercise of executive power. Its particular context, however, was the exercise of Crown prerogative over the environment. The Charter afforded rights to the people – of access to natural resources including the capacity to make a living – as well as a framework of good governance. It was no longer open to the Crown to take resources for its own purposes, to the exclusion of the people. The Charter establishes a principle of custodianship of nature – which could otherwise not speak for itself. The *Forest Charter* has stood alongside *Magna Carta* over centuries, informing the development of an English 'rule of law for nature'[17] and, as is argued here, of the principle of legality itself.

14 See e.g., C Cullinan, *Wild Law: A Manifesto for Earth Justice* (Siber Ink, 2002); P D Burdon, *Earth Jurisprudence: Private Property and the Environment* (Routledge, 2015); M Maloney and P D Burdon (eds), *Wild Law: In Practice* (Routledge, 2014).

15 J E Koons, 'Earth Jurisprudence: The Moral Value of Nature' (2008) 25(2) *Pace Environmental Law Review* 263.

16 G Van Bueren, 'More Magna than Magna Carta: Magna Carta's Sister – The Charter of the Forest' in R Hazell and J Melton (eds), *Magna Carta and its Modern Legacy* (Cambridge University Press, 2015) 194, 198.

17 N A Robinson, 'The charter of the forest: evolving human rights in nature' in D B Magraw, A Marinez and R E Wrownell (eds) *Magna Carta and the Rule of Law* (American Bar Association, 2014) 311, 315.

Unlike the techno-centrism of much contemporary environmental law, the *Forest Charter* dealt with the boundaries of executive authority over nature and in doing so evolved to uphold the ecological value of nature in its own right.[18] Charter provisions for the rights of subjects and the Crown to use resources in a sustainable way reflect what might be described as a 'stewardship' approach to environmental management, but the limits imposed on the Crown's domain within the forests prevented its destruction. The Charter's effect and its underlying premise was nature preservation for its own sake. In this respect the Charter represents a long history of the law affording power to the environment, treating it as the subject of rights and positioning humans as its custodian.

While *Magna Carta* represents anthropocentric concerns now incorporated within human rights, this judgment imagines the *Forest Charter* analogously providing a companion ecocentric bill of environmental rights. As such, just as common law human rights inform the principle of legality that guides statutory interpretation, so too does the principle of legality encompass principles of environmental custodianship and environmental standing as the subject of rights.

Aside from the *Charter* itself, a custodial approach to nature reflects the law and belief systems of indigenous peoples around the world.[19] Important inroads into environmental law have been made in recent years through adoption of these norms into contemporary legal frameworks. For example, in 2008, the Ecuadorian Constitution recognised rights for nature.[20] In 2012, the New Zealand government and the Whanganui Iwi entered into a Record of Understanding and Tūtohu Whakatupua, recognizing 'the Whanganui River as Te Awa Tupua, a living being and entity in its own right, and the unique status of the Whanganui River in relation to Te Awa Tupua and its governance'.[21]

There is a further contemporary context for incorporating principles of environmental custodianship within a reading of the law. The Brundtland Report[22] espoused at an international level the principles of environmentally sustainable development. It reflected the urgency of inevitable global environmental catastrophe in the face of the exclusion of the environment

---

18 Notably from the late nineteenth century to World War II. See ibid., 316.
19 See e.g., N J Turner, M Boelscher Ignace and R Ignace, 'Traditional Ecological Knowledge and Wisdom of Aboriginal Peoples in British Columbia' (2000) 10(5) *Ecological Applications* 1275.
20 Constitution of the Republic of Ecuador (2008) arts 71–4.
21 E C Hsaio, 'Whanganui River Agreement: Indigenous Rights and Rights of Nature' (2012) 42(6) *Environmental Policy and Law* 371, 373.
22 World Commission on Environment and Development, *Report of the World Commission on Environment and Development: Our Common Future* (1987).

from the paradigm of growth and development. As with the *Forest Charter*, it articulates principles of care for the environment. Responsibilities under the Report include intergenerational equity[23] – a reflection of the standing of nature itself as a subject, rather than an object of the law.

Principles of custodianship – of Earth jurisprudence – have informed the law of Queensland, enacted explicitly in the EPA and by implication also in the MRA. Without recognising the interrelationship of all of earth's systems, the principles enacted in these statutes have little force. In terms of Earth jurisprudence, giving meaning to these relationships requires interpreting statutory meaning through an ecological lens. This is a wild law approach to statutory interpretation.

Appreciating the interconnectedness of earth's systems, the pressures on land expressed within the evidence in *Hancock Coal* – impacts on ground-water, surface water, ecology, and climate change – are cumulative, amounting not just to discretely measurable instances but working together on an organism as 'death by a thousand cuts'.[24] Understanding global environmental interconnectedness requires that statutory construction accommodate the holistic environmental legal subject. To do so requires an insight into the nature of statutory interpretation not as simply being a 'neutral' conduit for parliamentary purpose or intention; but as the privileging of a worldview that omits the construction of 'nature' as itself.

This judgment in review seeks statutory intention in furtherance of the aims of wild law. In the same way that the courts fail to make explicit the anthropocentric or the economic ideological foundations of their decisions, so too does this judgment take as given the legitimacy of an Earth jurisprudence approach.

This is an application for statutory order of review under the *Judicial Review Act 1991* (Qld) ('JRA'), of recommendations made by the learned Member of the Land Court. Such recommendations are treated as administrative rather than judicial, invoking the JRA.[25]

The Member made alternative recommendations to the Honorable Ministers responsible respectively for the *Mineral Resources Act 1989* (Qld) ('MRA') and for the *Environmental Protection Act 1994* (Qld) ('EPA'). The alternatives were, first, rejection of MLA 70426 and refusal to the permit under the EPA, or alternatively, conditional approval in each case. The

---

23 Ibid., Annexe 1 289.
24 J Bradsen, 'Biodiversity Legislation: Species, Vegetation, Habitat' (1992) 9 *Environmental and Planning Law Journal* 171, 179.
25 *Dunn v Burtenshaw* (2010) 31 QLCR 156.

SUPREME COURT OF QUEENSLAND

CITATION: *Coast and Country Association of*
                                   Queensland Inc v Minister for Environment
                                   and Heritage Protection & Ors [2015] QSC
                                   260

PARTIES: In SC 4249 of 2014

                                   COAST AND COUNTRY ASSOCIATION
                                   OF QUEENSLAND INC (applicant)

                                   v

                                   PAUL ANTHONY SMITH, MEMBER OF THE
                                   LAND COURT OF QUEENSLAND
                                   (first respondent)
                                   HANCOCK COAL PTY LTD
                                   (second respondent)

FILE NO/S: SC No 4249 of 2014

DIVISION: Trial Division

PROCEEDING: Application for a statutory order of review

ORIGINATING COURT: Supreme Court at Brisbane

DELIVERED ON: 4 September 2015

DELIVERED AT: Brisbane

HEARING DATE: 22 and 23 April 2015

JUDGE: Galloway J

specific concern of the Member in each case rested upon what he found to
be inconclusive reports about groundwater.

Following the recommendations of the Land Court, the MRA Minister
approved the mining application, assuring the EPA Minister that a groundwater condition would be attached to the approval. The EPA Minister
approved the application subject to conditions.

The applicant seeks a review of these decisions on three bases: first, that
the Land Court prioritised possible economic benefits over the environmental and public interest considerations listed in the MRA; second, that the
Member failed to take into account scope three emissions when he should
have; and third, that the Member took into account irrelevant considerations in finding that there would be no net impact on climate change if the
mining activity were refused.

The power of the Land Court to make recommendations is derived
from sections 269 of the MRA and 190 of the EPA respectively. Each

statute provides for matters to be taken into consideration in making the recommendation.[26]

That the two Ministers are involved in considering the Member's recommendations and making a decision in respect of the same land and same activities highlights that the two statutes, the MRA and the EPA, are intimately related. Each forms part of the context for the other, and each is therefore relevant in interpreting the intention of the other. As pointed out in *CIC Insurance Ltd v Bankstown Football Club Ltd*:

> The modern approach to statutory interpretation (a) insists that the context be considered in the first instance, not merely at some later stage when ambiguity might be thought to arise, and (b) uses 'context' in its widest sense.[27]

This application concerns the way in which the Court is required to read these statutes in context, and harmoniously. In deciding whether the learned Member took into account irrelevant considerations, or failed to account for relevant considerations, the task is to ascertain not only the text of the relevant provisions of the MRA and EPA, but their context and purpose – including 'in light of the apprehended mischief sought to be overcome and the objects of the legislation'.[28]

A: Context

The heads of review in this case together concern the amenity of 'the environment' around the proposed Alpha Mine. As confirmed by ecological principles and reflected in Earth jurisprudence, environment is a holistic concept. As such it is greater than a composite of component parts such as water, landscape, vegetation, fauna, air, and noise.

Further, 'environment' is not limited by the legal concept of jurisdiction as geography. Thus mining activities on the land as a bounded spatial locale may have what are known as 'downstream' effects. This includes the so-called 'scope three emissions' of greenhouse gases: 'an optional reporting category that allows for the treatment of all other indirect emissions . . . a consequence of the activities of the company . . . [occurring] from sources not owned or controlled by the company'.[29] These might include in this case, the transportation of the coal, and its end use to produce energy.

---

26 MRA s 269(4) and EPA s 191.
27 (1997) 187 CLR 384, 408.
28 *Thomas v Mowbray* (2007) 81 ALJR 1414, 1439 (Gummow and Crennan JJ).
29 *Hancock Coal* [2014] QLC 12 [204].

The Court must therefore necessarily consider that globalised environmental issues arise from the interconnectedness of human activity from one sphere to another – even over considerable distance.[30] The contemporary context for this occurrence is climate change, considered in some depth in the decision in the Land Court,[31] and for example in *Xstrata v Friends of the Earth*.[32]

In contrast to an ecological conceptualisation of land, and the increasing understanding of the way in which human activity has global effect, Queensland statutes have been interpreted as providing separate enactments for compartmentalised components of land, and a geographically bounded concept both of environment, and of 'operations' affecting the environment. Thus the Land Court felt constrained to decide on the groundwater issue separately from other aspects of the objections, under the *Water Act 2000*. Further, the Member determined that the scope three emissions, while 'real and of concern',[33] could not properly be construed as part of the mining company's 'operations' under the MRA. Yet in determining the economics of the application, these same 'operations' were found to include the export earnings that necessarily involved the sale and transportation of the coal for use overseas.[34]

These examples illustrate the challenge in giving meaning to statutes to fulfill their ecocentric intention. Understanding the legislative context as environmental leads to the issue central to the three bases of appeal in this case – how to read these statutes harmoniously so as to make sense of their overarching environmental purpose.

B: Purpose

As has been observed, the statutes relevant to this decision cite ostensibly different purposes.[35] Further, their subject matter is apparently different. On the one hand, the MRA deals with minerals, located on and under the earth's surface. On the other hand, the EPA deals with 'the environment'.[36] However, the definition of 'environment' in the EPA specifically includes 'all natural and physical resources'. Minerals therefore form part of 'the

---

30 Evidenced by e.g., *United Nations Framework Convention on Climate Change*, opened for signature 9 May 1992, 1771 UNTS 107 (entered into force 21 March 1994).
31 *Hancock Coal* [2014] QLC 12.
32 [2012] QLC 13.
33 *Hancock Coal* [2014] QLC 12 [208].
34 *Hancock Coal* [2014] QLC 12 [258].
35 *Hancock Coal* [2014] QLC 12 [9].
36 EPA s 8.

environment'. This situates the MRA within the ambit of the EPA, its mineral 'object' not as a discrete object but part of an interconnected whole.

I must therefore resolve the relationship between the relative weight attached to 'environmental' issues[37] and the overarching statutory intention that interests of the environment itself remain at the forefront. To do so requires drawing attention to the way the statutes, with their differing purposes and ostensibly different objects, can be read harmoniously together.

To understand the alignment of differential purposes, in the Land Court, Member Smith cited from the judgment of Lord Wilberforce in *Associated Minerals Consolidated Ltd v Wyong Shire Council*. The statement is worth repeating:

> In its wider presentation the argument raises the issue, which frequently arises, of the interrelation in law of two statutes whose field of application is different, where the later statute does not expressly repeal or override the earlier. The problem is one of ascertaining the legislative intention: is it to leave the earlier statute intact, with autonomous application to its own subject-matter; is it to override the earlier statute in case of any inconsistency between the two; is it to add an additional layer of legislation on top of the pre-existing legislation, so that each may operate within its respective field?
>
> Discussion of such questions commonly starts from the use of the maxim *generalia specialibus non derogant* and with citation from the judgment of this Board in *Barker v Edger*:

>> The general maxim is, *Generalia specialibus non derogant*. When the Legislature has given its attention to a separate subject, and made provision for it, the presumption is that a subsequent general enactment is not intended to interfere with special provision unless it manifests that intention very clearly. Each enactment must be construed in that respect according to its own subject-matter and its own terms.

> The principle stated in this passage and others to a similar effect is, of course, unexceptionable, but cases are rarely so simple as this, for even where the earlier statute deals with a particular and limited subject-matter which is included within the general subject-matter with which the later statute is concerned, it is still a matter of legislative intention, which the courts endeavour to extract from all available indications, whether the former is left intact, or is superseded, and the cases in which the latter has been held are almost as numerous as the former.

---

37 Pursuant to MRA s 269(4)(i), (j) and (m) and EPA s 191(g).

Both Acts apply, or are capable of being applied, with complete generality to land in the State of New South Wales. Can they, in relation to
a given piece of land, coexist? In their Lordships' opinion they clearly
can, and do. The Acts have different purposes, each of which is capable
of being fulfilled.[38]

It is useful in this context to examine the respective purposes of the
relevant legislation to ascertain the scope of their application.
Section 2 of the MRA states:

The principal objectives of this Act are to –

(a) encourage and facilitate prospecting and exploring for and mining
    of minerals;
(b) enhance knowledge of the mineral resources of the State;
(c) minimise land use conflict with respect to prospecting, exploring
    and mining;
(d) encourage environmental responsibility in prospecting, exploring
    and mining;
(e) ensure an appropriate financial return to the State from mining;
(f) provide an administrative framework to expedite and regulate pros-
    pecting and exploring for and mining of minerals;
(g) encourage responsible land care management in prospecting, exploring
    and mining.

Section 3 of the EPA states:

The object of this Act is to protect Queensland's environment while
allowing for development that improves the total quality of life, both
now and in the future, in a way that maintains the ecological processes
on which life depends (ecologically sustainable development).

Each of these statutes provides for development *and* environmental protection. As with *Wyong*, '[both] Acts . . . are capable of being applied, with
complete generality to land in the State of [Queensland]'. However, 'can
they, in relation to a given piece of land, coexist?'[39]
The miner maintains that the environmental standards are met in relation to the mining activity, while the objectors maintain that they are not.
The Member has indicated that it is a question of taking into account and

---

38 *Associated Minerals Consolidated Ltd v Wyong Shire Council* [1975] AC 539, 553.
39 Ibid.

weighing the component environmental factors. However, this in itself does not address the ecological purpose of the statutes read as a legislative scheme. Instead, the real question is how to read the various purposes, expressed in terms of listed considerations,[40] so that they mesh as a cohesive regulatory framework.

For the EPA to be given effect, 'mining' under the MRA must be read in light of all of the Act's purposes. If this is the case, then the relationship between the purposes of each Act becomes iterative. Implicitly, and to ascertain the intention of both the Acts, I must integrate the concepts of land, resources, land use, environment, sustainability, and ecology that are repeated within the statutes' purposes, and which must be given meaning in the context of the proposed mining application. The interconnectedness of the natural world explains legislative intention through the use of these terms, and therefore guides the calibration of factors in assessing the application to undertake mining activity.

C: Weighing relevant factors

The EPA states that its purpose will be achieved through implementing an integrated management program in cyclical phases. Phase three is to 'integrate environmental values into land use planning and management of natural resources'.[41] Environmental values are defined in that Act, and thus inform our understanding of Parliament's intention as to such values where the environment is mentioned in the MRA. Both Acts are, after all, engaged in the management of the same object: the environment and its resources.

An environmental value is 'a quality or physical characteristic of the environment that is conducive to ecological health or public amenity or safety'.[42] 'Environment', under section 8, is framed widely but importantly, in ecological terms. Environmental values are informed by the well-known legal principles of ecologically sustainable development, the 'standard criteria' defined in Schedule 4 of the EPA – including the precautionary principle and inter-generational equity.

Parliament has, in expressly articulating these principles, enacted the framework of a wild law approach to land and resource management – including, implicitly, in the operation of the MRA. Contemporary science confirms the holistic principles of Earth jurisprudence and the notion of the interrelatedness of ecological and environmental systems. To the extent that these principles are reflected in the statutes, Parliament is giving expression

---

40 Pursuant to MRA s 269(4) and EPA s 191.
41 EPA s 4(6).
42 EPA s 9(a).

to Earth jurisprudence and to that extent, those principles must inform the Court's interpretation of the statutes.

In discussing presumptions to guide interpretation, Bankowski and MacCormick suggest that 'parliament is presumed not to trespass upon fundamental principles even if it is able to do so by provisions enacted in unequivocally clear terms'. Such presumptions, they suggest, 'operate in a quasi-constitutional way, expressing fundamental rights and liberties of citizens and regulating the relations between parliament, the executive and the courts'.[43]

An Earth jurisprudence approach adopts this same principle in terms of recognition of the environment. While the statutes themselves regulate human behaviour, they are designed to protect environmental values. Recognised within the framework offered in the legislation, it behooves the Courts to read the statutes against this presumption. Had Parliament intended to erode environmental values, or to subordinate environmental values to, say, untrammelled economic growth, it could have said so in unequivocally clear terms.

In fact, it did the opposite. It expressed that development might only go ahead if environmental values were upheld. The presumption to be invoked in interpreting the statutes is that the weighting of the elements in section 269(4) of the MRA is not to trespass upon the fundamental principles of wild law, embodied in the Earth jurisprudence approach of the legislative scheme.

D: Principle of legality

Another way of expressing this idea is through the analogous application of the principle of legality. An iteration of the rule of law, the principle of legality is considered to govern relations between Parliament, the executive and the courts. It operates as a presumption that legislation does not interfere with fundamental rights, freedoms and immunities.[44] It has also been described as a 'common law bill of rights'.[45]

The legislative scheme is not, of course, dealing with human rights. Instead, its intention is to deal with environmental rights. To the extent that

---

43 Z Bankowski and D N MacCormick, 'Statutory Interpretation in the United Kingdom' in D N MacCormick and R S Summers (eds), *Interpreting Statutes: A Comparative Study* (Routledge, 1991) 359, 391.

44 See e.g., *Coco v The Queen* (1994) 179 CLR 427, 437 (Mason CJ, Brennan, Gaudron and McHugh JJ), 446 (Deane and Dawson JJ).

45 J Spigelman, 'The Common Law Bill of Rights', paper presented at the 2008 McPherson Lectures: Statutory Interpretation and Human Rights, University of Queensland, 10 March 2008.

it invokes principles of Earth jurisprudence, it can be considered as treating the environment itself as the legal subject. Reflected in the text of the statutes are the principles of inter-generational equity, of ecological sustainability, and the precautionary principle.[46] These are the hallmarks of an otherwise anthropocentric legal system giving voice to the weak in the face of the strong: of articulating that the environment is central to executive decision-making. They represent an implied ecocentric bill of rights.

While not calling to mind *Magna Carta*-style rights, at the root of the interpretive problem in this case is the extent to which Parliament is engaging the long-held principle of stewardship. Originally embodied in the *Forest Charter* of 1217, this Royal Law evolved over centuries alongside *Magna Carta* implementing boundaries to executive decision-making, recognising the intrinsic value of 'the forest' (now 'the environment'), and articulating the relationship between the people and the Crown to care for and maintain the forest.

This is not to argue that the text of either *Magna Carta* or the *Forest Charter* is literally part of Australian law. The former can, however, be seen as an expression of common law.[47] In a similar way the latter has, over centuries in the common law tradition, informed government's relationship with nature. As Robinson points out:

> The Forest Charter's intangible legacies inhere in the principles that it – together with *Magna Carta* – forged to establish. These include establishing the rule of law and proclaiming the 'liberties of the forest' which shaped foundations for what has become sustainable natural resources law, and in particular regimes for protection of natural areas.[48]

The longevity of the *Forest Charter* is seen in contemporary English legislation but it can also be seen in the approach of the Queensland Parliament. As I have discussed above, the predilection of the common law is to compartmentalise the resource elements of the land as property – conceptually if not physically. However, in the Australian context this has occurred alongside resource regulation and environmental management and protection.[49] The structure of the legislative scheme itself, with its clearly stated environmental objectives, evidences the governmental tendency towards

---

46 EPA sch 4.
47 See e.g., *R v Wright, ex parte Klar* (1971) 1 SASR 103.
48 Robinson, above n 17, 314. Citations omitted.
49 See e.g., an overview of the history of land management principles in Queensland land grants, in S Christensen et al., 'Early Land Grants and Reservations: Any Lessons from the Queensland Experience for the Sustainability Challenge to Land Ownership' (2003) 15 *James Cook University Law Review* 42.

'sustainable natural resources law' – a contemporary iteration of the principles of the *Forest Charter*. Further, Australian governments' adherence to the principles of ecological sustainability derived from international law continues the historical tendency of government towards stewardship and environmental protection.

The ecocentric focus of the *Forest Charter* represents a system of governance under law that speaks for the environment. It iterates the principles of Earth jurisprudence that recognise our responsibility for the environment and the interconnectedness of all life. Like the principle of legality expressed in human rights terms, it stands for environmental rights and the presumption of their prioritisation in executive decision-making.

E: Conclusion and orders

It is clear with regard to the purpose and context of the statutes, that environmental considerations must be weighed holistically. They are to lie at the forefront of decision-making pursuant to the Acts, following wild law principles that are largely and pragmatically articulated in the 'standard criteria' of the EPA.[50]

I: Scope three emissions – relevant considerations

Considering scope three emissions in this context requires a holistic approach, recognising the interconnectedness of all things. It is entirely relevant therefore, to take account of the acknowledged 'downstream' effects of mining operations on climate change. It is a misinterpretation of the statutes' purpose and context to find that the scope three emissions are outside the mining 'operations' that are contemplated by the MRA.

An ecocentric understanding of the mining 'operations', including through application of the precautionary principle, contemplates the effect of all phases of mining: including scope one to scope three emissions. It is impossible to conceive of the operation of coal mining for a global market without including its transportation and sole end use. These effects must therefore fall within the consideration of the Member in terms of section 269(4) of the MRA. Further, they require consideration both under paragraph (j) – whether there will be 'any environmental impact caused by these operations' – and paragraph (k) – 'whether the public right and interest will be prejudiced'. It is noted that the Member found these scope three emissions to be 'both real and of concern'.[51] It is impossible to conceive of the

---

50 EPA s 191(g); sch 4.
51 *Hancock Coal* [2014] QLC 12 [209].

environmental impact of this mine without accounting for concerning emissions with grave environmental consequences. This is therefore a relevant consideration.

2: Scope three emissions – irrelevant consideration

Hancock Coal argued that in any event, environmental harm is negligible because if the Alpha Mine does not go ahead, then another coal mine overseas will fill its place and may do so using even dirtier coal. The Member accepted this argument, but I reject it as irrelevant.

It is relevant to consider global harm, because of the interconnectedness of the earth community. However according to law, the Court only has authority within its own jurisdiction. To achieve the overarching environmental purposes of the legislation, it must interpret the Acts and exercise its discretion over activities that do fall within its jurisdiction. In this case, these activities are the mining operations situated spatially at the Alpha Mine. These operations can – and indeed must – comply with the regulatory framework regardless of what may or may not occur overseas.

The Court can only make an order in respect of that which occurs within its jurisdiction. In this case, that is the mining operations that are the subject of this application. To the extent that they would cause environmental harm if they proceed, they fail to satisfy the requirements of the MRA or the EPA and should be refused. This is Queensland's contribution to protection of the environment, which is a global endeavour.

I note further that the Member took into account the export revenues forecast from the mining operation in considering the economic consequences of the approval.[52] As a question of consistency, the context of the mining operations cannot, for environmental purposes, be constrained to the immediate local 'operations' but for economic purposes be considered in terms of their broader 'downstream' effect.

3: Net benefit argument

While '[t]he Court must balance all of the relevant considerations, including the economic and associated benefits of the project proceeding',[53] to make the decision because of economic considerations implicitly places these ahead of the environmental concerns. While the Member accepted that 'the direct economic benefits from the mine are many',[54] and that there will be 'an

---

52 *Hancock Coal* [2014] QLC 12 [257].
53 *Xstrata v Friends of the Earth* [2012] QLC 67 [576].
54 *Hancock Coal* [2014] QLC 12 [256].

acceptable level of development and utilization of the mineral resource',[55] these findings simply address some of the elements to be considered. They do not address the environmental and public interest considerations enumerated within the relevant provisions in the legislation,[56] and that I find must inform the intention behind the decision-making process itself.

The learned Member's prioritisation of economic concerns over both unresolved groundwater issues and scope three emissions, fails also to accord with the statutory intention. Accounting for the context and purpose of the statutes, including the ecocentric principle of legality, requires the legislation to be read in terms of protection of the environment. Only if the environment is secure, taking account of the 'standard criteria', might the mining activity be approved.

For these reasons, the application is upheld.

---

55 *Hancock Coal* [2014] QLC 12 [260].
56 Enumerated in MRA s 269(4) and s 191 EPA (s 191).

Chapter 12

Exploring fundamental legal change through adjacent possibilities

The *Newcrest* mining case

Aidan Ricketts

Commentary

The case of *Newcrest Mining (WA) Ltd v Commonwealth*[1] (*Newcrest*) exemplifies the tensions between private property rights and environmental protection in rather grand terms. As a constitutional decision, it is of very great importance in relation to the future legislative competence of the Commonwealth Parliament to adequately respond to growing environmental crises, and highlights the tension between good environmental governance and competing anthropocentric property claims.

The case centres around the complex jurisprudence of section 51(xxxi) of the Commonwealth of Australia Constitution. This provides that:

> The Parliament shall, subject to this Constitution, have power to make laws for the peace, order, and good government of the Commonwealth with respect to:-
> (xxxi) The acquisition of property on just terms from any State or person for any purpose in respect of which the Parliament has power to make laws.

This particular section has been interpreted very broadly by High Court judges since Federation as a vehicle for entrenching property rights as against the legislative power of the Parliament.[2] Previously, this section was used to resist socialist programs such as bank nationalisation[3] but more recently, as

---

1 *Newcrest Mining (WA) Ltd v Commonwealth* (1997) 190 CLR 513.
2 S Joseph and M Castan, *Federal Constitutional Law: A Contemporary View* (Lawbook Company, 2010) 385.
3 *Bank of NSW v Commonwealth* (the Bank Nationalisation case) (1948) 76 CLR 1.

seen in the *Newcrest* case, has emerged at the cutting edge of the tension between environmental protection and property rights.

The case concerned attempts by the Commonwealth government to preserve environmentally significant land within an expanded Kakadu National Park and a challenge by a mining corporation to the loss of its rights to undertake certain activities under mining leases it held within the area. The outcome in the *Newcrest* case privileged commercial interests over and above the public interest in better environmental management and the decision expanded the concept of 'acquisition' in a way that seriously undermines the capacity of future Commonwealth governments to respond to environmental issues without facing a massive compensation bill payable to affected industries.

The case exemplifies the way in which property rights, a key value of political liberalism, have become deeply and often invisibly embedded in legal reasoning. The judgment of Justice Kirby in the *Newcrest* case actually features an explicit defence of private property rights as key fundamental rights in a 'civilised society'.[4] Justice Kirby's explicit articulation of the values of political liberalism in this case was unusual, as these values are usually represented in legal judgments *sub silento*, or behind the cloak of politically decontextualised legal reasoning. Such a spirited defence of property rights suggests that Justice Kirby also saw the case as a potential watershed on that issue.

It was for this reason that I chose to rewrite this judgment. My approach, however, has been to explore whether an Earth laws outcome could be achieved using existing, even arguably conservative forms of reasoning. This approach reflects my interest in complexity theory as an approach to achieving eventual fundamental change through seemingly small steps. At the end of my judgment I briefly canvass the Earth jurisprudential issues in the judgment but in a purely legal sense the judgment does not rely upon that discussion.

A: Property rights and Earth laws

Property rights are central to the ontology of classical and modern liberalism.[5] However, from an Earth laws perspective, property rights, particularly those related to land, represent a highly contentious anthropocentric and abstract projection onto the natural environment.[6] The negotiation of property interests between humans is one thing, but the much grander assertion

---

4 *Newcrest Mining (WA) Ltd v Commonwealth* (1997) 190 CLR 513, 659 (Kirby J).
5 N James and R Field, *The New Lawyer* (Wiley and Sons, 2013) 75.
6 N Graham, 'Owning the Earth' in P Burdon (ed.) *Exploring Wild Law: The Philosophy of Earth Jurisprudence* (Wakefield Press, 2011) 259.

of human ownership of the natural environment is another. The commodi-fication of land and of the natural environment conflicts sharply with an ecological or Earth laws perspective on the relationship of the human spe-cies to the environment within which it has co-evolved with the rest of the planetary biosphere.

Whilst many commentators call for fundamental shifts in the way that our legal system understands and responds to the natural world,[7] it remains challenging to postulate the practical ways in which such shifts could be initiated. Consequently it is useful to enquire into practical ways in which major transitions could be achieved by a series of smaller changes.

B: Complexity science and the idea of change through the 'adjacent possible'

Complexity science offers a useful perspective for understanding the way that eventual fundamental systemic change can occur as a result of numer-ous smaller changes that eventually contribute to the emergence of a new and radically different order.[8] Small changes in one part of a system are capable of setting initial conditions in train for massive transformations later, and the exploration of a relatively minor 'adjacent possible' (such as a modest shift in constitutional approach) can expand the field of possibilities for ongoing transition.[9] Whilst there is not sufficient space here for me to fully explore the way complexity theory can provide potential roadmaps for legal system transformation, there is a remarkable similarity between the way that legal theory develops under the doctrine of precedent and evolutionary biologist Stuart Kaufman's description of how adopting novel 'adjacent pos-sibles' can open up a whole new field of evolution in a system:

> When searching the space of possibilities . . . for . . . a different way of doing things, it is not possible to explore all possibilities. It may, how-ever, be possible to consider change one step away from what already exists. In this sense, exaptation may be considered an exploration of what is sometimes called the 'adjacent possible'. That is exploring one

7 P Burdon (ed.) *Exploring Wild Law: The Philosophy of Earth Jurisprudence* (Wakefield Press, 2011).
8 E Mitleton-Kelly, 'Ten principles of complexity and enabling infrastructures' in E Mitleton-Kelly (ed.), *Complex Systems and Evolutionary Perspectives on Organisations: The Application of Complexity Theory to Organisations* (Elsevier Science 2003) 23; J Urry 'The Complexity Turn' (2005) 22(5) *Theory, Culture and Society* 1.
9 A Pelizzon and A Ricketts, 'Beyond Anthropocentrism and Back Again: From Ontological to Normative Anthropocentrism' (2015) 18/2 *Australian Journal of Natural Resources Law and Policy* 119.

step away, using 'building blocks' already available, but put together in a novel way. According to Kauffman (2000: 22) the push into novelty in the molecular, morphological, behavioural, technological and organisational spheres, is persistent and happens through exploration of the adjacent possible. . . . Once discoveries have been realised in the current adjacent possible, a new adjacent possible, accessible from the enlarged actual that includes the novel discoveries from the former adjacent possible, becomes available.[10]

In the context of Australian law, initiating changes at the constitutional level, and for a liberalist system, initiating changes that disturb the primacy of property over nature, are likely to be fertile areas for sowing the seeds of ongoing change.

C: Newcrest: *just terms compensation as a fundamental right?*

The *Newcrest* case is a case in which our High Court, faced with the emerging tension between the old ontology of private property and the emerging impetus for improved environmental governance, chose in my view the wrong direction. The case builds upon the established jurisprudence of section 51(xxxi) by re-affirming prior decisions that treat the subsection as a form of fundamental guarantee to compensation in the event of property acquisition by the Commonwealth. However it also ushered in an expanded understanding of the concept of acquisition to include not only acquisition by transfer of title, but also the mere 'sterilisation' of prior proprietary rights through new land use restrictions.

The facts of the case also exemplify a very topical issue in Australia at present: the vast areas of the Australian continent that are covered by mining and petroleum licences, and efforts by civil society and in some instances government to provide better protection for these places.[11]

D: *Promoting radical change through conservative reasoning*

It is for this reason that this case provides a good starting point in the process of re-imagining constitutional jurisprudence from an Earth laws perspective. Despite the radical nature of attempting to partially dismantle the

---

10 Mitleton-Kelly, above n 8, 36–7.
11 Lock the Gate Australia, a national non-government organisation campaigning against expansion of the coal and gas industry, has published a map showing that coal and gas exploration titles now cover approximately two-thirds of the Australian land mass. <https://d3n8a8pro7vhmx.cloudfront.net/lockthegate/pages/2187/attachments/origi nal/1439522455/LTG-CallToCountry-Brochure-web.pdf?1439522455>.

primacy of property as a value in Australia's constitutional jurisprudence, I chose to approach the judgement from a relatively conservative form of legal analysis, primarily utilising traditional approaches to statutory interpretation. In the same way that liberalist property ontology has been inserted and reified in our constitutional jurisprudence since Federation, with scarce acknowledgment of the political and philosophical contentiousness of such an approach, I have adopted a conservative form of reasoning to achieve a radical outcome. My challenge was to see if property centredness could be dismantled with a similar form of positivistic stealth to that with which it has been constructed.

The issue of the meaning of the concept of acquisition was raised in the *Newcrest* case but only Justice McHugh upheld the argument put by the Commonwealth that as no property was transferred from Newcrest to the government, no acquisition had taken place.[12] In my judgment, the more limited view of the concept of acquisition proffered by Justice McHugh is adopted. Given the sufficiency of Justice McHugh's view on that point, my primary concern instead was to strategically disrupt the established jurisprudence, which has constructed the subsection as a shield from government action rather than simply as a head of power.

E: A head of power, not a defensive shield

This judgment aims to completely upend the property centred approach that has characterised 'just terms' cases in Australia but to do so by reference to well established legal reasoning, namely statutory interpretation principles. The argument that a fundamental guarantee of property rights and immunity to government action can be found in a head of power subsection is a glaring weakness in the reasoning that has produced cases like *Newcrest* and its predecessors.

The argument that a head of power can be transformed into a defensive shield is a vulnerable argument but, furthermore, the High Court's insistence on this approach has generated a confusing jurisprudence in which numerous exceptions have had to be made to the 'guarantee' for practical reasons.[13] The judgment of Chief Justice Mason in *Mutual Pools*[14] is cited in my judgment to illustrate the range of exceptions that have been necessitated.

---

12 *Newcrest Mining (WA) Ltd v Commonwealth* (1997) 190 CLR 513, 573–4 (McHugh J).
13 *Federal Commissioner Taxation v Clyne* (1958) 100 CLR 246; *Trade Practices Commission v Tooth Co Ltd* (1979) 142 CLR 397; *Attorney General v Schmidt* (1961) 105 CLR 361; *Nintendo Company Ltd v Centronics Systems Pty Ltd* (1994) 28 CLR 134.
14 *Mutual Pools And Staff Pty. Limited v The Commonwealth Of Australia* (1994) 179 CLR 155, 169–71.

Another interesting aspect of the *Newcrest* case is the judgment of Justice Kirby. To his credit, Justice Kirby does not hide behind the shield of neutrality and positivism in delivering his robust arguments in favour of the liberal political value of private property but instead elucidates his political position overtly.

He states, for example:

> There is one final consideration which reinforces the view to which I am driven by the foregoing reasons. Where the Constitution is ambiguous, this Court should adopt that meaning which conforms to the principles of fundamental rights rather than an interpretation which would involve a departure from such rights.[15]

Justice Kirby went on to cite numerous provisions from various constitutions, treaties and rights covenants that attested to the primacy of human private property rights. He summed this up by saying:

> In effect, the foregoing constitutional provisions do no more than reflect universal and fundamental rights by now recognised by customary international law. Ordinarily, in a civilised society, where private property rights are protected by law, the government, its agencies or those acting under authority of law may not deprive a person of such rights without a legal process which includes provision for just compensation. Whilst companies such as the appellants may not, as such, be entitled to the benefit of every fundamental human right, s 51(xxxi) of the Australian Constitution must be understood as it commonly applies to individuals entitled to the protection of basic rights. It must be given a meaning and operation which fully reflects that application. In this way, in Australian law, it extends to protect the basic rights of corporations as well as individuals.[16]

Justice Kirby's frankness in identifying the ontological foundations of property centred jurisprudence is refreshing because it at least facilitates critique. In my own judgment, however, I chose to adopt traditional statutory interpretation reasoning as a strategy to demonstrate how a radical view could be advanced even by using traditional approaches. Nevertheless, in the final paragraphs, there is some acknowledgement of the ecological values that underlie my approach.

---

15 *Newcrest Mining (WA) Ltd v Commonwealth* (1997) 190 CLR 513, 657–61 (Kirby J).
16 *Newcrest Mining (WA) Ltd v Commonwealth* (1997) 190 CLR 513, 660 (Kirby J).

F: Practical effect of re-imaging section 51(xxxi)

The practical effect of the judgment, if it were adopted in future as a majority position in the Australian High Court, would be to reverse the *Newcrest* decision and to overrule the pre-existing line of cases on section 51(xxxi) so that in future the Commonwealth government could, in the exercise of many of their legislative heads of power, acquire property rights without compensation being a constitutional requirement. I see this flexibility as key to repositioning the balance between property and economic interests with the ecological needs of the eco-sphere. It cannot be assumed that governments can, in future, always pay corporations for the adjustments to property rights that may be required to respond to the environmental crisis.

The interpretation of section 51(xxxi) as another head of power that supplements rather than obliterates the power of acquisition that is incidental to many other heads of power does not make the subsection redundant. There are conceivable situations in which the Parliament may not be able to point to any other head of power to justify a particular acquisition in which case section 51(xxxi) would still apply and would require just terms compensation.

G: Conclusion

A re-imagined jurisprudential approach to section 51(xxxi) on its own could not be expected to achieve the transformations needed to bring our legal system into alignment with a robust Earth laws perspective, but it certainly represents a practical shift away from current constitutional interpretations that privilege property rights over the growing need for substantial changes in environmental regulation, as well as an important symbolic challenge to the liberalist ontology that treats property rights as a fundamental value. The exploration of one adjacent possible in legal reasoning has the capacity to open up a wider field of future possibilities.

HIGH COURT OF AUSTRALIA
BRENNAN CJ,
DAWSON, TOOHEY, GAUDRON, McHUGH, GUMMOW, GAIA AND KIRBY JJ
NEWCREST MINING (WA) LIMITED
v
THE COMMONWEALTH OF AUSTRALIA and ors
14 August 1997

Gaia J (dissenting)

Two principal questions arise in this appeal. The first question is whether the Commonwealth, by making proclamations under section 7(2) of the

National Parks and Wildlife Conservation Act 1975 (Cth), acquired property of the appellants. The second question is whether, assuming the Commonwealth did so acquire property, the acquisitions are invalid because the Commonwealth failed to provide just terms. There is a further question of whether the exercise of power under section 122 of the Constitution – the territories power – is subject to section 51(xxxi) of the Constitution. I will address all three questions in turn.

This case concerns certain mining leases in the area of Coronation Hill in the Northern Territory held by the first appellant, Newcrest Mining (WA) Limited ('Newcrest'). The principal question in these proceedings concerns whether Newcrest's property in the mining leases was acquired by the Commonwealth otherwise than on just terms in breach of section 51(xxxi) of the Constitution. The relevant facts and legislation are set out in the judgment of Justice Gummow, but because of the view which I take of the scope and operation of section 51(xxxi), I have no need to refer to them in detail.

The lands over which Newcrest claims mining leases are included in Kakadu National Park. That park was proclaimed in three stages under section 7(2) of the *National Parks and Wildlife Conservation Act 1975* (Cth) ('the Act'). Stage 3 was extended by proclamation on 13 November 1989 and again on 21 June 1991. Under section 7 of the Act, the proclamations include the described areas of land including 1000 metres depth into the subsoil. It is those proclamations which, according to the appellants' contention, constitute an acquisition of property otherwise than upon just terms.

Under section 7(7) of the Act:

> Upon the declaration of a park or reserve, any interest held by the Commonwealth in respect of the land (including any sea-bed or any subsoil) within the park or reserve, but not in respect of any minerals, becomes, by force of this subsection, vested in the Director.

The Director is the Director of National Parks and Wildlife which is a corporation under section 15 of the Act. The Commonwealth accepts that Newcrest's mining leases, if they exist, are property within the meaning of section 51(xxxi) of the Constitution. However, it claims that any interest held by Newcrest under the leases was not vested in the Director pursuant to section 7(7) but continued to be held by it.

The Act was amended by the *National Parks and Wildlife Conservation Amendment Act 1987* (Cth) ('the Amending Act'). The Amending Act inserted in the definition section of the Act the following:

> 'Kakadu National Park' means the area for the time being declared under section 7 to be the park of that name.

The Amending Act also inserted section 10(1A) into the Act. That sub-section provides:

> No operations for the recovery of minerals shall be carried on in Kakadu National Park.

Section 7 of the Amending Act provides:

> Notwithstanding any law of the Commonwealth or of the Northern Territory, the Commonwealth is not liable to pay compensation to any person by reason of the enactment of this Act.

In asserting an acquisition of property, Newcrest relies upon the prohibition against the exploration of the mining tenements in question. Since section 10(1A) of the Act was inserted by the Amending Act, any acquisition of property was by reason of the enactment of the Amending Act.

It is sufficient for my purposes to assume that the proclamations dated 13 November 1989 and 21 June 1991, together with the statutory prohibition against mining operations in Kakadu National Park, constituted the purported acquisition to which the appellants object.

The Commonwealth argues that the proclamations did not acquire any property in any of the mining leases but that section 10(1A) of the Act merely sterilised the use of the land for mining purposes when that section declared that '[n]o operations for the recovery of minerals shall be carried on in Kakadu National Park'. The Commonwealth points out that the pro-hibition in section 10(1A) was universal in its application. Since it applied to the Commonwealth and the Director as well as to Newcrest, the prohibi-tion produced no benefit of a proprietary nature for the Commonwealth or the Director. The proclamations may have prevented Newcrest from mining the land but neither the Commonwealth nor the Director received any corresponding advantage. That being so, the Commonwealth contends that there was no acquisition within the meaning of section 51(xxxi) of the Constitution.

In relation to this question I agree with the judgement of Justice McHugh, which I have had the benefit of reading. The effect of the proclamations was merely to impinge on Newcrest's rights to exploit its interests in the mining leases. Even accepting that this amounted to an extinguishment of all or part of Newcrest's interests, there was no gain by the Commonwealth (or the Director). There is an important difference between an acquisition in which property changes hands, and laws which merely impinge upon an individual's previously held proprietary rights. It would unduly hamper the process of government and legislation in its entirety if each and every new law that impinged on an individual's pre-existing property rights required compensation at the public expense.

In view of the position taken by the majority in relation to the above question, however, I will now turn to consider whether, in the event that the Commonwealth were to be found to have acquired property, the acquisitions are invalid because the Commonwealth failed to provide just terms.

This question requires consideration of the entire scope and meaning of section 51(xxxi). In my opinion this question must be resolved by reference to ordinary principles of statutory interpretation as they apply to constitutional interpretation.

Section 51(xxxi) of the Constitution provides that:

> The Parliament shall, subject to this Constitution, have power to make laws for the peace, order, and good government of the Commonwealth with respect to:-
> (xxxi) The acquisition of property on just terms from any State or person for any purpose in respect of which the Parliament has power to make laws.

Section 51 resides in Part V of the Constitution. That Part bears the title 'Powers of the Parliament' and section 51 itself bears the title 'Legislative powers of the Parliament'. The whole of section 51 is dedicated to enumerating specific heads of legislative power, of which the power contained in section 51(xxxi) is but one. As a subsection granting legislative power to the Parliament, it plainly provides the Parliament with a power, separate and distinct from any other head of power, to acquire property on just terms for any purpose in respect of which the Parliament may make laws.

If subsection (xxxi) were not included in section 51, then the power to acquire property would have needed to be justified in all cases where it was exercised by the Parliament by reference to powers incidental to the exercise of one of the other heads of power. I have no doubt that many of the other heads of power include the power to acquire property as an incident of the subject matter of the power conferred.

The question that has occupied so much of this Court's time since the enactment of section 51(xxxi) is whether that particular head of power also acts as a limitation upon the powers that otherwise are granted in other subsections in section 51. From the outset it would be unusual for a head of power to operate as a fetter on all of the other heads of power, but in the interpretation of provisions such an outcome is not impossible.

I acknowledge that there is a significant line of cases that affirm the status of section 51(xxxi) as a de facto constitutional guarantee, designed to protect citizens from being deprived of their property except on just terms: *Minister* of State for the Army v Dalziel;[17] Clunies-Ross v The Commonwealth;[18] Australian Tape Manufacturers Association Ltd v The Commonwealth.[19]

The language of referring to a grant of power as a guarantee of immunity from certain kinds of legislative action is, however, highly problematic.

Chief Justice Mason in *Mutual Pools and Staff Pty Limited v the Commonwealth Of Australia*[20] (*Mutual Pools*) criticises the description of the subsection as a constitutional guarantee and points out that such a description:

> (d)oes not reveal completely its true character and its relationship with the other legislative powers conferred upon the Parliament by the Constitution. The true position is, as Dixon J pointed out in *Grace Bros Pty Ltd v The Commonwealth*, that s 51(xxxi) was introduced into the Constitution as a specific power: 'not, like the Fifth Amendment for the purpose of protecting the subject or citizen, but primarily to make certain that the Commonwealth possessed a power compulsorily to acquire property, particularly from the States. The condition "on just terms" was included to prevent arbitrary exercises of the power at the expense of a State or the subject.'
>
> The words 'for any purpose in respect of which the Parliament has power to make laws' are, in the context of a grant of legislative power, words of limitation. They confine the exercise of the power to an implementation of a purpose within the field of Commonwealth legislative power. They are not to be read as an exclusive and exhaustive statement of the Parliament's powers to deal with or provide for the involuntary disposition of or transfer of title to an interest in property.

The precise relationship between section 51(xxxi) and the other heads of power is not simple. The Chief Justice in *Mutual Pools* referred to many examples where the very nature of a specific power may indicate that it is not constrained by section 51(xxxi):

> Hence, the effect of s 51(xxxi) when read in conjunction with the other legislative powers of the Parliament is that, subject to any contrary intention, it forbids the making of laws with respect to the acquisition

17 *Minister of State for the Army v Dalziel* (1944) 68 CLR 261, 276 (Latham CJ), 284–5 (Rich J).
18 *Clunies-Ross v Commonwealth* (1984) 155 CLR 193, 201–2 (Gibbs CJ, Mason, Wilson, Brennan, Deane and Dawson JJ).
19 *Australian Tape Manufacturers Association Ltd v Commonwealth* (1993) 176 CLR 480, 509.
20 *Mutual Pools And Staff Pty Limited v the Commonwealth Of Australia* (1994) 179 CLR 155, 168–9 (Mason CJ).

of property from any State or person for a relevant purpose on terms that are not just. . . . An indication of contrary intention may be provided by the express terms in which a specific power is conferred or by the very nature of the subject-matter of a specific power or what is included within it.[21]

His Honour noted bankruptcy laws, taxation laws, various criminal laws and importation laws as such examples. In conclusion, his Honour noted that:

Whether a law falls within s 51(xxxi) and must comply with the requirement ultimately depends upon the characterization of the law and upon the context of the expression 'acquisition of property' within the meaning of s 51(xxxi).[22]

The view that section 51(xxxi) is itself a head of power, rather than a grand constitutional guarantee or immunity, accords entirely with its placement in section 51 and in Part V of the Constitution. The notion that it operates somehow to provide a general immunity to various kinds of legislative action is misplaced and has generated a complex jurisprudence in which the Court has needed to explain why, notwithstanding the operation of section 51(xxxi), some particular heads of powers allow for the acquisition of property other than on just terms. This has arisen specifically with regard to taxation in *Federal Commissioner Taxation v Clyne*[23] with Chief Justice Dixon finding that taxation did not amount to an acquisition of property despite the creation of debt owed to the Commonwealth. In *Trade Practices Commission v Tooth Co Ltd*,[24] in which Justice Aikin discusses several notable exceptions to the restrictive application of section 51(xxxi),[25] including taxation, forfeiture of prohibited imports[26] and seizure of enemy property during warfare as notable examples. More recently in *Nintendo Company Ltd v Centronics Systems Pty Ltd*,[27] the conferral and removal of intellectual property rights by legislation has been questioned in relation to section 51(xxxi) and found again to be outside of its operation. It is entirely in accord with established

---

21 *Mutual Pools And Staff Pty Limited v the Commonwealth Of Australia* (1994) 179 CLR 155, 169–70 (Mason CJ).
22 *Mutual Pools And Staff Pty Limited v the Commonwealth Of Australia* (1994) 179 CLR 155, 172 (Mason CJ).
23 *Federal Commissioner Taxation v Clyne* (1958) 100 CLR 246, 263.
24 *Trade Practices Commission v Tooth Co Ltd* (1979) 142 CLR 397.
25 *Trade Practices Commission v Tooth Co Ltd* (1979) 142 CLR 397, 453–5 (Aikin J).
26 *Burton v Honan* (1952) 86 CLR 169; *Attorney General v Schmidt* (1961) 105 CLR 361.
27 *Nintendo Company Ltd v Centronics Systems Pty Ltd* (1994) 28 CLR 134.

constitutional principles to take the approach, as Chief Justice Mason did in *Mutual Pools*, that the question of whether a particular 'acquisition' is an acquisition that relies upon the grant of power contained in section 51(xxxi) is a question of statutory interpretation in each instance.

In view of the confusing jurisprudence that has followed attempts to define section 51(xxxi) as a fundamental guarantee that places limits upon all of its sibling heads of power under section 51, it is pertinent to return to interpretative principles to enquire whether a simpler and more coherent interpretation may be available.

Chief Justice Mason stated in *Mutual Pools* that:

> it is a well-accepted principle of interpretation that, when a power is conferred and some qualification or restriction is attached to its exercise, other powers should be construed, absent any indication of contrary intention, so as not to authorize an exercise of the power free from the qualification or restriction.[28]

This may be so, but it would be a mistake to take that particular interpretative approach to the extreme position where a single head of power was enabled to operate as a fetter on every other head of power.

The question of the relationship of one head of power to other heads of power is complex, and in general, particularly in relation to constitutional interpretation, a grant of power should be read broadly and liberally and given its full amplitude and plain meaning: *Jumbunna Coal Mine NL v Victorian Coal Miners Association*; *Australian National Airways v Commonwealth*; *R v Coldham*.[29] It also follows where a number of heads of power are enumerated in the same section as occurs in section 51 that each head of power should be read broadly and that it should not be assumed that a single head of power is intended to act as a fetter upon its siblings.

Looked at in its plain context, section 51(xxxi) is simply and no more than a distinct grant of power to the Commonwealth Parliament to acquire property on just terms for any purpose in respect of which the Parliament may make laws. In keeping with its location as a head of power, its nature is to expand and enhance the powers of Parliament, not to detract from them. Each and every head of power should be interpreted liberally. The plain meaning of the provision is to provide to the Parliament the power, in case

---

28 *Mutual Pools And Staff Pty Limited v the Commonwealth Of Australia* (1994) 179 CLR 155, 169 (Mason CJ).
29 *Jumbunna Coal Mine NL v Victorian Coal Miners Association* (1908) 6 CLR 309, 368; *Australian National Airways v Commonwealth* (1945) 71 CLR 29, 81; *R v Coldham* (1983) 153 CLR 297, 312.

of any doubt, to acquire property for any purpose in respect of which the Parliament may make laws. It has this effect whether or not such a power was seen to be also implicit in any other head of power or not.

It is contrary to the principle of broad interpretation of grants of legislative power to lend to a grant of power an operation that could limit the exercise of every other head of power. If section 51(xxxi) were ever intended to operate as a limitation upon all other heads of power, it would in my opinion need to have been contained in a separate section, outside section 51, outside Part V in the same way that, for instance, sections 92 and 116 were drafted.

It is true that interpretative principles often compete in their application, and the limitation to the principle that powers should be given a liberal interpretation is that in doing so, every head of power should be meaningful and able to achieve its purpose. In avoiding the problematic consequence of section 51(xxxi) fettering all other powers, it is also important to adopt an interpretation that gives section 51(xxxi) some meaningful efficacy.

To give section 51(xxxi) an operation in which it adds to the powers implicit in other heads of power, but does not operate to fetter all of the other powers does not render the subsection meaningless nor redundant. It simply means that, when on a proper construction of any other head of power that power includes an incidental or implicit power of acquisition, then the occasion for the application of section 51(xxxi) simply does not arise. However, in all other cases where the Parliament exercises a power to acquire, and cannot point to an acquisition power elsewhere in the Constitution, it will need to do so consistently with section 51(xxxi).

This is not to say that as a matter of public policy the Parliament should not give due consideration to the provision of compensation for acquisition, particularly where issues of human rights are concerned, but simply that under the Commonwealth Constitution the position is similar to the Australian States, where the question of compensation rightly belongs to the Parliament itself.

As a result of the view that I have taken in relation to the operation of section 51(xxxi) generally I also find that section 122 of the Constitution is not constrained in any way by section 51(xxxi).

A case such as the present one exemplifies the danger of treating section 51(xxxi) as a form of general constitutional guarantee rather than as a mere head of power. Governments over the years issue numerous rights in the form of leases and licences to mining companies to the extent that such interests cover vast swathes of the Australian continent. To contend that governments cannot at any future point in time, in the exercise of their legislative powers and their obligations to protect places of international conservation significance, place restrictions on land use without a corresponding need to compensate all parties affected by such restrictions would unduly hamper the power of the Parliament to make laws for the peace, order and good government of the nation and to take actions to redress the very grave environmental concerns that have become the pressing issues of our times.

Constitutions are lasting documents, and in interpreting them it is important for the courts to adopt an ambulatory interpretation that allows the meaning of particular provisions to move with changing circumstances. On the view I have taken of the proper meaning and effect of section 51(xxxi), no issue arises as to the governments' competence to affect the rights of the mining company in question. But beyond issues of interpretation there remains substantive reasons why governments, in attempting to address issues of environmental management should not be unduly constrained by political sensitivities concerning property rights.

The absolute nature of private property rights is a key historical cornerstone of political liberalism, a philosophy that has historically been and continues to be highly influential in our legal system. But a reified and absolute defence of individual property rights had its origins in a time when the capacity of humans to irreparably damage the natural environment on a global and potentially catastrophic scale was not seriously anticipated. The alarming scale of anthropogenic environmental destruction has become one of the most pressing issues for governments across the globe. In facing the global ecological crisis, we as humans are forced to confront not only our impacts but, more deeply, the flawed nature of the worldviews that have previously underpinned our systems of law and governance. The anthropocentric, individualistic and property centred worldview inherited from various strands of European and American liberalism may have served our societies well in various ways in the past, but as the evidence of our current global environmental predicament and the empiric scientific knowledge about the interconnectedness of all life continue to build, we need to be prepared to shift our underlying jurisprudential and philosophical approaches. As judges and courts, we cannot ignore that we are necessarily confronting our own Copernican moment.

The growing scientific evidence of the need to re-integrate human economic activities within a framework that recognises and works within observable ecological constraints means that we can no longer consider property rights to be an absolute value unconstrained by ecological considerations.

Developing a jurisprudence that concedes some significant and growing humility on the part of human beings in relation to the ecological principles upon which all life depends is a pressing and solemn responsibility. Whilst this sense of awe and responsibility guides my judgment, it is not in a purely legal sense essential to the reasoning upon which my judgment in this particular case relies.

The appeal should be dismissed.

Chapter 13

Metgasco Limited v Minister for Resources and Energy

Cristy Clark

Commentary

A: Background

In February 2014, an unusual camp was established on a farm in Bentley, a small community near Lismore in Northern New South Wales. The farm was next door to Metgasco's Rosella drill site – the location of a proposed exploratory drilling project to determine the potential of PEL 16 – one of three gas exploration licences held by the oil and gas company over a significant area of land in the Northern Rivers region of New South Wales. Depending on the results of these proposed drilling operations, Metgasco was hoping to establish a large gas field in the region.

By April, the daily community protest against the proposed drilling had swelled to over 2,000 people from all walks of life. In addition to a desire not to live in a gas field, one of the key issues motivating the community resistance, which came to be known as the Bentley Blockade, was the risk posed by coal seam gas (CGS) and other forms of unconventional gas drilling to local water resources and the environment. In order to extract methane gas, CSG drilling brings large volumes of salty water up to the surface. The theory is that this heavily salinated water will be safely disposed of, but this has not always been the reality. In July 2012, for example, the New South Wales Environment Protection Authority found both Metgasco and Eastern Star Gas guilty of illegally dumping wastewater on more than one occasion.[1]

---

1 S Harlum, 'Metgasco in hot water', *Northern Star* (online), 13 June 2012 <www.north ernstar.com.au/news/metgasco-in-hot-water/1415322>; C Adams, D Eaton and J Wagner, 'Metgasco's wastewater fine', *Northern Star* (online), 19 June 2012 <www.northernstar.com. au/news/metgascos-wastewater-fine/1422760>; NSW Environment Protection Authority, 'Eastern Star Gas fined for pollution in the Pilliga' Media release, 6 July 2012) <www.epa. nsw.gov.au/epamedia/epamedia12070601.htm>.

The Bentley Blockade came to a head in May 2014, when thousands of people gathered at the site to prevent Metgasco from delivering its drilling equipment. The New South Wales government was reported to be ordering around 700 to 800 State police to the site to break up the blockade and a strong community campaign of lobbying politicians was initiated in response.[2]

B: The two suspensions

At the eleventh hour, the government changed tack and suspended Metgasco's Activity Approval (to drill at the Rosella site) citing a failure to comply with 'Condition 8' of its licence. Condition 8 required that the company engage in 'effective community consultation'.[3] Metgasco appealed the suspension and submitted documentation to the government to support its argument that it had engaged in adequate consultation.

After considering the evidence, the Minister for Resources and Energy for the State of New South Wales (the Minister) made a second decision through his delegate (the Delegate) confirming the first decision to suspend the Activity Approval.[4] In doing so, the Delegate cited Metgasco's failure to adequately identify all relevant stakeholders and failure to engage in effective consultation.[5] The Delegate emphasised the need to comply with the *Guideline for community consultation requirements for the exploration of coal and petroleum, including coal seam gas* ('the Guideline'), which states that 'community consultation is about involving people in the making of decisions that affect them [and] aims to understand and address issues in the early stages of a project to ensure that the interests of the community are considered during the planning process'.[6] Metgasco's consultation activities had been primarily confined to information provision and it had chosen not to engage with community members who were opposed to CSG drilling – instead characterising them as being misinformed about the nature of its operations.[7]

2 P J Henderson, 'Managing Director's Address and Presentation to 2014 AGM' (Metgasco, 28 November 2014) <www.metgasco.com.au/asx-announcements/managing-directors-address-and-presentation-2014-agm>.
3 *Metgasco Limited v Minister for Resources and Energy* [2015] NSWSC 453 [21–23].
4 Ibid. [30–32].
5 Ibid. [30–32].
6 *Guideline for community consultation requirements for the exploration of coal and petroleum, including coal seam gas* (NSW Trade & Investment, 2012) p.1.
7 Henderson, above n 2.

C: The judgment

Metgasco sought judicial review of the Delegate's two decisions to suspend its Activity Approval before the Supreme Court of New South Wales. In his judgment, Justice Button found the suspension was invalid because the Delegate had failed to give adequate notice of the first suspension and failed to 'give the holder of the title a reasonable opportunity to make representations with respect to the proposed . . . suspension, and to take any such representations into consideration', as required by section 22(6) of the Act.[8] He then found that the second suspension (which took place after this procedural justice process had been complied with) was invalid because it purported to confirm the first (invalid) suspension.[9]

Justice Button also found that section 22(3A) of the *Petroleum (Onshore) Act 1991* (NSW) only permitted the suspension of a licence for the breach of a condition that has been 'identified as a condition related to environmental management in the title, or in any notice of the imposition or variation of the condition given to the title holder'.[10] Condition 8 had not been identified as an environmental management condition and so Justice Button found that it was not a 'suspendable condition' under the Act.[11]

In my rewritten judgment, I do not reconsider these aspects of Justice Button's judgment, primarily because they relate more to questions of statutory interpretation than wild law. Instead, my judgment reconsiders the final section of Justice Button's judgment, in which he considers Metgasco's alternative argument that the Delegate's decision was invalid because it took into account a number of irrelevant considerations 'in the sense discussed in the well-known decision of *Minister for Aboriginal Affairs v Peko Wallsend Ltd* [1986] HCA 40; 162 CLR 24'.[12]

The irrelevant considerations in question, identified by Metgasco and upheld by Justice Button, were that in concluding that Metgasco had failed to engage in 'effective consultation' as mandated by Condition 8, the Delegate had appeared to place substantial weight on the persuasive effect of the consultations upon those who were being consulted, instead of only focusing on the inherent nature of the consultation. This distinction between the inherent nature of the consultation process from the persuasive effect of the process on those being consulted can only be supported if the concept of 'consultation' is reduced to a process of information provision. If, instead, 'effective consultation' necessarily involves community participation in the

---

8 *Metgasco Limited v Minister for Resources and Energy* [2015] NSWSC 453 [33–42].
9 Ibid. [43–48].
10 Ibid. [51–62].
11 Ibid.
12 Ibid. [64].

decision-making process, then the process cannot be distinguished neatly from the outcome.

D: Contribution of a wild law perspective

In rewriting Justice Button's judgment, I take a different approach to the interpretation of 'effective community consultation' by drawing on the right to a healthy environment and the rights of nature. The procedural rights associated with these rights include the right to actively participate in decisions affecting the environment – a core right in empowering the community to protect the environment from destructive development.[13]

Across the world, communities have been demanding the right to meaningfully participate in decisions that affect the environment, and this right has been reflected in international instruments such as Principle 10 of the *Rio Declaration* and the *Aarhus Convention*.[14] As the International Law Association (a non-governmental organisation comprised of international legal experts) noted in 2004, 'in contemporary society, legitimacy largely depends on the consent of the governed, and hence on the sense that the governed have a voice through direct participation, representation, deliberation, or other methods'.[15]

It is significant that this right of participation has emerged from within international environmental law, because it demonstrates the link between participation and environmental justice. If the rights of nature are going to become a more central consideration within our anthropocentric legal system and policy making processes, it will be essential that communities are empowered to take on a custodianship role over the environment. Although human rights have been criticized for being inherently anthropocentric, new conceptions of the right to a healthy environment (and human rights broadly) have questioned this understanding.

13 S Atapattu, 'The Right to a Healthy Life or the Right to Die Polluted?: The Emergence of a Human Right to a Healthy Environment Under International Law' 16 (2002–3) 65 *Tulane Environmental Law Journal* 90; R Cantley-Smith, 'A Human Right to a Healthy Environment' in P Gerber and M Castan (eds), *Contemporary Perspectives on Human Rights Law in Australia* (Thomson Reuters, 2013) 447, 470.

14 *Rio Declaration: Report of the United Nations Conference on Environment and Development*, UN Doc A/CONF 151/26/Rev 1 (1992), Principle 10; *Convention on Access to Information, Public Participation in Decision-making and Access to Justice in Environmental Matters (Aarhus Convention)*, opened for signature 25 June 1998, United Nations. Treaty Series, vol 2161, 447, entered into force on 30 October 2001.

15 J W Dellapenna, 'The Berlin Rules on Water Resources' (Report of the 71st Conference, International Law Association, 2004) art 18.

The Constitutions of Ecuador and Bolivia, for example, have recognised a range of human rights (including the right to a healthy environment) along-side of the rights of nature. Article 83 of the Constitution of Ecuador provides that Ecuadorians have the duty and obligation '[t]o respect the rights of nature, preserve a healthy environment and use natural resources rationally, sustainably and durably'.[16] Similarly, Bolivia's Constitution adopts an Earth-centric approach to human rights by enshrining the 'right to a healthy, protected, and balanced environment . . . [for] individuals and collectives of present and future generations, *as well as to other living things*, so they may develop in a normal and permanent way'.[17] Article 108(16) also highlights the duty of Bolivians to 'protect and defend an environment suitable for the development of living beings'.[18]

What is notable in both Constitutions is that human rights and the rights of nature are not treated as separate agendas. This natural link between human rights and earth rights has also been recognised in the Preamble to the proposed *Declaration on the Rights of Mother Earth*: 'We, the peoples and nations of Earth: considering that we are all part of Mother Earth, an indivisible, living community of interrelated and interdependent beings with a common destiny'.[19] As activists at the 2015 Paris climate summit articulated so clearly, 'We are not defending nature – we are nature defending itself'.[20]

Local communities have a clear interest in promoting sustainable approaches to development in contrast to corporations that have a history of extracting resources and walking away from environmental damage. The Bentley Blockade, for example, was spearheaded by a social movement calling itself the Gasfield Free Northern Rivers movement. Established in 2010, its stated aims are 'to protect the biodiversity, water resources, agricultural lands and sustainable industries of the Northern Rivers, and the livelihoods and well-being of the people who live here, from the impacts of coal seam gas (CSG) and other forms of unconventional gas mining'.[21]

When responding to the government's decision to suspend Metgasco's Activity Approval, Managing Director Peter Henderson was dismissive of

16 *Constitución Política del Ecuador 2008* art 83(6).
17 *Bolivia (Plurinational State Of)'s Constitution 2009* art 33 (emphasis added).
18 Ibid. art 108(16).
19 *Proposed Declaration on the Rights of Mother Earth*, World People's Conference on Climate Change and the Rights of Mother Earth (2010) <https://pwccc.wordpress.com/programa>.
20 N Bloch, 'COP21 actions go ahead: 'We are not defending nature – we are nature defending itself'', *Ecologist* (online), 28 November 2015 <www.theecologist.org/News/news_analysis/2986467/cop21_actions_go_ahead_we_are_not_defending_nature_we_are_nature_defending_itself.html>.
21 Gasfield Free Northern Rivers, Homepage (2015) <http://csgfreenorthernrivers.org>.

the community opposition to Metgasco's operations, arguing that it repre-sented only a minority of people, mostly from outside Metgasco's licence areas.[22] He also opined that these people were from '"life style" and "alter-native" areas represented by Nimbin and Byron Bay' rather than being from farming or industrial areas.[23] These comments are not supported by local government surveys, which found 87 per cent of residents were opposed to CSG exploration and production in the region,[24] or by the fact that over 7,000 people marched against CSG on the streets of Lismore in 2012.[25] The implication that farming communities supported Metgasco's operations is also undermined by the key role played in the Bentley Blockade by the Lock the Gate Alliance – a national organisation with over 40,000 supporters, including many farmers, 'who are concerned about unsafe coal and gas mining'.[26]

The significance of these social movements, and the Bentley Blockade itself, is that they represent examples of communities successfully mobilising to protect their local environment and to demand a more sustainable and earth-centred approach to governance and development. These demands were initially ignored by the New South Wales government, which continued to approve and renew gas licences in the region despite the overwhelming oppo-sition of the community. It was not until the community made it impossible to continue to ignore their demands that the government started to backpeddle – first by suspending the Activity Approval and eventually by announcing a buy back of all of the gas licences in the Northern Rivers Region.[27]

The argument that community control might serve to promote sustainable development and protect the rights of the earth is contrary to the parable of the *tragedy of the commons* made popular by Garrett Hardin.[28] In his 1968 paper of the same title, Hardin argued that whenever a valuable resource is treated as part of the commons, and (crucially) not subject to private prop-erty rights, that resource will be quickly depleted, as there will be nothing to deter individuals from externalising the negative consequences of overuse.[29]

22 Henderson, above n 2.
23 Ibid.
24 NSW Electoral Commission, *Report on the 2012 Local Government Elections* (2012) 448.
25 Gasfield Free Northern Rivers, *About the Gasfield Free Campaign* (2015) <http://csg freenorthernrivers.org/about-the-csg-free-campaign>.
26 Lock the Gate Alliance, *About Us* (2015) <www.lockthegate.org.au/about_us>.
27 NSW Government, *NSW Gas Plan – Protecting what's valuable, securing our future* (December 2014) <www.nsw.gov.au/sites/default/files/miscellaneous/sc000218_nsw_gas_ plan_announcement_web.pdf>.
28 G Hardin, 'The tragedy of the Commons' (1968) 162 *Science* 1243.
29 Ibid.

Hardin's thesis was very influential. For example, Mansuri reports that Hardin's thesis, supported by property rights theorists like Demsetz[30] and North,[31] generated scepticism in 'the World Bank and other multi-laterals about the viability of any local collective action in the provision of public goods, and created . . . an emphasis on the development of private property rights'.[32] However, Hardin's thesis was challenged by Nobel laureate Elinor Ostrom's research into the commons, which demonstrated that cooperative institutions to manage common property resources were both widespread and often highly successful.[33]

Ostrom's work established that the inexorable depletion argument only really applies to open access or global commons, such as the high seas or the atmosphere.[34] Most local commons have historically been subject to 'well-defined rules of access and use' that have been established in response to local conditions and have continued to be adapted to remain in tune with local culture and traditions without resorting to enclosure through private property rights.[35] As the United Nations Development Programme has pointed out, drawing on examples of local water resource management from Senegal, 'customary law often involves strict controls on water use, with water rights structured to balance claims based on inheritance, social need and sustainability. Institutional cooperation is common'.[36] Not only is cooperation at the community level common, community involvement in rural water supply projects and forest management has been found to increase the local sense of ownership, willingness to contribute, and (importantly) concern for preservation.[37]

If the rights of the earth are to receive greater protection in our legal and political system, we are going to need to move away from the priority currently

---

30 H Demsetz, 'The private production of public goods' (1970) 13(3) *Journal of Law and Economics* 293.
31 D North, *Institutions, Institutional Change and Economic Performance* (Cambridge University Press, 1990).
32 G Mansuri and V Rao, *Evaluating Community-Based and Community-Driven Development: A Critical Review of the Evidence* (World Bank, 2003).
33 E Ostrom, *Governing the Commons: The Evolution of Institutions for Collective Action* (Cambridge University Press, 1990); E Ostrom and R Gardner, 'Coping with asymmetries in the commons: self-governing irrigation systems can work' (1993) 7(4) *Journal of Economic Perspectives* 93.
34 Ibid. See also S Osmani, 'Participatory governance: an overview of issues and evidence' in UNDESA (ed.), *Participatory Governance and the Millennium Development Goals (MDGs)* (UNDESA, 2008) 16–17.
35 See ibid. See also Mansuri and Rao, above n 32, 7.
36 United Nations Development Programme, *Human Development Report – Beyond scarcity: power, poverty and the global water crisis* (UNDP, 2006) 185.
37 See World Bank, *World Development Report: Infrastructure for Development* (World Bank, 1994); Osmani, above n 34, 15–18.

granted to private property rights and growth-driven economic development. In order to shift towards a more sustainable and just approach to development, it will be necessary to protect the right to a healthy environment and to empower communities to carry out their 'solemn responsibility to protect and improve the environment for present and future generations'.[38] The first step to achieving this goal will be to recognise and protect the right to participate in decisions that affect the environment, and that is what this case is all about.

	Supreme Court
	New South Wales
Case Name:	Metgasco Limited v Minister for Resources and Energy
Medium Neutral Citation:	[2015] NSWSC 453
Hearing Date(s):	20–21 October 2014
Decision Date:	24 April 2015
Jurisdiction:	Common Law
Before:	Clark J
Legislation Cited:	Petroleum (Onshore) Act 1991 (NSW) 22(3B)(b)
Case Cited:	Minister for Aboriginal Affairs v Peko Wallsend Ltd [1986] HCA 40; 162 CLR 24
Parties:	Metgasco Limited (Plaintiff)
	Minister for Resources and Energy (First Defendant) NSW Department of Trade and Investment, Office of Coal Seam Gas (Second Defendant)
File Number:	2014/165970

Judgment

A: Background

The applicant, Metgasco Limited (Metgasco), has applied for judicial review of the decision of the Minister for Resources and Energy for the State of New South Wales (the Minister) by his delegate (the Delegate) to suspend specified operations approved under Petroleum Exploration Licence No 16 (PEL 16) at the Rosella E01 conventional gas exploration well

---

38 United Nations Conference on the Human Environment, Stockholm, 16 June 1972, UN doc A/CONF 48/14, Principle 1.

(the Rosella well), pursuant to section 22(3A) of the *Petroleum (Onshore) Act 1991* (NSW) (the Act).

The Delegate's decision to suspend the Activity Approval (the construction of the Rosella well at the Bentley site) was based on a finding that Metgasco had failed to undertake 'effective community consultation' in breach of Condition 8 of the licence, which read:

> The licence holder must engage with the community in relation to the planning for and conduct of prospecting operations authorised under this exploration licence.
>
> The consultation must be undertaken in accordance with the *Guideline for community consultation requirements for the exploration of coal and petroleum, including coal seam gas* (NSW Trade & Investment, 2012) as amended or replaced from time to time.
>
> An annual report on Community Consultation must be submitted to the Department within 28 days of the anniversary of this licence being granted, together with evidence that community consultation has been undertaken in accordance with the Guideline.

Metgasco submitted that the reasons for the suspension show that a number of irrelevant considerations were taken into account by the Delegate in the sense discussed in the well-known decision of *Minister for Aboriginal Affairs v Peko Wallsend Ltd* [1986] 162 CLR 24.

Specifically, Metgasco submitted that the Delegate erred in her assessment of its compliance with the requirement to engage in 'effective community consultation' by focusing on the persuasive effect of its consultation upon those being consulted with, rather than restricting her assessment to the inherent nature of the consultation itself.

B: Effective consultation

In considering the submissions of the applicant, it is necessary to first determine the meaning of the phrase 'effective consultation' as imported into Condition 8 of the licence.

The first paragraph of Condition 8 imposes a requirement to 'engage with the community in relation to the planning for and conduct of prospecting operations authorised under this exploration licence'. From this wording it is apparent that the applicant was expected to not only inform the community of its proposed operations, but to involve them in the planning process itself.

The second and third paragraphs of Condition 8 refer to the *Guideline for community consultation requirements for the exploration of coal and petroleum, including coal seam gas* (NSW Trade and Investment, 2012) ('the Guideline'), which the applicant is required to adhere to. While the Guideline focuses primarily on information provision, it also states that 'community consultation is about involving people in making decisions that affect them' and requires that licence holders provide 'feedback to the community on how their input has influenced decisions'.[39]

In 1969, Sherry Arnstein wrote what has come to be seen as a seminal article entitled, 'A ladder of citizen participation',[40] in which she set out the eight levels of participation from (1) Manipulation and (2) Therapy (which she described as 'nonparticipation'), through to (3) Informing, (4) Consultation, and (5) Placation (which she described as 'tokenism'), to (6) Partnership, (7) Delegated Power, and (8) Citizen Control (which she described as 'citizen power').[41] She introduces this ladder by arguing that '[t]here is a critical difference between going through the empty ritual of participation and having the real power needed to affect the outcome of the process'.[42]

The Guideline emphasises the importance of community members being included from the very beginning of the project identification process, and continually involved throughout the planning, implementation, monitoring, and evaluation processes, and of steps being taken to address any concerns raised by the community, because it recognises that effective consultation involves moving beyond information-sharing or token consultations into more empowering forms of participation.[43]

C: International law

Arnstein's definition of participation is reflected in emerging international norms around community participation in development and the right to a healthy environment. The World Bank, for example, defines participation as 'a process through which stakeholders influence and share control over development initiatives and the decisions and resources which affect them'.[44] Similarly, the United Nations defines 'effective participation' as:

> one in which all the relevant stakeholders take part in decision-making processes and are also able to influence the decisions in the sense that at the end of the decision-making process all parties feel that their views

---

39 *Guideline for community consultation requirements for the exploration of coal and petroleum, including coal seam gas* (NSW Trade & Investment, 2012) 1–2.

40 S R Arnstein, 'A ladder of citizen participation' (1969) 35(4) *Journal of the American Institute of Planners* 216.

41 Ibid. 4.

42 Ibid. 1.

43 Ibid; M Aycrigg, 'Participation and the World Bank: success, constraints, and responses' (Paper presented at the International Conference on Upscaling and Mainstreaming Participation: of Primary Stakeholders: Lessons Learned and Ways Forward, Washington DC, November 1998) 4.

44 See World Bank, *The World Bank participation sourcebook* (World Bank, 1996).

and interests have been given due consideration even if they are not always able to have their way.[45]

Historically the right to political participation in international human rights law focused on representative democracy, as reflected in articles 19 and 20 of the non-binding *Universal Declaration of Human Rights*[46] and articles 19 and 25 of the *International Covenant on Civil and Political Rights* ('ICCPR').[47]

In recent decades, a right to a more direct and active form of participation has started to be incorporated into the right to water and the right to a healthy environment.[48] In articulating the normative content of both of these rights, it has been acknowledged that the procedural rights to information, participation and legal redress are essential prerequisites to enabling communities to protect and secure their rights for the future.[49] Relevantly to the facts before us, the link between these rights and 'the rights of nature in the context of sustainable development' has also been acknowledged.[50]

The strongest recognition of participatory rights in the context of the right to a healthy environment comes from the *Aarhus Convention* (1998), which has the stated objective of contributing:

> to the protection of the right of every person of present and future generations to live in an environment adequate to his or her health and well-being [by guaranteeing] the right of access to information, public participation in decision-making, and access to justice in environmental matters in accordance with the provisions of this Convention.[51]

These procedural rights build on Principle 10 of the *Rio Declaration* (1992), which states:

> Environmental issues are best handled with participation of all concerned citizens, at the relevant level. . . . States shall facilitate and encourage public awareness and participation by making information

45 Osmani, above n 34, 1 (n 4).
46 *Universal Declaration of Human Rights*, UN Doc A/RES/ 217A(III) (1948) arts 19, 20.
47 *International Covenant on Civil and Political Rights*, opened for signature 16 December 1966, GA Res 2200A (XXI) arts 19, 25.
48 Resolution on human rights and access to safe drinking water and sanitation, 15th HRC sess, UN Doc A/HRC/15/L.14 (2010); The Future We Want, 66th UNGA sess, UN Doc A/RES/66/288 (2012) [43].
49 Ibid.
50 The Future We Want, 66th UNGA sess, UN Doc A/RES/66/288 (2012) [39].
51 *Convention on Access to Information, Public Participation in Decision-making and Access to Justice in Environmental Matters (Aarhus Convention)*, opened for signature 25 June 1998, United Nations, Treaty Series, vol 2161, 447, entered into force on 30 October 2001, art 1.

widely available. Effective access to judicial and administrative proceedings, including redress and remedy, shall be provided.[52]

Although it is a regional agreement, having been developed by UNECE, article 19(3) provides that the *Aarhus Convention* is open to accession by any United Nations Member State ('upon approval by the Meeting of the Parties').[53] It forms part of a growing body of international instruments that demonstrate an emerging consensus on the existence of the right to a healthy environment and, relevantly, the procedural rights that accompany it.

After examining a wide range of international instruments, the International Law Association (ILA) has argued for the existence of a general principle that people should have the right to participate in all decisions that affect their lives.[54] The ILA asserts, 'there can be little doubt that a right of public participation has now become a general rule of international law regarding environmental management even beyond the specific provisions of these agreements'.[55]

Finally, the emerging Rights of Nature movement acknowledges that humans are not separate from nature itself, and embeds these rights to participation within a broader understanding of the right and responsibility of communities to protect the Earth for its own sake.[56] This embedding of human rights within a broader conception of the Rights of Nature has been given clear expression in the Constitutions of Ecuador and Bolivia,[57] which both adopt an Earth-centric understanding of human rights and specifically empower and call on citizens to 'protect and defend an environment suitable for the development of living beings'.[58]

D: Interpretation

After considering the wording of Condition 8, the Guideline and emerging international norms around the meaning of community participation, it is clear that it would not be appropriate to expect the Delegate to separate any assessment of the inherent nature of the consultation from a consideration

---

52 *Rio Declaration: Report of the United Nations Conference on Environment and Development*, UN Doc A/CONF 151/26/Rev 1 (1992), principle 10.
53 As of 26 September 2012 there were 46 Parties to the *Aarhus Convention*.
54 Dellapenna, above n 15.
55 Ibid.
56 *Proposed Declaration on the Rights of Mother Earth*, World People's Conference on Climate Change and the Rights of Mother Earth (2010) <https://pwccc.wordpress.com/programa>.
57 *Constitución Política del Ecuador 2008* art 83(6); *Bolivia (Plurinational State Of)'s Constitution 2009* arts 33, 108(16).
58 *Bolivia (Plurinational State Of)'s Constitution 2009* art 108(16).

of the effect of the consultation on those being consulted with. Indeed, any attempt to engage in such a separation would undermine the rights of the community to be involved in the planning process itself.

If 'effective consultation' requires that community members are actively involved in the decision-making process, the effect of the consultation process on the community is a relevant consideration in determining its effectiveness. Although the expectation is not that all community members get their way, the outcome that is relevant to consider is whether the process was sufficiently participatory so as to ensure the community felt their views and interests (including the interests of the environment itself) were given due consideration.

The comments by the Delegate in cancelling the operations at the Rosella Well reflect a finding that the community did not feel as though their views and interests had been given due consideration. If over 80 per cent of the community had clearly expressed a desire that no drilling take place within their region, due consideration of these views would necessarily require that a good proportion of such a significant majority be convinced to change their opinion or that the proposed operation be cancelled. The expectation of the community that Metgasco and the New South Wales government would obtain a 'social licence' prior to drilling for gas is consistent with emerging international norms.

E: Conclusion

The Delegate's second decision to suspend Metgasco's Activity Approval was not separately invalid because, in asserting that a condition requiring 'effective consultation' had been breached, it took into account an irrelevant consideration, that being the results of the consultation, rather than focusing upon the attributes of the consultation itself. This submission from the applicant relies on a mischaracterisation of the nature of 'effective consultation'.

F: Orders

(1) It is declared that the decision of the Minister, by his delegate, on 14 May 2014 (the First Decision) to suspend specified operations under Petroleum Exploration Licence No 16 (PEL 16), being the construction of the Rosella E01 conventional gas exploration well approved by the Minister's delegate on 6 February 2014 (the Approved Activity), was not made according to law.

(2) It is declared that the decision of the Minister, by his delegate, on 26 June 2014 (the Second Decision) to confirm the First Decision was made according to law.

(3) The First Decision is quashed.

(4) The Second Decision is upheld.

Part III

First Nations law

Chapter 14

Aboriginal laws of the land

Surviving fracking, golf courses and drains among other extractive industries

Irene Watson

I: Introduction

A shark swims the channel and returns some time later with a seal in its mouth; the old people watch – all the time knowing the shark's next move. They know there is a balance, there is reciprocity, and this is law, the law of the land. When the law of the land is then layered by colonialism there is a shift and another world emerges, that of the colonial. In the world of the colonial the shark moves around unobserved and the balance shifts.[1]

This chapter is not a judgement but rather a talking back to colonialism, and a singing up of the decolonial. In this chapter I draw upon First Nations ontologies in my critique of the colonial foundation and practices of Australian law. Its purpose is positioning another way of knowing law: the laws of the First Nations of this place we now call Australia. Instead of judgement we dream back the song lines which have always been and remain the core of law in this land. In doing so, the focus of this chapter is on the authority held by First Nations to speak on the ongoing sustainability and health of our natural world. I write about that which I know and am actively involved with; I speak with the land of my ancestors the Tanganekald, Meintangk and Boandik Peoples of the South East of South Australia and against some of the latest proposals for 'development', which will damage it. This year's agenda includes proposals for a new golf course and tourist development in coastal sand dunes,[2] natural

---

1 I Watson, *Aboriginal Peoples, Colonialism and International Law: Raw Law* (Routledge, 2015).

2 I Watson, Submission Kungari Association, in response to the *Major Project Public Environmental Report – Proposed Nora Creina Golf Course and Tourism Resort, Development* (2016) <www.facebook.com/savenoracreina>. The Kungari Association was established by the elders in the 1980s and has a membership comprising the Tanganekald, Meintangk and Boandik First Nations People; its objectives are to protect land and culture.

gas fracking across the whole region[3] and the construction of new drains.[4] These proposals have such potential to damage our natural world that we are called to stand up in the laws of the land, and say: 'NO, you cannot do this'. The state has power derived from its original violent colonial foundation to go ahead but it does not have the law-full[5] authority. The state does not have the consent of First Nations and even if consent was given by individuals, it is important to note that there are situations in which consent could never be given, where consent would be against law. There is no law-full authority held to consent to destruction of the land, for that is the law.

With the advent of the colonial world the terra nullius doctrine was used to ignore our existence and had the effect of silencing and burying First Nations ontologies and epistemologies, even while we, in its face, continue to exist. We exist beyond the impact of terra nullius, even though in contemporary times First Nations laws have to compete with the loudness of the Australian state. We are Peoples living in two worlds; our laws compete with colonial power which frequently enables harmful development of our territories. While the Australian legal system runs so too does First Nations laws of the land. The business of colonialism is not yet finished, indeed, it rolls on, business as usual. This fact presents us with certain challenges. For Australia, the legal limit in dealing with the fact of First Nations Peoples' existence was reiterated and again set by the High Court in *Mabo (No 2)*.[6] The court determined that the extent of recognition of First Nations cannot

3 I Watson, 'Submission 29 January, 2015', *Natural Resources Committee Inquiry into Unconventional Gas (Fracking), pursuant to section 16 (1) (a) of the Parliamentary Committee Act 1991: Potential Risks and Impacts in the Use of Hydraulic Fracture Stimulation to Produce Gas in the South East of South Australia.*

4 On 13 December 2014, I emailed the South Australian Minister of Environment, Ian Hunter, advising him against the proposed construction of drains across the lands of the Tanganekald, Meintangk and Boandik Peoples for the purpose of diverting water flow into the Southern Coorong. What was highlighted in that email was the lack of information and evidence that the drains would be a solution to the problems of climate change impact upon the Coorong. The Minister was also advised on the Department's failure to consult with our Peoples, in accordance with the principles of free, prior and informed consent. The ministerial response was to fob off complaints, directing them to the government's own advisory committee, the South East Focus Group, (SEFG), and the Ngarrindjeri Regional Authority, (NRA). And while a number of First Nations individuals are members of the SEFG, their position in any process of obtaining free, prior and informed consent is conflicted, due to their role as advisors to the Minister, and not representatives of First Nations. Similarly the NRA does not have the authority to represent the First Nations of the South East.

5 Law-full is intended to acknowledge the lawfulness of First Nations Peoples in contrast to the legal history of terra nullius.

6 *Mabo v Queensland* (No 2) (1992) 175 CLR 1.

move beyond the 'skeletal' principle which is the foundation of the colonial state.[7] This position was defined further by the Australian state in its native title law.[8]

II: Obligations – sustainable lands for future generations

Australia holds records for animal and plant extinctions and the ecocide of our territories continues with each new and damaging development proposal.[9] First Nations' authority and obligations to maintain sustainable relationships and futures is both conflicted and limited by the power of the colonial state. Its foundations pass as law, but it is military power and colonial violence which remain the foundations of the colonising project in Australia. The urgencies of our time call for the continuity of Aboriginal laws, laws which have encoded our obligation to keep our natural worlds living, and whether they mean struggles must be maintained to protect the lands of my ancestors in the South East, or to stop oil-drilling in the Great Australian Bight[10]; or prevent the building of nuclear waste dumps on the lands of the Adnyamathanha Peoples.[11] Across Aboriginal Australia, First

---

7 Ibid., 30.
8 *Native Title Act 1993* (Cth), for a discussion on limitations, see I Watson, 'Sovereign Spaces, Caring for Country and the Homeless Position of Aboriginal Peoples' (2009) 108 *South Atlantic Quarterly* 27; Watson, above n 1.
9 For example, Australia has the highest rate of mammalian extinction in the world. Over the past 230 years, one in three of Earth's mammal extinctions have occurred in Australia. A 2015 review of the literature indicates that since 1788, 28 species of Australian land mammals have become extinct. This is to be contrasted with post-invasion North America, in which only one mammal species has become extinct in more than 400 years. See J Woinarski, A Burbidge and P Harrison, 'Ongoing Unravelling of a Continental Fauna: Decline and Extinction of Australian Mammals since European Settlement' (2015) 112 *Proceedings of the National Academy of Sciences* 4531. For information on flora both extinct and threatened, see <www.environment.gov.au/cgi-bin/sprat/public/publicthreat enedlist.pl?wanted=flora>; and a comparison with international data on Australia, 'Fact check: does Australia have one of the "highest loss of species anywhere in the world"?' *ABC News*, 4 March 2016 <www.abc.net.au/news/2015–08–19/fact-check-does-australia-have-one-of-the-highest-extinction/6691026>.
10 N Pelle, 'Six reasons to keep oil out of the Great Australian Bight' *Greenpeace General*, 27 June 2016 <www.greenpeace.org.au/blog/six-reasons-to-keep-oil-out-of-the-great-australian-bight/>.
11 The commonwealth government of Australia has proposed to create a nuclear waste dump in an area which borders the Yappala Indigenous Protected area without the consultation of the Adnyamathanha People. Jane Norman, 'Nuclear Dump: Barndioota Station in SA Earmarked as Site of Waste Facility', *ABC News*, 29 April 2016 <www.abc.net.au/news/2016–04–29/nuclear-waste-dump-expected-south-australian-cattle-station/7369346>.

Nations are standing with our ancient laws and obligations to ensure our lands remain alive and well for future generations.

The argument and concern here is not for the reform of colonial law, but for the re-emergence of the ancient laws of this continent now called Australia.[12] As First Nations living in two worlds we understand that our ancient laws will ensure the future health of our lands. The laws of the land are as old as the continent itself; they continue to exist. The laws of the land cannot be finished, other than perhaps in the minds of those who proclaim their ending. Law continues just as the natural world continues, regardless of how it may be denied by humans. For that is the law. Law is what cares for country, for that is the law.

Kombu-Merri and Waka Waka philosopher Mary Graham argues that First Nations have managed this country forever and that we still have the authority to do so today. And the sources of that authority is the land:

> The two most important kinds of relationship in life are, firstly, those between land and people and, secondly, those amongst people themselves, the second being always contingent upon the first. The land, and how we treat it, is what determines our human-ness. Because land is sacred and must be looked after, the relation between people and land becomes the template for society and social relations. Therefore all meaning comes from land.[13]

While we do not have the power to determine what happens to the land in terms of colonial developments, we hold the lawful authority to carry out ancient obligations to care for country. Many First Nations share the obligation to care for country; it is in our laws to ensure that future generations – not only human – but those of all species, including our ngaitji[14] relations, have a sustainable future. The challenge for the non-Indigenous is to see First Nations Peoples and the laws of the land from another horizon, one of law outside and beyond the terra nullius-colonial project. It is to see beyond the commodification of land and the constructions of Aboriginality as backward and savage, as beings without law. The challenge is also to understand that a debt is accruing – a debt is owed to First Nations and our territories. It's a debt measurable in the displacement of First Nations

---

12 For a further discussion see Watson, above n 1.
13 M Graham, 'Some thoughts about the philosophical underpinnings of Aboriginal world views', (2008) 45 *Australian Humanities Review* 181, 181–182 <www.australianhuman itiesreview.org/archive/Issue-November-2008/graham.html>.
14 Our ngaitji represents the relationship or kinship we share with our surrounding natural world. It is a relationship which teaches us about the unity we share with all natural things.

governance,[15] a foundational sovereign debt,[16] and one which is almost inestimable in dollar terms. In respect of the environment, the debt continues to mount, and many would agree with this. However the debt owed to First Nations is rarely, if ever, considered.

This continuance of the current global trends of 'progress and development' is to ensure the decline of life on Earth, and the extinction of many more species. Possibilities for our survival are enmeshed in our relationship with all life forms. First Nations laws' core philosophies are relational; they are diminished by fracking and other destructive developments on the lands to which we are related and obliged to care for. All life is reliant upon these connections: we are all dependent upon air and water, and our airs and waters flow together and are connected. The maintenance and centring of First Nations' epistemologies is probably essential to survival and if we are not positioned to reframe the dialogue (as we are not currently), then what is the alternative? Coloniality will continue and with that its same old progress-development agenda, at any cost to our natural world, towards its inevitable disastrous end. What is really essential to the survival of all species is to progress a horizontal dialogue between colonialist interests and First Nations-centred epistemologies. It is essential because currently the fullness of Indigenous epistemologies is ignored and remains largely unfathomable to the non-Indigenous world. The critical need for another way of being law-full is not appreciated or is beyond the colonial comprehension. The urgency for all to open eyes wide and to see and know law beyond a colonialist foundation is critical. But standing in the way of seeing First Nations law is a colonial legal system still sourced in the myths of terra nullius, and the lies that First Nations' laws do not exist. The colonial project posits and constructs rules which overlay the laws of the land. They reject First Nations' laws as being no more than oral stories, mere myths and fables. This continues while many of our stories and songs are only now being cited by science (as noted in a recent example concerning the rising waters which resulted in the Great Barrier Reef).[17] They are too often fobbed off as childlike or primitive, and hence remain unexamined for their ancient coded knowledge of the co-dependence between humans and our environments.

On behalf of our community organisation Kungari, I presented a submission to the South Australian Parliamentary Committee Natural Resources

15 S Motha, 'The debt crisis as crisis of democracy' (2012) 8 *Law, Culture and the Humanities* 390.

16 M Giannacopoulos, 'Sovereign debts: global colonialism, austerity and neo-liberal assimilation' (2015) 19 *Law Text Culture* 166.

17 N Reid and P Nunn, 'Aboriginal memories of inundation of the Australian coast dating from more than 7000 years ago' (2016) 47 *Australian Geographer* 11.

Committee Inquiry into Unconventional Gas (fracking), where it was argued that the state does not have the consent of our Peoples to go ahead and frack the unceded territories of the Meintangk and Boandik Peoples of lands now known as the South East of South Australia.[18] Our never-ceded territories, which prior to invasion and colonisation were pristine and provided a sustainable lifestyle from time immemorial, continue to sustain our People who carry the law-full authority and obligations to ensure they remain alive and well for future generations. We do not have a mandate to agree to their mass-damage and desolation. The state does have an obligation to consult and obtain our free, prior and informed consent regarding any proposals to develop our lands,[19] and it is important to note, that taking a First Nations' ontology, we could never consent to the destruction of our territories. To do so would be to act against our own First Nations' laws. First Nations Peoples know that without the land we are without food and water, and the future of humanity, along with other species, is threatened. Our mandate thus carries with it the obligation to ensure that our territories continue to sustain future generations.

III: Colonial rules and regulations – Aboriginal futures and law

First Nations' survival within the confines of colonialism has meant a forced march to assimilate and fit within the colonial project. In contemporary times there has been much hype in respect of reconciliation,[20] recognition,[21] and native title. Often the intent of these state-initiated projects is to carry the appearance of transformation, new beginnings and openings, but the hype is misplaced. Reconciliation movements continue as ad-hoc initiatives across Australia, while at the same time the battle against high incarceration rates, poor health, homelessness and poverty is being lost, with worsening levels of disadvantage.[22] Now we have the recognition movement,

---

18 Watson, above n 3.

19 *United Nations Declaration on the Rights of Indigenous Peoples*, GA Res 61/68, UN Doc A/RES/61/295 (September 13, 2007) (hereinafter *Declaration*) art 32 (2) outlines the minimum standard regarding negotiations with Indigenous Peoples.

20 P Patton, 'Reconciliation, Aboriginal rights and constitutional paradox in Australia' (2001) 15 *Australian Feminist Law Journal* 25.

21 Watson, above n 1, 2–5, 7–8 for a critique on the assimilatory effects of recognition.

22 Caro Meldrum-Hanna compiled a recent expose of how disadvantage translates into high incarceration and inhumane treatment while in detention, and has shown First Nations children being treated in a cruel and inhumane way in the Northern Territory: see C Meldrum-Hanna, 'Australia's Shame', *ABC 4 Corners*, 25 July 2016 <www.abc. net.au/4corners/stories/2016/07/25/4504895.htm>.

out that remains vague in defining what it is that is to be recognised. Is it our humanness – our human right to no longer be treated as the barbarian outside the colonial invasion/settlement? Certainly there's little discussion around Australian parliaments concerning our rights to take care of country as we have done since time immemorial.

The much-discussed case *Mabo No 2* enabled the recognition of our right to hold a native title, and native title is frequently proclaimed as having put an end to terra nullius, but this is not what actually occurred.[23] This 'right' fails to secure our obligations to land and law, and we can only treat or negotiate with the state within the confines of land as property and not as it stands in First Nations Law. The long-held conceptualisation within the colonial state is that land, or the natural world as a whole, is a resource for appropriation for economic growth, and an endless resource for the development of the state. Accumulation is seen as a natural activity of all human societies and is the premise of the natural-resource industries, but this is in opposition to an Indigenous ontological way of knowing.[24] The colonial failings to see the law of the land, to see the natural world, pits law against colonial power. The colonial process of characterising the natural world as a piece of property, and to divide it and parcel it up for sale and profit and or exploitation, is an act against law, First Nations law. It is a breach of our laws, but this most basic of colonial positions remains an unacknowledged breach.

Trying to move on from the void of state laws' failings to see Aboriginal laws, I will consider the possibilities that international law might provide a reprieve from the dichotomy of the two worlds we live in, and its concomitant genocide and ecocide of First Nations Peoples. Even though international law is itself a formation of colonialism,[25] I will consider recent international standards which have been hailed as recognition of First Nations' rights to justice and how those standards mirror, or improve upon state initiatives in the field of recognition.

More recently First Nations have accused the colonial states of failing to properly consult with them over development proposals on our lands. The government of Queensland failed to consult with the Mithaka Peoples over proposals to frack their lands in 2014.[26] Proscribed minimum standards

23 Watson, above n 1, 88–89.
24 An approach that occurs also in Bolivia, see R Merino, 'The politics of extractive governance: Indigenous peoples and socio-environmental conflicts' (2015) 2 *The Extractive Industries and Society* 85.
25 A Anghie, *Imperialism, Sovereignty and the Making of International Law* (Cambridge University Press, 2008).
26 The Mithaka Peoples made a submission to the UN Special Rapporteur on Indigenous Peoples in 2014, on the failure of the Queensland government to consult in accord with

set out in the United Nations' *Declaration on the Rights of Indigenous Peoples* (UNDRIP),[27] which set out as a minimum the expectation that Indigenous Peoples be meaningfully consulted and involved in any decisions about proposals to exploit resources on our territories, were ignored by the Queensland government.

While the standards within the UNDRIP are non-binding, there remains an expectation that both the state and industries involved in the extraction of non-renewable resources would comply with these international law standards. UNDRIP Article 32 (2) requires that states undertake good faith consultations in order to obtain free and informed consent to any large-scale projects and states thus:

> States shall consult and cooperate in good faith with the indigenous peoples concerned through their own representative institutions in order to obtain their free and informed consent prior to the approval of any project affecting their lands or territories and other resources, particularly in connection with the development, utilization or exploitation of mineral, water or other resources.

However, remaining a declaration, the UNDRIP does not of itself invoke an enforceable right, because it is not binding on states. However, it does provide a minimum standard and the expectation that states might exercise goodwill and comply with the Declaration in future political action which might be taken in relation to extractive industries, Indigenous Peoples and human rights, is a reasonable one. The principles outlined above have become customary international law, and given the general acceptance they have received, they should be considered so. They should be instruments of balance given the limitations in Australian law regarding the inherent rights of First Nations Peoples, and should ensure proper consultation processes which include free, prior and informed consent.

While consultations and the process of free, prior and informed consent should be fundamental protocols in any dealings with First Nations Peoples and our territories, it is also argued, in accord with First Nations ontologies and laws, that proposed developments that would impact so as to damage the natural world should be rejected. So while free, prior and informed consent has relevance to those situations, it is of no consequence if the First Nations have no power to enable a rejection or alteration. The option of

UNDRIP Article 32(2) minimum standards. <https://mithaka.files.wordpress.com/2014/11/submission-to-the-special-rapporteur-by-the-mithaka-people-03–12–2014.pdf.>

27 Ibid., see also the *Declaration on the Rights of Indigenous Peoples*, above n 19.

a complete veto should be held by First Nations where the land and natural world have the right to say 'no'.

Where I am situated it is clear that the state has failed to properly consult with the First Nations Peoples Boandik-Meintangk. In the eyes of the state we are deemed insignificant Peoples with no real rights to speak about our territories. However, more importantly, the proposal, for example, to frack on our territories presents such a high risk of fundamental damage to our natural environment and a conflict with our ancient laws, that we should have a right to exercise our authority as carriers of Aboriginal law and say 'no', regardless of whether or not there was a process of free, prior and informed consent. But of course, the state continues to hold power, underpinned by military force, to determine the outcome of extractive process projects regardless of any First Nations concerns or opposition.

IV: Conclusion – what will happen if the state remains blind to the laws of First Nations?

Critical sociologist and legal theorist Boaventura de Sousa Santos reminds us of our obligations beyond critical discourse and of the need for talk about the unspeakable when he warns of the risk of an epistemicide – the murder of knowledge – occurring if the exclusion of 'different' voices continues.[28] Although, as I have said above, the law is the law, it just is and it cannot be extinguished, the potential for a juricide looms while Aboriginal laws continue to be ignored. Perhaps it is more accurate to say that there is the potential for a death of knowledge in its connection to First Nations' Peoples as the carriers of law, notwithstanding that the law goes on as it has always done.

There is a myopia and there is also no dialogue or space in which the voices of First Nations speaking in the law of the land can be heard. So is there a possibility for a dialogue on critical concerns beyond the dominance of a western-centric universalism and incommensurability between cultures?[29] Can we imagine an equality of power relations, that is, relations of shared authority? It's possible, but the necessary precursors are intercultural translations, since only then can reciprocity amongst First Nations and states be obtained.[30] In thinking through this terrain, First Nations' critical approaches to vague and loaded concepts such as 'equality' remain essential. They are essential to the possibility of First Nations' voices being registered,

---

28 B de Sousa Santos, *Epistemologies of the South: Justice Against Epistemicide* (Paradigm Publishers, 2014).
29 Ibid., 212.
30 Ibid., 214.

heard and understood.[31] In opening up the possibility for a horizontal dialogue between colonialist states and peoples, Santos raises a series of concerns which I have found helpful when considering the possibility that First Nations' voices may be heard by dominant colonial states. I have accounted those concerns in raising the following questions: how do the power relations between First Nations and states translate into western law?; what place or space is there to speak of coloniality, for where is the world free of colonialism?; and can 'proper' dialogue only flourish where there is a commitment to decolonising power relations? So then, what are the possibilities for creating decolonial spaces where we can begin the dialogue, and begin also to see other ways of being law-full?

What are the possibilities for First Nations' laws? (The task before us is to re-centre them with the natural world). The enabling of an inter-states and First Nations' dialogue should include First Nations' perspectives on authority and power; this should be central to any dialogue, along with the understanding that the natural world holds authority.

We need to listen to the natural world all the time; now it is changing, howling, raining and drying up. We need to continually monitor dangerous extractive industries which might damage our natural ecosystems.

Under the pressure of humanity, our natural world is in crisis and this makes the need to see law truly ever more urgent.[32] The west has reached the end-point of its project progress, and does not have the solutions to the crisis it has constructed. It has no other lands to invade and colonise beyond leaving our mother Earth and searching for other planets. Current regimes of recognition and protection don't work. We are on the brink of sacrificing our waters, oceans, lands, skies and fellow creatures which still provide for an over-populated planet. Laws recognising First Nations come in the form of Native Title,[33] Aboriginal Heritage protection,[34] and other-named environmental laws,[35] and none of them have the capacity to protect the environments which are vital to our survival. We are on a trajectory which, it appears, could sacrifice all life forms, but we still have the capacity for ongoing life. Cycles do return, to begin again. Aboriginal law is an ongoing cycle, it is the law. Within the cycles of law there is the capacity to return to an ongoing cycle of life on Earth.

---

31 M Fricker, *Epistemic Injustice, Power and the Ethics of Knowing* (Oxford University Press, 2007).
32 de Sousa Santos, above n 28, 233.
33 *Native Title Act 1993* (Cth), *Native Title Amendment Act 1998* (Cth).
34 *Aboriginal and Torres Strait Islander Heritage Protection Act 1984* (Cth).
35 *Environment Protection and Biodiversity Conservation Act 1999* (Cth), among other Australian State and Commonwealth laws.

Chapter 15

Reimagining Aboriginal land rights: Crown, Country and custodians

Mabo v Queensland (No 2)

Stephen Summerhayes

Commentary

Mabo v Queensland (No 2) (*'Mabo'*)[1] recognised native title in Australia, overturning terra nullius, a common law fiction legitimising the dispossession of Aboriginals from their ancestral lands. My judgment further reimagines and redefines land tenure and the rights of Australia's traditional inhabitants. I depart from the Court's finding by vesting rights, 'ecological title', in Country (an Aboriginal concept similar to but broader than the Western notion of nature) and empowering traditional inhabitants as its custodians.

Australian Aboriginal culture, one of the oldest continuous living traditions, is tightly interwoven with Country, an embodiment of the physical, biological, spiritual, cultural and experiential elements of landscapes such as land, sea and sky.[2] My handing down of judgment on Australia Day is symbolic. Whilst to some, Australia Day represents the birth of a nation, to others, particularly Aboriginals, it is a poignant reminder of the injustice and other rights infringements that colonisation has brought upon Aboriginals and Country.[3]

1 (1992) 175 CLR 1.
2 See, e.g., D B Rose, *Nourishing Terrains. Australian Aboriginal Views of Landscape and Wilderness* (Australian Heritage Commission, 1996); P Burgess and J Morrison, 'Country' in B Carson et al. (eds), *Social Determinants of Indigenous Health* (Allen & Unwin, 2007) 177.
3 The first colonialists 'systematically dispossessed, murdered, raped and incarcerated the original owners': A Moreton-Robinson, 'I Still Call Australia Home: Indigenous Belonging and Place in a White Postcolonising Society' in S Ahmed (ed.), *Uprootings/Regroundings: Questions of Home and Migration* (Berg Publishing, 2003) 31. Genocide of Aboriginal communities was considered in *Coe v Commonwealth* [1993] HCA 42.

In *Mabo*, the High Court, the ultimate court in the Australian judicial system, rejected 200 years of traditional colonial legal theory,[4] rewrote the general law's treatment of Aboriginal interests in land and disrupted the power imbalance between Aboriginals and colonial-based interests. For communities with an ongoing connection to ancestral land, the Court recognised a native title that survived white settlement.

The Court's seven sitting judges delivered five judgments. Six of the seven members held that the Plaintiffs, on behalf of the indigenous Meriam people of the Murray Islands, possessed by virtue of their continuous occupation, and in accordance with their laws and customs, collective land rights that survived English sovereignty and the annexation of the Islands by the Colony of Queensland.[5] The Court abandoned the common law doctrine of terra nullius that provides that a sovereign state could, through conquest, cession or occupation, acquire territory over uninhabited or unsettled land.[6] Under the doctrine, land was also regarded as unsettled where the social system of the inhabitants could not 'be reconciled with the institutions or the legal ideas of civilized society'.[7] The Court found that the Islands were not terra nullius and rejected any notion of Aboriginals being low in social organisation.[8]

Before *Mabo*, in the only other Australian case to have considered an Aboriginal claim to traditional land, it was held that the 'doctrine of communal native title . . . does not form, and never has formed, part of the law of any part of Australia'.[9] It was said therefore that Aboriginal title to land could only originate from statute or a specific Crown grant.[10]

In some quarters, *Mabo* was hailed as a 'judicial revolution'[11] that expounded a credible doctrine protecting Aboriginal interests and supporting their culture[12], and 'one of the most necessary, important and overdue

---

4 L Strelein, *Compromised Jurisprudence: Native Title Cases since Mabo* (Aboriginal Studies Press, 2009) 10.

5 *Mabo* 15 (Mason CJ and McHugh J).

6 *Cooper v Stuart* (1889) 14 App Cas 286, 291 (Lord Watson).

7 *In re Southern Rhodesia* (178) (1919) AC 233.

8 See, e.g., *Mabo* 41–42, 58 (Brennan J).

9 *Milirrpum v Nabalco Pty Ltd* (1971) 17 FLR 141, 143 (Blackburn J).

10 See, e.g., U Secher, 'The Doctrine of Tenure in Australia Post-Mabo: Replacing the 'Feudal Fiction' with the 'Mere Radical Title Fiction' – Part 1' (2006) 13 *Australian Property Law Journal* 107.

11 See, e.g., M Stephenson and S Ratnapala (eds), *Mabo: A Judicial Revolution: The Aboriginal Land Rights Decision and Its Impact on Australian Law* (University of Queensland Press, 1993).

12 Strelein, above n 4, 23.

debates in Australia's history'.[13] It was also said to constitute a re-evaluation of colonisation against 'contemporary values of equality and social justice, of the status of Aboriginal "social organisation and customs", specifically in relation to the status of their connection to the land'.[14] Further, the case has been expressed to have had 'a profound impact on the legal, social and political reality of Indigenous–non-Indigenous relations in Australia'[15] that required the development of new legislative, policy and administrative approaches.[16]

Despite the enthusiasm surrounding the judgment, the extent of the 'revolution' has not been without debate.[17] Ritter remarked that '[a]lthough the doctrine of terra nullius may have been rejected, the existing super-structure of power-relations in Australia was dramatically re-affirmed'.[18] Mansell too observed that '[t]he Court did not overturn anything of substance, but merely propounded white domination and superiority over Aborigines by recognising such a meagre Aboriginal form of rights over land'.[19] Subsequent discourse also labelled the doctrine of terra nullius a 'legal fiction',[20] 'convenient scapegoat',[21] and a 'rhetorical explanation for why Aboriginal land rights had historically not been recognised',[22] It has also been said to have legitimised discrimination against Aboriginals based upon their 'nonconformity to the dominant culture'.[23] More fundamentally, commentators have argued that since no previous Australian or English case had decided

13 J Gardiner-Garden, 'The Mabo Debate – A Chronology' (Background Paper No 23, Parliamentary Research Service, Parliament of Australia 1993) 1.

14 M Donaldson and W Jonas, 'Native Title: The Promise of Acceptance' in G Doeker-Mach and K A. Ziegert (eds), *Law and Legal Culture in Comparative Perspective* (Franz Steiner Verlag, 2004) 357.

15 Strelein, above n 4, 1.

16 Parliament of Australia, *Mabo: Ten Years On*, E-Brief (28 December 2015) <www.aph. gov.au/About_Parliament/Parliamentary_Departments/Parliamentary_Library/Publica tions_Archive/archive/mabo>; Richard Bartlett, 'Political and Legislative Responses to Mabo' (1993) 23 *Western Australian Law Review* 352.

17 See, e.g., R Bartlett, 'Review of M A Stephenson and Suri Ratnapala (eds), *Mabo: A Judicial Revolution*' (1993) 23 *Western Australian Law Review* 383.

18 D Ritter, 'The "Rejection of Terra Nullius" in Mabo: A Critical Analysis' (1996) 18(5) *Sydney Law Review* 32.

19 M Mansell, 'The Court Gives an Inch but Takes Another Mile' (1992) 2(57) *Aboriginal Law Bulletin* 4.

20 A Howe, 'A Poststructuralist Consideration of Property as Thin Air – Mabo, A Case Study' (1995) 2(1) *Murdoch University Electronic Journal of Law*.

21 Ritter, above n 18, 7.

22 Ibid., 33.

23 Ibid., 12.

that Australia was terra nullius,[24] the doctrine was not even a common law concept.[25]

I deliver a sixth judgment of the High Court, installing myself as a fictitious eighth member. I write as if sitting alongside the judges in 1992. I identify, interpret and apply the law as it was back then, but reposition and address concepts of justice within contemporary frameworks of Earth jurisprudence, biocentrism and ecological justice.

The Court embraced a holistic and broadminded approach in its 'politically charged and emotionally resonant'[26] judgment. In a similar progressive spirit, I adopt an alternative jurisprudential trajectory and novel conceptualisation of Australian land rights – declaring Country as an entity capable of enjoying legal rights. Expression is given to Earth jurisprudence by rejecting the traditional model of land law in which human and corporate actors hold a unique, unilateral right of ownership and reframing it from a nature and ecosystems perspective.[27] I draw upon and extend models of indigenous custodianship of land and ecosystems, as well as intra-generational equity and social justice. My decision is premised on Country and people as interconnected, interdependent and interacting; with human systems functioning within Country rather than vice versa.

I conclude that rights in land did not accrue to the Meriam people pursuant to native title, but as custodians and constructive trustees of a unique 'ecological title' held by Country. My decision closely aligns land law with Aboriginal traditional laws and customs rather than moulding or interpreting Aboriginal rights within inappropriate Western legal models.

To paint a clear historical picture of the significant justice issues under consideration within a complete and discreet narrative, I spend time canvassing the factual elements of the case. I also introduce elements of the ecology and geology of the Murray Islands to subtly underpin their uniqueness and significance, especially in the context of Country. I deliberately formulate a pragmatic decision that readers can readily imagine and which proposes an iterative step and scaffold in the development of ecological jurisprudence.

24 Ibid., 9.
25 R Bartlett, 'The Mabo Decision' (1994) 1(3) *Australian Property Law Journal* 240.
26 G Simpson, 'Mabo, International Law, Terra Nullius and the Stories of Settlement: An Unresolved Jurisprudence' (1993) 19 *Melbourne University Law Review* 196.
27 See generally J Murray, 'Earth Jurisprudence, Wild Law, Emergent Law: The Emerging Field of Ecology and Law—Part 1' (2014) 35 *Liverpool Law Review* 217; J Murray, 'Earth Jurisprudence, Wild Law, Emergent Law: The Emerging Field of Ecology and Law—Part 2' (2014) 36 *Liverpool Law Review* 105.

HIGH COURT OF AUSTRALIA

Mason CJ, Brennan, Deane, Dawson, Toohey, Gaudron, McHugh and Summerhayes JJ

Mabo v Queensland (No 2) (1992) 175 CLR 1

Summerhayes J

A: Introduction

My judgment departs from that of my learned colleagues. I find that the Aboriginal[28] inhabitants, the Meriam people, are not native titleholders but custodians of Country and, pursuant to the terms of the custodianship, are entitled to occupation, use and enjoyment of the subject land, the Meriam Islands.

B: The plaintiffs' claim

There are two declarations sought by the plaintiffs on behalf of the Meriam people that are relevant to my findings:

a)   That the Murray Islands are not Crown land.
b)   That the Meriam people hold native title to the Islands and therefore are entitled as against the whole world to possession, occupation, use and enjoyment thereof.

C: The Murray Islands

The Murray Islands comprise three islands between Australia and New Guinea in the Torres Strait: Mer (also known as Murray Island), Dauer (also spelt Dauar and Dawar) and Waier. The Islands possess a rich and unique geology and ecology. For example, Mer Island, the largest of the three, has a landscape dominated by a volcanic vent, the only known Australian example. The Meriam people identify the vent as 'Gelam', the creator of the dugong. The island is also characterised by six natural vegetation communities as well as two regional ecosystems that do not appear elsewhere in Queensland. Biodiversity comprises 299 floral and 115 faunal species.

---

28 I use the term 'Aboriginal' rather than 'indigenous' to refer to native inhabitants such as the Meriam people, acknowledging however that mainland and islander Aboriginal cultures are distinct.

D: The Meriam people

The Meriam people inhabit the Murray Islands. They are descendants of Melanesian immigrants from Papua New Guinea and have been in continuous occupation of the Islands since before European contact. The Meriam people have a deep, fundamental and vital interconnection and identity with the Islands evidenced, in part, by their care and cultivation of the land for sustenance.

E: Annexation of the Murray Islands

When the Colony of New South Wales ('NSW') was proclaimed on 7 February 1788, as much of the common and statute law of England became domestic law as was applicable to the local context.[29] The Crown acquired the absolute beneficial ownership of the land pursuant to a radical title (ultimate ownership). In practical terms, colonisation facilitated the introduction of the English system of land ownership and the related conveyancing process. This acquisition of sovereignty is an act of State that cannot be challenged before the courts of that State, no matter how fair or unfair.[30] Notwithstanding, section 24 of the *Australian Courts Act 1828*[31] confirmed the inheritance by NSW of the laws of England.

Queensland originally formed part of the Colony of NSW. When the Colony of Queensland separated from NSW in 1859, it inherited the laws of NSW including those that NSW had inherited from England. On 1 August 1879, the Colony of Queensland annexed the Murray Islands. Sovereignty over the Islands was confirmed by the *Colonial Boundaries Act 1895*[32] which retrospectively incorporated the Islands into Queensland. Queensland law then became the law of the Murray Islands and the Meriam people became entitled to such rights and privileges and subject to such liabilities as the inherited law provided.

F: Terra nullius

The doctrine of terra nullius provides that a sovereign State can acquire uninhabited land as a new territory, a radical title, through conquest, cession or occupation. The doctrine was later expanded to legitimise the colonisation of occupied territories where the occupants were without 'settled

---

29 *Cooper v Stuart* (1889) 14 App Cas ('Cooper').
30 *New South Wales v The Commonwealth* (1975) 135 CLR 388 (Gibbs J) ('*the Seas and Submerged Lands Case*').
31 (Imp) (9 GEO IV c.83).
32 (Imp) (58 and 59 Vict c 34).

law'[33] or, in the words of Lord Sumner of the Privy Council, *In re Southern Rhodesia*:

> are so low in the scale of social organization that their usages and conceptions of rights and duties are not to be reconciled with the institutions or the legal ideas of civilized society.[34]

Thus, although Aboriginals inhabited Australia at the time of colonisation, the doctrine justified their dispossession resulting in their estrangement from Country and their way of life.

Early colonial government communications provide an insight into the attitude towards and justification for the acquisition of sovereignty over land inhabited by Aboriginals. For example, the despatch of Lord John Russell, English Colonial Secretary, to Sir George Gipps, Governor of NSW, dated 21 December 1839 refers to the Crown's 'sacred duty . . . to impart to the former occupiers of NSW the blessings of Christianity, or the knowledge of the Arts and advantages of civilised life'.[35] It is, however, difficult to identify any tangible benefits flowing to the Meriam people from such purported blessings, knowledge and advantages.

I disagree that Australia was terra nullius at the time of British sovereignty. It is implausible to suggest that the Aboriginals were of low social organisation. To the contrary, Justice Moynihan of the Supreme Court of Queensland in the present case found that the Meriam people had an 'elaborate and complex social organisation'.[36] They had, however, no formal scheme of land ownership and registration, but that was because the Western legal construct of real property had no place in their social system. I agree with the sentiments of Justice Brennan in his judgment in this case, that an application of terra nullius in Australia is 'a discriminatory denigration of indigenous inhabitants, their social organization and customs'.

G: *Native title*

Native title is *sui generis* and cannot be readily compartmentalised within the common law.[37] It confers upon Aboriginal communities in occupation of land the right to use and enjoy that land in accordance with their

---

33 *Cooper* 291 (Lord Watson).
34 *In re Southern Rhodesia* (1919) AC 233.
35 F Watson (ed.), *Historical Records of Australia. Series 1. Governors' Despatches to and from England Volume XX February 1839–September 1840* (The Library Committee of the Commonwealth Parliament, 1924) 440.
36 Vol 1 155–156.
37 *Guerin v The Queen* (1984) 13 DLR (4th) 321, 339 (Dickson J).

traditional laws and customs. Dealings with native title rights must be under those traditional laws or customs, as they evolve over time, and cannot be transferred or assigned outside of the particular system. Community members are descendants of those who have traditionally occupied the land, identify as members and are accepted as such by the community.

Native title survives sovereignty, as a burden on the Crown's radical title, but will be extinguished in the following circumstances, whereupon the Crown's radical title expands to a full beneficial title:

1   Voluntary surrender to the Crown.
2   Death of the last surviving community member.
3   Community laws and customs are no longer observed by the community.
4   Cessation of the community's connection with the land.
5   Clear and plain legislation on the part of the Crown.
6   Alienation of land by the Crown which is wholly or partially inconsistent with the continued right to enjoy native title (extinguishment is to the extent of the inconsistency).[38]

In *Milirrpum v Nabalco Pty Ltd* ('*Milirrpum*'),[39] the only prior reported Australian decision examining Aboriginal native title, Justice Blackburn stated that the doctrine of native title had no place in a settled colony. I respectfully disagree with His Honour. Whilst there are Australian cases supporting his contention,[40] these can be distinguished on the basis that they were either misguided in considering Australia as relevantly unoccupied, or the judicial statements were made *obiter dicta*, without compelling reasons. The true nature of the Aboriginals' occupation and relationship with the land are, as outlined below, key determinants of my decision.

H: Country

The footprints of the Grassland Rodent, *Melomys burtoni*, delicately impressed in fertile basaltic soil gently erode over time. The salty coastal winds cause the rusty, golden-coloured hairs on the new growth of the small, endangered tree, *lectryon repandodentatus* to bristle. The moving

38 *In re Southern Rhodesia* (1919) AC 233; *Amodu Tijani v Secretary, Southern Nigeria* (1921) 2 AC 399 ('*Amodu Tijani*').
39 *Milirrpum v Nabalco Pty Ltd* (1971) 17 FLR 141 ('*Milirrpum*').
40 *Attorney-General v Brown* (1847) 1 Legge 312; *Cooper v Stuart* (1889) 14 App Cas; *Williams v Attorney-General for New South Wales* [1913] HCA 33; *Randwick Corporation v Rutledge* [1959] HCA 63.

reflection of billowing clouds can be seen in the short-lived water pools that are formed after tropical drenching rains. These are characteristics of the Meriam Islands and captured by the Aboriginal English word Country, a culturally relativist term that embodies the physical, biological, spiritual, cultural and experiential elements of landscapes such as land, sea and sky.

Deborah Bird Rose in her 1986 work on behalf of the Australian Heritage Commission defines Country in the following terms:

> Country in Aboriginal English is not only a common noun but also a proper noun. People talk about country in the same way that they would talk about a person: they speak to country, sing to country, visit country, worry about country, feel sorry for country, and long for country. People say that country knows, hears, smells, takes notice, takes care, is sorry or happy. Country is not a generalised or undifferentiated type of place, such as one might indicate with terms like 'spending a day in the country' or 'going up the country'. Rather, country is a living entity with a yesterday, today and tomorrow, with a consciousness, and a will toward life. Because of this richness, country is home, and peace; nourishment for body, mind, and spirit; heart's ease.[41]

In somewhat blander terms, Garnett and Sithole describe Country as the lands with which Aboriginal people 'have a traditional attachment'.[42] Country is also a concept well-known to the Australian Government who present a slightly more nuanced interpretation:

> Land is fundamental to the wellbeing of Aboriginal people. The land is not just soil or rocks or minerals, but a whole environment that sustains and is sustained by people and culture. For Indigenous Australians, the land is the core of all spirituality and this relationship and the spirit of 'country' is central to the issues that are important to Indigenous people today.[43]

---

41 D B Rose, *Nourishing Terrains. Australian Aboriginal Views of Landscape and Wilderness* (Australian Heritage Commission, 1996) 7.

42 S Garnett and B Sithole, *Sustainable Northern Landscapes and the Nexus with Indigenous Health: Healthy Country, Healthy People* (Australian Government, 2007).

43 Australian Government, *Australian Indigenous Cultural Heritage. Land – at the Core of Belief* (1 January 2016) <www.australia.gov.au/about-australia/australian-story/austn-indigenous-cultural-heritage>.

Country resonates with the Latin American concept of Pachamama or Mother Earth:

> Pachamama is earth as a mother, Mother Earth. She is the mother of the mountains, of life that is born and grows, of all people. It is because of her that fruit ripens, animals multiply, and women have babies. It is she who controls frosts and rains. If she gets angry, she sends thunder and storms.[44]

Pachamama has been recognised as a living being, afforded rights under Bolivian Law,[45] and echoed in the Principles of Environmental Justice[46] developed at the First National People of Color Environmental Leadership Summit held in 1991. Principle 1 affirms the 'sacredness of Mother Earth, ecological unity, and the interdependence of all species'.

I accept that the Meriam people have a complex and deeply intertwined relationship with Country. Indeed, Justice Moynihan in the present case found that they possess 'a strong sense of relationship to their Islands and the land and seas of the islands which persists from the time prior to European contact'.[47] To my mind, the legal rights of the Meriam people over the Murray Islands derive from this fundamental interconnection and interdependence.

I: Country distinguished from nature and the natural environment

There is considerable discourse on the philosophical and ideological concepts of nature. For example, nature has been posited as an anthropocentric construct, a commodity, a set of ecosystem services, or a matriarchal nurturing figure.[48] For the purpose of this case, I define nature as 'the physical world collectively, including biota, the landscape, and other features which exist independently of humans'. For the natural environment, I adopt the

---

44 P Martin and M R MacDonald, *Pachamama Tales: Folklore from Argentina, Bolivia, Chile, Paraguay, Peru, and Uruguay* (ABC-CLIO, 2014) xiv.

45 Bolivian Government, *Ley de Derechos de la Madre Tierra* (21 December 2010) <www. ine.gob.bo/indicadoresddhh/archivos/alimentacion/nal/Ley%20N%C2%BA%20071. Pdf>.

46 First National People of Color Environmental Leadership Summit, *Principles of Environmental Justice* (October 1991) <www.ejnet.org/ej/principles.html>.

47 Ibid.

48 See, e.g., L Tulloch, 'Is Emile in the Garden of Eden? Western Ideologies of Nature' (2015) 13(1) *Policy Futures in Education* 20.

definition of Johnson et al. as an environment that is relatively changed, modified, disturbed, or created by human cultural activities.[49]

This case examines interdependent human and non-human systems and therefore, the terms nature and natural environment are too narrow. The term Country is apposite, encapsulating the Islands, the Meriam people's relationship with them, and the real property over which they claim an interest.

J: Land, Country and custodianship

For Aboriginals, land is inseparable from Country and a mere legal construct. Brazenor et al. express the Aboriginal concept of land in the following manner:

> The importance of land within Aboriginal culture transcends generations and involves a spiritual and material connection to land, where land is not measurable in mathematical terms, nor delineated through the mapping or textual mediums of a western society. Traditional interests in land are measured and quantified through the traditions and customs as held by the Aboriginal peoples of Australia. These traditions and customs are demonstrated through paintings, song, dance, symbolic totems and oral traditions, and used to reinforce the rights, restrictions and responsibilities for managing and protecting land. The land tenure system as held by Aboriginal peoples, defines cultural identity and is held in a communal manner.[50]

For living memory, Aboriginals have existed in harmony with Country. They consider Country as an inextricable part of their community's functioning. Country is deeply embedded in their beliefs, values, cultural practices and spirituality, and is thus firmly allied to their sense of belonging and physical, social and emotional wellbeing. This is fundamentally different to the Judaeo-Christian approach that subordinates Country.[51] It also contrasts with the cadastral land system inherited from England that commodifies, privatises and commercialises land, derogating from the

49 D L Johnson et al., 'Meanings of Environmental Terms' (1997) 26(3) *Journal of Environmental Quality* 582.
50 C Brazenor, C Ogleby and I Williamson, 'The Spatial Dimension of Aboriginal Land Tenure' (Paper presented at the Proceedings of the 6th South East Asian Surveyors Congress, Fremantle, Western Australia, 1999).
51 C C Williams and A C Millington, 'The Diverse and Contested Meanings of Sustainable Development' (2004) 170(2) *The Geographical Journal* 99; Lynn White, 'The Historical Roots of Our Ecologic Crisis' (1967) 155 *Science, New Series* 3767.

intrinsic value of Country and obviating the establishment or recognition of custodial duties over it.

In the words of Purdie et al.:

> Land is fundamental to Indigenous people, both individually and collectively. Concepts of Indigenous land ownership were and are different from European legal systems. Boundaries were fixed and validated by the Dreaming creation stories. Each individual belonged to certain territories within the family group and had spiritual connections and obligations to particular country. *Hence land was not owned; one belonged to the land.*[52] (emphasis added.)

Justice Blackburn in *Milirrpum* acknowledged that 'the Aboriginals have a more cogent feeling of obligation to the land than of ownership . . . it seems easier, on the evidence, to say that the clan belongs to the land than that the land belongs to the clan'.[53]

As mentioned earlier, native title confers upon Aboriginal communities the right to use and enjoy land in accordance with their traditional laws and customs. Based upon the foregoing, because ownership of land is inconsistent with Aboriginal laws and customs, native title is an inappropriate vehicle to recognise their interests in it.

The Aboriginal ontology of a physical, cultural and spiritual connection with and duty of care for Country,[54] rather than ownership of it, more properly establishes them as custodians of land and constructive trustees as opposed to landowners. A number of authors avert to Aboriginal custodianship of land. Donnelly says:

> Custodianship of land is important in Aboriginal culture. It is held on behalf of the ancestors and must be adequately managed. The current generations of Indigenous Australians recognise they have a responsibility to leave land and water resources so future generations can be

52 P Dudgeon et al., 'The Social, Cultural and Historical Context of Aboriginal and Torres Strait Islander Australians', in N Purdie et al. (eds), *Working Together: Aboriginal and Torres Strait Islander Mental Health and Wellbeing Principles and Practice* (Australian Government Department of Health and Ageing, 2010) 26.

53 *Milirrpum* 270–271.

54 A Voigt and N Drury, *Wisdom from the Earth: The Living Legacy of the Aboriginal Dreamtime* (Shambhala, 1998); A Moreton-Robinson, 'I Still Call Australia Home: Indigenous Belonging and Place in a White Postcolonising Society' in S Ahmed (ed.), *Uprootings/Regroundings: Questions of Home and Migration* (Berg Publishing, 2003) 31.

sustained. Innate respect for the land and resources afforded to them have been emphasised in traditional Aboriginal communities.[55]

Bayet remarks:

Aboriginal people perceive the Australian landscape as their cultural domain. It was their traditional duty to be custodians of the land.[56]

Further, the Australian Government identifies Aboriginals as traditional custodians. For example, the President of the Senate on taking the chair each day reads the Lord's prayer and acknowledges the 'Ngunnawal and Ngambri peoples who are the traditional custodians of the Canberra area and pay[s] respect to the elders, past and present, of all Australia's Indigenous peoples'.[57]

K: Ecological sustainability and justice

For millennia, Aboriginals, discharging their role as custodians, have sustainably managed the health, integrity, stability and beauty of Country. Rose opined that Aboriginals adopt an ecocentric approach to land management based upon knowledge and understanding[58]; a position supported by Ross et al. who conclude that Aboriginal ecological knowledge is derived from tens of thousands of years of stewardship of the country's resources.[59] This stewardship of Country facilitates Aboriginal self-determination and is a necessary condition of justice to future generations: it constitutes a species of intergenerational equity because it supports Country's continued provision of benefits that are vital to life,[60] and affords equal opportunity for present

---

55 K Donnelly, 'The Challenge: Shared Responsibility – Report on the Building Cultural Competencies Between Indigenous and Non-Indigenous Australians' (2007) 84 *Culturescope* 7.

56 F Bayet, 'Overturning the Doctrine: Indigenous People and Wilderness – Being Aboriginal in the Environmental Movement' in J Baird Callicott and M P Nelson (eds), *The Great New Wilderness Debate* (University of Georgia Press, 1998) 316.

57 Parliament of Australia, *Chapter 8 – Sittings, Quorum and Adjournment of the Senate, Annotated Standing Orders of the Australian Senate* (1 January 2016) <www.aph.gov.au/About_Parliament/Senate/Powers_practice_n_procedures/aso/so050>.

58 Rose, above n 41.

59 A Ross et al., *Indigenous Peoples and the Collaborative Stewardship of Nature: Knowledge. Knowledge Binds and Institutional Conflicts* (Left Coast Press Inc, 1991).

60 D Schlosberg, *Defining Environmental Justice: Theories, Movements, and Nature* (Oxford University Press, 2007).

and future generations to enjoy Country's benefits.[61] It is capable of formal recognition by the law.

Some commentators have questioned the ecological credentials of Aboriginal land management, asserting that environmental sustainability is merely an unintended consequence of the Aboriginals' limited technological development.[62] However, there is cogent historical evidence that sustainability is intentional and consistent with kinship of Country: Aboriginals nurture Country, maintaining its ecological structure and function. And, accepting that all organisms modify their environment in some way, Aboriginal land management is low impact and, as far as reasonably practicable, in harmony with it.

Whether the Aboriginal custodial role gives rise to rights and obligations recognised in law turns upon whether Country can possess a distinct and independent existence.

L: Country as an entity

The law recognises many non-human entities. For example, society is familiar and comfortable dealing with corporations. Corporations are abstract entities afforded legal personality by the legislature[63] and characterised as property that is owned by humans or other corporations. Like the human form, Country is a system of interconnecting, interacting and interdependent physical, chemical, biological and spiritual components. Similarities between Country and humans also extend to such things as sentience, agency, integrity and capacity.[64] A corporation does not share these attributes and thus it can be reasonably argued that legal personality is better suited to Country than to corporations.

The law recognises the legal rights of animals, enforceable by a human legal representative.[65] The *United Nations Convention on the Rights of the Child* identifies the need to protect unborn children.[66] The *Nasciturus* doctrine also grants rights to unborn children,[67] and this doctrine has been

61 Ibid.
62 See e.g., P Ehrlich, 'Human Natures, Nature Conservation, and Environmental Ethics' (2002) 2(1) *Bioscience*.
63 *Corporations Act 2001* (Cth) s 1.5.1.
64 Schlosberg, above n 60.
65 See e.g., *Prevention of Cruelty to Animals Act 1979* (NSW).
66 *Convention on the Rights of the Child*, opened for signature 20 November 1989, 1577 UNTS 3 (entered into force 2 September 1990).
67 The doctrine is an exception to the common law rule that the unborn child has no legal personality. A testamentary gift to a child *in utero* but conceived before the testator's death is entitled to take under the Will.

codified in State legislation.[68] The principle of intergenerational equity also recognises future generations, that is, humans who have not yet been born. Following this line of reasoning, Country can be a recipient of justice and capable of possessing rights as do infants or those under a disability.[69]

The common law has also seen fit to vest native title rights in entities other than individuals. In *Amodu Tijani*, Viscount Haldane recognised that interests in land can vest in a community and transcend property systems and processes. He observed:

> The title, such as it is, may not be that of the individual, as in this country it nearly always is in some form, but may be that of a community. . . . The introduction of the system of Crown grants which was made subsequently must be regarded as having been brought about mainly, if not exclusively, for conveyancing purposes, and not with a view to altering substantive title already existing.[70]

A fortiori, if the common law can recognise land rights in a community, it can vest land rights in Country.

Country recognises the intrinsic value of the Earth and all its elements (that is, value independent of any demonstrable human benefits), whilst still allowing resource use conditional upon the maintenance of socio-ecological integrity. Country explicitly acknowledges that humans are an essential part of the system and vice versa. The relationship between a healthy Country and a healthy community is axiomatic. Further, as I averred to earlier, the maintenance of the integrity of Country is an essential precondition for ecological justice (doing justice to nature,[71] a right to exist and persist)[72] and Aboriginal justice. It would be inimical to justice if the courts characterised Country as too amorphous or abstract to possess rights and interests in land.

The fact that courts have not previously recognised Country is no obstacle to doing so in this case. Judges often decide novel questions like the ones we have been asked to consider. The Privy Council in *Adeyinka Oyekan v Musendiku Adele* held that rights in land can exist (in that case, native title rights) 'even though those interests are of a kind unknown to English

---

68 See e.g., *Succession Act 2006* (NSW) s 3(2).
69 See e.g., Schlosberg, above n 60.
70 *Amodu Tijani* (1921) 2 AC 403–404, 407–408.
71 See, e.g., D Schlosberg and L Collins, 'From Environmental to Climate Justice: Climate Change and the Discourse of Environmental Justice' (2014) 3 *Wiley Interdisciplinary Reviews: Climate Change* 359.
72 J Murray, 'Earth Jurisprudence, Wild Law, Emergent Law: The Emerging Field of Ecology and Law – Part 2' (2015) 36 *Liverpool Law Review* 105.

law'.[73] In *Cattanach v Melchior*[74] the High Court held that a doctor could be responsible for the costs of raising and maintaining a child conceived as a result of the doctor's negligence. Judges are also at liberty to overturn decisions that do not reflect contemporary societal values. For example, in *R v L*, it was found that 'whatever may have been the position in the past, the institution of marriage in its present form provides no foundation for a presumption which has the effect of denying that consent to intercourse in marriage can, expressly or impliedly, be withdrawn'.[75]

It may be argued that conceptualising Country as capable of enjoying legal rights is tantamount to making new law. Justice Dixon advanced the declaratory theory whereby judges are to declare what the law is and always has been based upon strict legal reasoning.[76] However, the common law is concerned with outcomes rather than strict adherence to an unwavering interpretation of historical case law. The common law is dynamic, evolving, ideally in step with the world and new knowledge and understanding, looking toward *lex ferenda* rather than *lex lata*. Justice Windeyer in *Skelton v Collins* said:

> [T]he common law is a body of principles capable of application to new situations, and in some degree of change by development.[77]

His Honour then quoted Lord Reid in the House of Lords' case *Myers v Director of Public Prosecutions*:

> The common law must be developed to meet changing economic conditions and habits of thought, and I would not be deterred by expressions of opinion in this House in old cases.[78]

In *Corkill v Forestry Commission of New South Wales*, Justice Stein challenged the traditional declaratory theory of judicial decision-making. In relation to a submission that the precautionary principle should be incorporated into domestic law he said:

> It seems to me unnecessary to enter into this debate. In my opinion the precautionary principle is a statement of common sense and has already been applied by decision-makers in appropriate circumstances prior to the principle being spelt out.[79]

---

73 *Adeyinka Oyekan v Musendiku Adele* (1957) 2 All ER 788.
74 *Cattanach v Melchior* [2003] HCA 38.
75 *R v L* (1991) 174 CLR 379 405 (Dawson J).
76 See e.g., O Dixon, 'Concerning Judicial Method' (1956) 29 *Australian Law Journal* 470.
77 *Skelton v Collins* (1966) 115 CLR 94.
78 *Myers v Director of Public Prosecutions* (1965) AC 1001 1021.
79 *Corkill v Forestry Commission of New South Wales* (No 2) (1991) 73 LGRA 126.

Justice Dixon did, however, acknowledge that the common law was flexible enough to apply to new factual situations, to new conclusions or to the logical development of a principle:

> It is one thing for a court to seek to extend the application of accepted principles to new cases or to reason from the more fundamental of settled legal principles to new conclusions or to decide that a category is not closed against unforseen instances which in reason might be subsumed thereunder. It is an entirely different thing for a judge, who is discontented with a result held to flow from a long accepted legal principle, deliberately to abandon the principle in the name of justice or of social necessity or of social convenience.[80]

Elevating the status of Country is in tune with society's increased preoccupation with environmental issues. Legal recognition of Country reflects contemporary international notions and development of environmental law. Emmenegger and Tschentscher distinguish three stages in the development of international environmental instruments:

a)   environmental protection;
b)   recognition of future generations;
c)   recognition of nature's intrinsic value.[81]

There are international environmental instruments evidencing a transition to the third stage. An example is the *United Nations Convention on Biological Diversity*. Australia is a party. The preamble states that the parties are:

> [c]onscious of the intrinsic value of biological diversity and of the ecological, genetic, social, economic, scientific, educational, cultural, recreational and aesthetic values of biological diversity and its components.[82]

Moreover, the 1982 *United Nations World Charter for Nature*, proclaims five 'principles of conservation by which all human conduct affecting nature is to be guided and judged'.[83] Principle one depicts nature as an entity, stipulating that it 'shall be respected and its essential processes shall not be impaired'.

---

80 Above n 76, 472.
81 S Emmenegger and A Tschentscher, 'Taking Nature's Rights Seriously: The Long Way to Biocentrism in Environmental Law' (1994) 6 *Georgetown International Environmental Law Review* 550.
82 *Convention on the Biological Diversity*, opened for signature 5 June 1992, 1760 UNTS 79 (entered into force 29 December 1993).
83 *World Charter for Nature*, UN GAOR, GA Res 37/7, 48th plenary mtg, UN Doc A/RES/37/7 (28 October 1982).

Affording Country rights does not mean that it must take on all those that humans enjoy,[84] but rather only those relevant to the context. Stone proffered:

> *In re Gault* gave 15-year-olds certain rights in juvenile proceedings, but it did not give them the right to vote. Thus, to say that the environment should have rights is not to say that it should have every right we can imagine, or even the same body of rights as human beings have.[85]

It is clear that Country is readily ascertainable and comprehendible as a subject rather than an object, and should be acknowledged in the eyes of the law. Any conceptual difficulty in Country possessing legal identity is merely illusory. My view does not depart from settled legal principles but is a logical and necessary development in respect of a new situation, and not so fanciful as to impugn our legal system. The case falls within Justice Dixon's exception. Echoing the words of Justice Brennan in this case, my position can be taken 'without fracturing the skeleton which gives our land law its shape and consistency'. In fact, my decision fortifies the skeleton, reinstating Country and Aboriginals to their rightful position.

M: Declarations

I declare that:

1   the land in the Murray Islands is not Crown land for the purposes of the *Land Act 1962* (Qld);
2   a *sui generis* 'ecological' title in respect to the land in the Murray Islands is vested in Country;
3   the Meriam people hold the Murray Islands as custodians and constructive trustees for Country; and
4   the Meriam people are entitled to the occupation, use and enjoyment of the Murray Islands in accordance with their traditional laws and customs pursuant to the constructive trust, specifically, it is a term of the trust that they maintain the Island's ecological integrity (structure, composition, and function, operating within the bounds of natural or historic range of variation and change).

---

84 Schlosberg, above n 60.
85 C Stone, 'Should Trees Have Standing? – Toward Legal Rights for Natural Objects' (1972) 45 *Southern California Law Review* 450.

Chapter 16

Nuclear waste dump

Sovereignty and the Muckaty mob

Greta Bird and Jo Bird

Commentary

The quest to find a permanent nuclear waste dump in Australia has been ongoing for decades. At present 'temporary' facilities exist for radioactive waste, for example, the Lucas Heights research reactor. In 1998, land in South Australia was selected as a site for a permanent national repository. However, there was strong opposition from both the Aboriginal owners of the land and the South Australian government. A legal battle ensued. The case was won against the compulsory acquisition by an action in the federal court in 2004 on constitutional law and native title grounds.[1]

The search for a potential site then moved to the Northern Territory, in part due to its constitutional weakness in relation to States. Initially, the Commonwealth had identified defence land as a suitable repository. However, during the drafting of the *Commonwealth Radioactive Waste Management Act 2005*, the Northern Land Council negotiated a nomination process for Aboriginal land in the Northern Territory that would lead to the government acquiring all property interests in the chosen site for the dump in perpetuity. The land to be nominated was made more attractive because under the current Act,[2] Commonwealth legislation was to be suspended for the site. Laws such as the *Native Title Act 1993* (Cth), *Aboriginal and Torres Strait Islander Heritage and Protection Act 1984* (Cth) and the *Environment Protection and Biodiversity Conservation Act 1999* (Cth) are deemed not to apply.

In 2007, Muckaty Station in the Northern Territory was selected as a potential site by the Howard government after negotiations with the Northern Land Council. Again, Aboriginal land had been earmarked for the

---

1 *State of South Australia v Honourable Peter Slipper* [2004] FCAFC 164.
2 The first *Radioactive Waste Management Act* (Cth) was passed in 2005 and was repealed by the 2012 Act of the same name which is much more expansive.

dump. The *United Nations Declaration on the Rights of Indigenous Peoples* provides that states must 'take effective measures to ensure that no storage or disposal of hazardous materials shall take place in the lands or territories of Indigenous peoples without their free, prior and informed consent'.[3] The question of consent therefore loomed large.

In effect, the Commonwealth was asking the traditional owners to cede their land to the Commonwealth without a treaty. There was no need for just compensation pursuant to section 51(xxxi) of the Constitution because this is not a compulsory acquisition by the Commonwealth, in that the traditional owners would be consenting to the destruction of their title. Under the *Aboriginal Land Rights Act 1976* the Northern Land Council could, after consultation, give the Commonwealth an assurance that the traditional owners had consented to the nomination. The Northern Land Council signed off on the necessary consent.

A: *Environmental concerns*

The site selection process has largely been conducted for political purposes rather than environmental reasons. An argument has been put forward that the waste dump should be in a 'remote' area. Julie Bishop, then federal science Minster, stated that the NT sites were 'some distance from any form of civilization'.[4] The idea is that desert areas are 'sacrifice zones'.

For those concerned about the dangers of radioactive waste storage it is vital that the site be environmentally stable. However, Muckaty does not fit this criteria. Scientists have warned that, environmentally, the area is subject to earthquakes.[5] In 1988, three earthquakes measuring between 6.3 and 6.7 on the Richter scale were recorded at Tennant Creek, which lies 110 kilometres from Muckaty Station. This resulted in $2.5 million in damage to buildings and a gas pipeline, and the aftershocks were felt as far away as Adelaide and Perth.[6]

In terms of radioactivity, more than 90 per cent of the waste from Lucas Heights proposed to be stored at Muckaty is from spent nuclear fuel reprocessing. Only a small percentage comes from nuclear medicine. Waste from other locations includes 2000 cubic metres of contaminated soil which is

3 *United Nations Declaration on the Rights of Indigenous Peoples*, adopted by the General Assembly, 2 October 2007, A/RES/61/295 art 8.
4 N Wasley, 'Protecting Manuwangku: radioactive wrongs or Indigenous rights' (2012) 7 *Indigenous Law Bulletin* 16.
5 See website of Geoscience Australia <www.ga.gov.au/earthquakes>; 'Proposed nuclear dump in quake hot spot', *The Sydney Morning Herald* (online), 2 July 2007 <www.smh. com.au/national/proposed-nuclear-dump-in-quake-hot-spot-20070701-li9.html>.
6 J Bowman, 'The 1988 Tennant Creek, Northern Territory earthquakes: a synthesis' (1992) 39 *Australian Journal of Earth Sciences* 651.

currently in storage on the land at Woomera. The Woomera test sites from British nuclear weapon tests in the 1950s is an environmental disaster zone.[7] A concern has been expressed that 'requiring ANSTO [Australian Nuclear Science and Technology Organisation] to store its own waste is the best and perhaps the only way of focussing the organisation's mind on waste minimisation'.

However, this option has not been pursued. Indeed, former Nationals leader Tim Fischer told the *Weekend Australian* that when Cabinet was deliberating in 1999 about investing in the Adelaide to Darwin rail line, it took into account the possibility of using the line as part of an international nuclear storage industry.[8] As we write, Prime Minister Malcolm Turnbull is supportive of Australia having a commercial role in the nuclear fuel cycle, and accepting high-level waste where currently Australia stores only low and medium level waste. He says 'we could bring spent fuel rods back to Australia for storage. We have got stable, very stable geology in remote locations and a stable political environment'.[9]

Following the Muckaty case, the South Australian government held a Royal Commission into the nuclear fuel cycle. The commission has to date only made tentative findings.[10] What is required is a national independent inquiry into radioactive waste management based on Indigenous, environmental and scientific considerations.[11]

There is a need to respect the First Nations peoples of Muckaty and their sovereignty and custodianship of the land. With the recognition of sovereignty, First Nations peoples are able to follow their law and care for country. Even without this recognition, there have been successful environmental protests against mining. For example, the Mirrar people opposed the establishment of a uranium mine at Jabiluka. They were concerned

---

7 F Walker, *Maralinga: The Chilling Exposé of our Secret Nuclear Shame and Betrayal of our Troops and Country* (Hachette Publications, 2014).

8 P Cleary, 'Genesee and Wyoming's transcontinental nuclear strategy on track', *The Australian* (online), 30–31 January 2016 <www.theaustralian.com.au/business/genesee—wyomings-transcontinental-nuclear-strategy-on-track/news-story/a3bfd548cdda30f5e1333c8763af5e58>.

9 M Kenny, 'An Australian nuclear industry still on the table for Malcolm Turnbull', *The Sydney Morning Herald* (online), 29 October 2015 <www.smh.com.au/federal-politics/political-news/an-australian-nuclear-industry-still-on-the-table-for-malcolm-turnbull-20151028-gkl3zj.html>.

10 South Australian Government, *Nuclear Fuel Cycle Royal Commission: Tentative Findings*, February 2016 <http://nuclearrc.sa.gov.au/tentative-findings>. The 2016 Royal Commission addresses Indigenous rights to land and protection of sacred sites in a few paragraphs of a 48 page document. The Commissioner is Rear Admiral the Honourable Kevin Scarce. None of the advisory committees have any First Nations members.

11 Wasley, above n 4, 19.

at the environmental damage that mining would cause, including damage to the wetlands at Kakadu. They were successful in their campaign and Rio Tinto entered into an agreement with the Mirrar people in 2003 that the mine would not be opened without the consent of the traditional owners. The mine shaft was filled to protect the sacred knob tailed gecko dreaming. Australian First Nations peoples have often been joined by environmentalists in actions to protect country.

B: *The original hearing*

A number of the clans on Muckaty Station were dismayed at the nomination of a site on their country and decided to take action in the Federal Court. They decided to challenge the decision and the case came before the Federal Court in June 2014.

In the proceedings, issued against the Northern Land Council, the Commonwealth and the Minister for Resources, there were a number of causes of action. These included unlawful conduct by the Minister and a breach of statutory duty by the Northern Land Council. Consent was an issue at the heart of the proceedings. Under the *Aboriginal Land Rights Act 1976* (Cth), section 23(3) provides that:

> In carrying out its functions with respect to any Aboriginal land in its area, a Land Council shall have regard to the interests of, and shall consult with, the traditional Aboriginal owners (if any) of the land and any other Aboriginals interested in the land and, in particular, shall not take any action, including, but not limited to, the giving of consent or the withholding of consent, in any matter in connexion with land held by a Land Trust, unless the Land Council is satisfied that:
>
> (a) the traditional Aboriginal owners (if any) of that land understand the nature and purpose of the proposed action and, as a group, consent to it; and
>
> (b) any Aboriginal community or group that may be affected by the proposed action has been consulted and has had adequate opportunity to express its view to the Land Council.

At the hearing, the determination of who the relevant traditional owners were and the matter of their consent to the site nomination was challenged. The Commonwealth argued that the Northern Land Council had the statutory power to determine who the traditional owners were and whether they had given their consent to the site nomination. The Commonwealth said that it could rely on the advice of the Northern Land Council. Even if the Land Council had been negligent in performing its statutory duty, the Commonwealth could, in law, act on the advice of the Northern Land Council.

The Northern Land Council identified a sub group of the Ngapa dreaming group as the traditional owners of the proposed site. This meant that the other six clans and the other sub groups of the Ngapa clan had no power to prevent the nomination of the site from going ahead. The nomination resulted in an upfront payment by the Commonwealth to the Ngapa sub group.

Another $11 million was to be paid into a charitable trust once the site was accepted for the dump. The purposes and objects of the trust were ill defined. However, the money was to be shared amongst all the traditional owners affected by the site nomination.

Lizzie O'Shea, a pro bono lawyer working for Maurice Blackburn who represented the applicants, said 'hopefully this case can mark the beginning of a new chapter of Australian legal and political history'.[12] Unfortunately, the Commonwealth avoided this possibility by withdrawing from the case midway once the traditional owners spoke their mind at the on country hearing. From our viewpoint, the case did provide the opportunity for a new chapter in legal history because it brought into focus the inherent sovereignty of First Nations peoples.

The sovereignty issue was raised for the first time before the High Court in the *Mabo* case.[13] Here, the High Court considered an application by traditional owners in Mer to have the title to their lands recognised.

The authorities prior to this date had asserted that First Nations peoples did not have any rights in property that deserved recognition under British laws. Their laws, Justice Blackburn said in the landmark *Gove* land rights case, were strong, but did not contain rights to land. His final decision was to deny land rights. He concluded 'there is so little resemblance between property, as our law, or what I know of any other law, understands that term . . . that I must hold that these claims are not in the nature of proprietary interests'.[14]

The *Mabo* case overturned the *Gove* land rights case. Six of the seven High Court judges found that the doctrine that had legitimated the taking of the continent by Britain, the doctrine of terra nullius, was a 'legal fiction'. They then held that the common law recognised a form of native title. The claim to the continent was declared an 'act of state' which could not be challenged in the municipal courts.

We assert that one reason that the Commonwealth chose not to continue with the Muckaty hearing was the spectre of a future High Court challenge which may have raised constitutional issues not addressed in the *Gove* and

---

12 L O'Shea, 'Muckaty and nuclear waste', Issue 118 *Precedent*, September/October 2013, 35.
13 *Mabo v Queensland* (1992) 175 CLR 1.
14 *Milirrpum v Nabalco Pty Ltd and the Commonwealth of Australia* (1971) 17 FLR 141, 272.

Mabo cases such as how an 'act of state' doctrine could legally found a nation and destroy the sovereignty of Australia's First Nations peoples.[15]

C: Sovereignty

In the United States of America, the sovereignty of First Nations peoples has to some extent been recognised.[16] This has enhanced their ability to care for their land and to prevent environmental destruction. In Canada, the Nunavut peoples have been granted province status over traditional land. This limited form of sovereignty enables them to negotiate on equal terms with the federal government over developments that threaten the environment such as oil pipelines. In New Zealand, the *Treaty of Waitangi* was signed with the Maori peoples in 1840 and has led to a measure of Maori authority in environmental matters.

For us, as we explore in our judgment, the legal context of the Muckaty case involves a reworking of the *Mabo* case. In *Mabo*, the racist doctrine of terra nullius, the idea that the British discovered and claimed absolute ownership over the entire continent due to its status as empty, uninhabited land was rejected by the High Court. However, in its place, there ought to have been a treaty, which to this day has been denied. The National Congress of Australia's First Peoples 2016 survey found that 68.8 per cent of members want a treaty.[17]

D: Treaty

A treaty has been called for since the early days of colonisation in Australia. John Batman entered into a treaty with First Nations elders on the banks of the Merri Creek in Northcote in 1835. The terms of the treaty included a sale of a large tract of land for an annual fee, akin to a rent. The treaty was

---

15 The official explanation from the Commonwealth government, as reported in the media, was that it was withdrawing from legal action on the advice of the Northern Land Council. Mr Joe Morrison, CEO of the Northern Land Council advised the Commonwealth not to act on the nomination 'because of divisions within the Aboriginal community'; see P Akerman, 'Muckaty waste dump abandoned' *The Australian* (online), 19 June 2014 <www.theaustralian.com.au/news/nation/northern-territory-muckaty-waste-dump-plan-abandoned/story-e6frg6nf-1226959705496>.

16 The First Nations peoples of the United States are regarded as 'domestic dependent nations'. See, for example, *Cherokee Nation v State of Georgia* 30 US 1 (1831). For a critique of this limited form of sovereignty see R Williams, 'An algebra of Federal Indian Law: the hard trail of decolonizing and Americanizing the white man's Indian jurisprudence' (1986) *Wisconsin Law Review* 219.

17 Australian Broadcasting Corporation, 'Interview with Nayuku Gorrie', *Lateline*, 27 April 2016.

proclaimed to be unlawful on the basis of the Crown's absolute beneficial ownership.[18] However, the *Mabo* case found that the Crown did not have absolute, beneficial ownership; this leaves the status of the Batman treaty in legal limbo. More recently, in 1988, then Prime Minister Bob Hawke agreed to a treaty in the Barunga Statement. John Howard immediately declared such a 'treaty' impossible and the matter did not go forward. As Aboriginal rock group Yothu Yindu sing in *Treaty*, in response to the Australian people:

> This land was never given up.
> This land was never bought and sold.
> The planting of the union Jack
> Never changed our law at all.[19]

The matter is unfinished legal business and may require the assistance of the International Court of Justice.

E: First Nations jurisprudence

The matter of the legality of the British claim to Australia has often been raised by First Nations peoples. In 1993 Aboriginal activist Noel Pearson published an article dealing with the claim to Aboriginal sovereignty put forward by Michael Mansell, an Aboriginal lawyer and scholar. Mansell argued for 'Indigenous Law sovereignty inhering in Aboriginal people who would be an independent nation as understood in international law'. He looked to the 'International Court of Justice as a means of determining the issue'. Pearson pointed out the obstacles to this course of action; however, he acknowledged that notions of 'sovereignty are widespread throughout the Aboriginal community. . . . [A]nd as a matter of international legal theory, the validity of the acquisition of sovereignty over Australia by the British Crown is a moot point. Aboriginal people are perfectly entitled to agitate the issue where they can'.[20] Irene Watson speaks, in 2015, of: 'First Nations status as independent and sovereign peoples'.[21]

One of the applicants in the Muckaty Federal Court case, Mark Lane Jangala argues 'we are going to challenge them in court – then through our

18 *Williams v AG NSW* (1913) 16 CLR 404.
19 Yothu Yindi, 'Treaty', *Tribal Voice* (Mushroom Records, 1991).
20 N Pearson, 'To be or not to be – separate Aboriginal Nationhood or Aboriginal self-determination and self-government within the Australian nation?' 3(61) *Aboriginal Law Bulletin*, April 1993, reprinted in (2001) 5(11) *Indigenous Law Bulletin* 24.
21 I Watson, *Aboriginal Peoples, Colonialism and International Law: Raw Law* (Routledge, 2015) 145.

court – Aboriginal law and culture with the dot paintings on our body. Both sides have law'.[22] This statement clarifies the position in Australian jurisprudence with regards to sovereignty. First Nations peoples are exercising their sovereignty each day as they go about living their law on country. This is an ancient and unshakeable law. 'White fellas' are claiming sovereignty for their law, through a line of authority which when traced to its very recent origins, reveals nothing more than ink on paper and the force of the gun.[23]

We have taken some licence with the project of rewriting an environmental judgment. The original case, with Mark Lane Jangala as applicant, took place in the Federal Court in Melbourne and on country. We sat in on the whole of the Melbourne hearings. The Commonwealth, on the advice of the Northern Land Council withdrew from the case following the on country evidence.

In the judgment below we have imagined a further case brought to the High Court by the clans on Walmanpa who oppose the nuclear waste dump. We have done this because the unexamined issue at the heart of this matter is whether the sovereignty of the clans, and their right to care for country is continuing. First Nations peoples' law is formed in the earth, their sacred beings reside there – theirs is an Earth jurisprudence.

HIGH COURT OF AUSTRALIA

Mark Lane Jangala and Ors

v

Commonwealth and the Northern Land Council

KEYWORDS

First Nations Peoples – Sovereignty – Whether extinguished by annexation by Crown – Reception of common law in Australia – Terra nullius – Act of state doctrine

Constitutional Law – Reception of common law in British colony – Effect on sovereignty of Indigenous people – Terra nullius

Nuclear waste dump – First Nations consent – Environmental Law Post *Mabo* jurisprudence

Bird J and Bird J

---

22 Mark Lane Jangala cited by the Beyond Nuclear Initiative <http://beyondnuclearinitiative.com/muckaty>.
23 We agree with Jacques Derrida's analysis in 'The force of law: the "mystical foundation of authority"' in Drucilla Cornell et al. (eds), *Deconstruction and the Possibility of Justice* (Routledge, 1992).

A: The nature of the proceedings

We sit on country among the spinifex grasslands of Warlmanpa, an Indigenous name for the country, also known by its white name Muckaty Station. Sovereignty over country has been exercised for 40,000 plus years by the seven clans, Milwayi, Ngapa, Ngarrka, Wirntibu, Kurrakurraja, Walanypirri and Yapayapa and their ancestors. Members of the clans sat down with us to share their knowledge of country and have informed the judgment offered here.[24]

The judgment raises important issues for the nation, involving as it does an examination of the legal basis of Crown sovereignty and the concept of the Crown's radical title. There has not to date been a treaty between First Nations peoples and the Australian people. This failure to engage with the various First Nations as sovereign peoples is a glaring lacuna in Australian jurisprudence. The Warlmanpa clans are asking for a declaration that the nomination of their country for a radioactive waste dump is invalid.

At an earlier hearing in the Federal Court in central Melbourne reference was made by the applicants to the 'white fella' nature of the proceedings. Reflecting on that hearing we acknowledge the dominance of the Judeo-Christian culture. The language of the court was English, the architecture sleek, technocratic and hierarchical, with the judge elevated and counsel and legal advisers at a front table. Some traditional owners attended and sat at the back of the court room. We are happy to sit now on country and acknowledge the continuing sovereignty of the traditional owners in this judicial space, for the reasons set out in the judgment below.

B: Whose law?

The 'white fella' nature of the proceedings in Melbourne was further embedded in the law being applied. The legal principles at the hearing were derived from the colonial British system and the common law and statutes developed through the Australian judicial and legislative organs. The only incursion of First Nations jurisprudence was through its limited recognition by the white system. The common law the High Court in the *Mabo* case (Mabo) found was said to recognise a form of native title, which was not however commensurate with white title.[25] Indeed the destruction of 'native title' was easily accomplished and its content was uncertain, beyond rights of foraging and hunting.

---

24 We have also drawn on the scholarship of First Nations philosophers, such as Professor Irene Watson.
25 *Mabo v Queensland (No 2) (1992)* 175 CLR 1.

The common law was later enacted in legislation as the *Native Title Act 1993* (Cth). This act did not recognise sovereignty or full ownership in country and was referred to at the time by First Nations scholar, Gary Martin and white academic, Greta Bird, as the *Native Title Extinguishment Act* or the *White Title Validation Act*.[26] The Act only recognises 'traditional connection' to land and where a court refuses to find 'tradition', a First Nations plaintiff will be refused remedy. We are mindful here of cases such as the Yorta Yorta where the plaintiffs were found to have 'revived' their culture, and not been sufficiently 'traditional' to satisfy the legislation, according to white judgment.[27] As First Nations philosopher and member of the Tanganekald and Meintangk Peoples, Professor of Law Irene Watson asserts:

> The vulnerability of native title lies in its recognition as 'merely a personal right unsupported by any prior actual or presumed crown grant of any estate or interest in the land' and where it is susceptible of being extinguished by an unqualified grant by the crown of an estate in fee or some lesser estate. [28]

However, Professor Watson further writes 'while the First Nations ways remain alive in the land and the bodies of its peoples, they can be dreamed and visioned back into the present'.[29]

The Warlmanpa clans applied to have white fella recognition of their country as 'native title'. The Northern Land Council was the body charged under statute to assist in the claim. From 1991 they began the task of gathering the evidence necessary to prove, according to white law, 'traditional connection' to country. In 1997, Justice Gray made the recommendation relying on the work of consultants – white anthropologists – and the 'native title' determination was made. There were seven descent groups that passed through the Warlmanpa country.

As part of the search for a radioactive waste dump the federal government passed the *Commonwealth Radioactive Waste Management Act 2005*[30]. The Northern Land Council nominated sites for the dump on Muckaty Station under the provisions of the Act, accompanied by an anthropologists' report, which differed in some respects from that of Justice Gray. The Northern Land Council's report identified one sub group of the Ngapa clan as the

26 Australian Broadcasting Corporation, North Coast Radio, *Interview with Gary Martin and Greta Bird*, February 1994.
27 *Members of the Yorta Yorta Community v Victoria* (2002) 214 CLR 422.
28 Watson, above n 21, 148.
29 Ibid.
30 Later repealed and replaced with the *Commonwealth Radioactive Waste Management Act (Cth)* 2012.

exclusive traditional owners for the site. This decision was 'controversial' and meant that only the consent of this sub group was required for a site nomination to be validly made. The Commonwealth relied on the 'consent' communicated to it by the Northern Land Council. However, other traditional owners disputed the authority of that sub group to 'talk straight for country' and initiated a legal challenge to the site nomination and the charitable trust that was to hold the compensation in an action in the Federal Court. The compensation for having a toxic waste dump on the land was money for basic services such as health and education to be held in the trust. The applicants, representatives of various dreamings in the Muckaty Land Trust area, issued proceedings in the Federal Court against the Northern Land Council, the Commonwealth Government and the Resources Minister. They sought a declaration that the nomination was invalid and an injunction to restrain the Minister from acting on the nomination.

The legal talk in the earlier case was complex to those uninitiated in white law, with its feudal and indeed pre-feudal legacy, in the areas of property law and equity. There were arguments put about the nature of the charitable trust designed to hold the Commonwealth compensation for allowing the radioactive waste dump. The question of the connection between the site nomination and the charitable trust was unclear. Would the site nomination have been agreed to in the absence of the charitable trust? Could the site nomination stand if the charitable trust was invalid in some respects? The definition of 'traditional owner' for the purposes of the agreement was debatable. Also the Northern Land Council may have acted without 'due care' in the exercise of its statutory powers under the *Aboriginal Land Rights (Northern Territory) Act 1976* (NT) ('*Aboriginal Land Rights Act*') to identify the 'traditional owners' of the nominated site and obtain their consent. If negligence of this type were proved, would the agreement be invalid? The consent that had been obtained by the Northern Land Council might not satisfy the requirements of the *Aboriginal Land Rights Act*. Over days the applicants, who were a number of traditional owners of Warlmanpa and the defendants, the Commonwealth Government and the Northern Land Council, through their legal counsel, put forward arguments on these complicated matters. They did not directly raise the question of sovereignty. The decision in the *Mabo* case, that sovereignty was not justiciable, would appear an insurmountable hurdle to a judicial inquiry into this aspect of colonisation and the establishment of the Australian nation. Following *Mabo*, and the public hysteria that surrounded the case, Chief Justice Mason dismissed an application by Isabel Coe for the recognition of the continuing sovereignty of the Wiradjuri Nation.[31]

---

31 *Isabel Coe v Commonwealth* (1993) 214 CLR 422.

We have decided on the basis of the evidence given on country that the traditional owners identified for the site nomination by the Northern Land Council do not represent all the dreamings for the site. The site nomination is therefore invalid. We have also decided that the charitable trust cannot stand with the failure of the nomination. We could conclude our judgment at this point. However, justice demands that we look at the broader question of connection to land first addressed 25 years ago in *Mabo*. That case was a step towards justice, however it is now outdated. The question of First Nations sovereignty must be examined.

We reiterate the First Nations peoples of Warlmanpa have never ceded their sovereignty over country. They are living their law as they have over millennia. The dreaming spirits are strong and moving through and across the land. They continue to care for country and do not want a radioactive waste dump on their land. The Commonwealth government has been searching since the 1980s for a site to dump this toxic waste, most importantly from the residue of nuclear medicine. There has been a reluctance to offer up land for these purposes given the long life of high-level radioactive material. The Commonwealth has land available to deposit this material. For example, there is land set aside for defence purposes and there is the prohibited area of Woomera, which is already contaminated as a result of nuclear bomb testing. However, the federal government has chosen not to take up those options. Instead it has encouraged First Nations people to offer up their traditional country. This encouragement has included the withdrawal of funding for essential services.

C: The consent of the 'traditional owners'

A Nomination Agreement was signed by a few traditional owners identified by the Northern Land Council under their statutory powers as the owners whose consent was required. Other traditional owners, who opposed the waste dump, applied to the court to have the site nomination declared invalid, on the grounds, *inter alia* of lack of due diligence by the Northern Land Council. In particular there were concerns about the identity of the traditional owners or as a witness put it: 'Who could talk straight for country?'

The traditional owners, applicants in this case, stated through their legal counsel that they would not accept any amount of monetary compensation in return for their consent to a waste dump on country. They opposed the whole nature of the project, which they perceived as incompatible with their responsibility to care for country according to their law. In effect they were arguing that their continuing custodianship of country was equivalent to sovereignty and not compatible with a nuclear waste dump. They wanted the court to recognise this.

D: Rights to sovereignty over land

In light of this claim by traditional owners we must decide the nature of their rights in the relevant land. Until the 1970s there was little case law on First Nations legal rights. Indeed history shows a sorry denial even of the humanity of First Nations peoples. In the first land rights case, *Milirrpum v Nabalco*, the *Gove* case, the judge found the applicants had '"a government of laws and not of men" . . . shown in the evidence before me'.[32] However, their system of law, his Honour found, did not include any property rights. The decision that binds us most tightly is the *Mabo* case. This decision, although not widely recognised as such, is a constitutional law case.

E: Terra nullius: a legal fiction

There is a glaring lacuna at the heart of Australian jurisprudence. What Justices Deane and Gaudron referred to as 'the darkest aspect of our history' remains.[33] Majority judges Justice Brennan, Chief Justice Mason and Justice McHugh, while finding that the common law cannot be 'frozen in an age of racial discrimination',[34] recognised a form of title in Aboriginal and Torres Strait Islander peoples that was not reflective of their law, with its inherent sovereignty. The right to land recognised by the court was a racist form, called a 'step above bare possession' by First Nations lawyer, Michael Mansell.[35] Furthermore, the traditional rights were liable to extinguishment where the 'tide of history', in the form of the gun, the stolen children, the genocidal practices, has 'washed away' connection to the land.[36] This 'tide of history' is a benign phrase, calling to mind a natural process that cannot be resisted, concealing the violence of the foundation of the nation. It also conceals the complexities in the relationship of First Nations peoples with the land, a relationship that is not 'washed away' by removal at the barrel of a gun or by hands bearing a bible.

We point out, however, that the connections that the First Nations people of Warlmanpa have with country have not been 'washed away by the tide of

---

32 *Milirrpum v Nabalco Ptd Ltd* (1971) 17 FLR 141.
33 *Mabo v Queensland (No 2)* (1992) 175 CLR 1, 109 (Deane and Gaudron JJ).
34 *Mabo v Queensland (No 2)* (1992) 107 175 CLR 1, 42 (Brennan J).
35 Quoted in G Bird and N Rogers, 'Talking to judges about the art of judging: an annotated performance text' (2009) 3 *Public Space: The Journal of Law and Social Justice*, 15.
36 Justice Brennan, in one of the majority judgments in *Mabo* said that 'where the tide of history has washed away . . . traditional law . . . the foundation of native title has disappeared'; *Mabo v Queensland (No 2)* (1992) 107 175 CLR 1, 60.

history'. The connections remain as they have over 40,000 plus years and, according to the *Mabo* case, their rights must be recognised. However, the High Court, viewed as progressive at the time, recognised only a watered down version of First Nations laws. The court constructed British sovereignty as incompatible with First Nations sovereignty. However, in other parts of the British Empire the sovereign rights of First Nations peoples were, to some extent recognised and found to co-exist. This occurred through treaties in Aotearoa (New Zealand), Canada, the United States and so on.

Australia was the only nation to create a 'legal fiction' and deny the humanity of its First Nations people. As Justice Brennan in *Mabo* recounts:

> the doctrines of the common law which depend on the notion that native peoples may be 'so low in the scale of social organisation' that it is 'idle to impute to such people some shadow of the rights known to our law' can hardly be retained. If it were permissible in past centuries to keep the common law in step with international law, it is imperative in today's world that the common law should neither be nor be seen to be frozen in an age of racial discrimination.[37]

With the passing of the *United Nations Declaration on the Rights of Indigenous Peoples*,[38] it becomes more pressing today to resolve this conundrum.

With the 'legal fiction' of terra nullius dismissed by the High Court in *Mabo*, the question of sovereignty appears in full view. Sovereignty could lawfully be acquired at the time by conquest, cession or terra nullius. Absent all three, as was the case in the Australian colonies, the 'act of state' doctrine was conjured up by the judges in *Mabo* to hide the legal black hole; to cover up the illegality. The doctrine was basically a story telling device to declare 'impermissible'[39] the questioning of the legitimacy of the political and legal framework set up by the colonisers. They set up a system of courts to hand out justice and then defined what that justice consisted of. They drafted a Constitution, an Act of the British Parliament, which created a fiction that First Nations peoples did not exist. It spoke of 'we the people' when the people of the Constitution were white men. 'Since the advent of colonisation First Nations Peoples have been asking the question of the invaders: "by what law have you come to our lands, and breached our laws?"'[40]

---

37 *Mabo v Queensland (No 2)* (1992) 107 175 CLR 1, 41–2 (Brennan J).
38 *United Nations Declaration on the Rights of Indigenous Peoples*, adopted by the General Assembly, 2 October 2007, A/RES/61/295.
39 *Coe v Commonwealth* (1979) 24 ALR 118.
40 Watson, above n 21, 18.

The majority judges in *Mabo*, by declaring 'terra nullius' a legal fiction, grounded the sovereignty of Britain, and later Australia, on the 'act of state' doctrine.[41] The doctrine emerged from the colonial project whereby Britain became an Empire. It has no place in a modern democracy committed to the 'rule of law'. It is the role of the judicial branch to prevent the executive exceeding its powers. The Constitution is an Act of the British Parliament whereby the Australian nation was created in a racist denial of First Nations sovereignty. The 'act of state' doctrine has rarely been used to establish sovereignty over colonised lands. We, following Professor Garnett,[42] would argue for its removal. We are left then with the only argument in favour of white sovereignty over Warlmanpa country being one of non-justiciability. In other words, legitimacy resides in the convention that a court will not examine the acts of the executive.

This is entirely unsatisfactory for a modern nation looking to be respected by the international community. We are left with the undeniable fact that the 40,000 plus years of sovereignty of the Warlmanpa over country has not been affected by the claims of the British and Australian governments. As justices of this court, sworn to uphold the law 'without fear or favour' we come to this conclusion in favour of the applicants. They are a sovereign peoples, living their law.

F: A judicial revolution

We are mindful that some may criticise this judgment as a radical break from 'settler' law. We may, like the majority in *Mabo*, be vilified as 'an activist court'.[43] However, we would point out that our judgment cannot be rooted in an era of racial discrimination. We have, like the majority in *Mabo*, relied on international law at the time of British colonisation and have accepted that the doctrine of 'terra nullius' was a 'legal fiction', derived from the imperial enterprise. The only legal basis for assuming sovereignty was the 'act of state' doctrine which, as demonstrated above, is nebulous and discredited. The majority judges in *Mabo* said that as a municipal court they could not inquire into the legality of the exercise of power that established

---

41 See e.g., *Mabo v Queensland (No 2)* (1992) 107 175 CLR 1, 78–9 (Deane and Gaudron JJ).
42 R Garnett, 'Foreign states and Australian courts' (2005) 29 *Melbourne University Law Review* 704.
43 The High Court under Chief Justice Mason and his successor, Chief Justice Brennan, was described as an 'unfaithful servant of the constitution', a 'pathetic . . . self-appointed [group of] Kings and Queens' and 'gripped . . . in a mania for progressivism', and said to be guilty of 'plunging Australia into the abyss'. See G Williams, 'When the umpire takes a stand', *The Sydney Morning Herald*, 12 November 2011.

the court.[44] We take the position that the law as stated in *Mabo* must be followed, the law cannot be frozen in 'racial discrimination'. It is now 25 years since the *Mabo* decision was handed down. Justice for First Nations peoples is increasing, albeit slowly. For example, there is the Canadian agreement with the Nunavut peoples, there are long standing treaties in the United States and Aotearoa.[45] Australia must move towards justice. This leads to our conclusion that a lawful basis for the Australian state requires a treaty with First Nations peoples. As Irene Watson asserts, we need 'thinking in terms of a new international law, which is "disconnected from its own imperial sources and history"'.[46] Following on from this will be constitutional reform, not of the modest kind, but a major reworking.

G: Conclusion

The racist 'act of state' doctrine at the foundation of the Australian legal system can no longer receive judicial affirmation. The applicant clans have retained their connection to the land and their continuing custodianship may prevent environmental damage. The extinguishment through the 'act of state' doctrine and the 'tide of history' narrative cannot, in our judgment, wash away the truth of the continuing sovereignty of Milwayi, Ngapa, Ngarrka, Wirntibu, Kurrakurraja, Walanypirri and Yapayapa, the First Nations people of Warlmanpa. Such a foundation cannot in this era of racial equality provide legitimacy for the Australian state.

We realise that this decision may cause concern among many citizens. However, we cannot base a judgment on feudal principles[47] of the radical title of the Crown and a racist, and legally flawed, acquisition of sovereignty. We echo the words of Justices Deane and Gaudron in the *Mabo* case: 'the lands of this continent were not terra nullius or "practically unoccupied" in 1788. . . . The nation as a whole must remain diminished unless and until there is an acknowledgment of, and retreat from, those past injustices'.[48]

---

44 Justice Brennan said that 'the recognition is accorded simply on the footing that such a prerogative act is an act of state the validity of which is not justiciable in the municipal courts'; *Mabo v Queensland (No 2)* (1992) 107 175 CLR 1, 31 (Brennan J).

45 *Nunavut Land Claims Agreement* (1993) Canada; *Treaty of Waitangi* (1840) New Zealand; in the United States there were over 30 treaties with First Nations peoples between 1778 and 1868. The original texts of these treaties are available through The Avalon Project, Yale Law School <http://avalon.law.yale.edu>.

46 Watson, above n 21, 149.

47 The rejection of the feudal inheritance of legal concepts is explored by Justice Michael Kirby in his judgment in *Wik Peoples v State of Queensland* (1996) 167 CLR 1.

48 *Mabo v Queensland (No 2)* (1992) 107 175 CLR 1, 109 (Deane and Gaudron JJ).

We are also mindful of the great environmental issues raised by the case. The applicant clans have cared for country over millennia and bring this action to enable that care to continue. They have destocked the area to enable rehabilitation to occur. They wish to protect the land from radioactive waste. They will not give up any of their land for monetary or other compensation. We thank them for their patient teaching.

Appeal allowed.

In the result, we make the following orders:

1   We recognise the continuing sovereignty of the First Nations peoples of Warlmanpa.
2   Strike out the Site Nomination deed and the Charitable Trust.
3   The respondents to pay the applicant's costs.

Part IV

International law

Chapter 17

Whaling in the Antarctic

(Australia v Japan:
New Zealand intervening)

Hope Johnson, Bridget Lewis and Rowena Maguire

Commentary

I: Introduction

The *Whaling in Antarctic (Australia v Japan: New Zealand Intervening)*[1] decision by the International Court of Justice (ICJ) found that Japan's permits for whaling were not in compliance with the *International Convention for the Regulation of Whaling 1946* (ICRW).[2] This convention allows a State party to permit the use of lethal methods on whales 'for the purposes of scientific research' under Article VIII(1). The majority of the ICJ ruled that Japan's purpose in providing the permits was 'broadly scientific', but the manner in which the research was being carried out was not 'for the purposes of scientific research'. This decision left open the possibility of whaling being legal as long as it was carried out for the purposes of science. Despite the ICJ's ruling, Japan subsequently launched a new research programme, known as NEWREP-A, which killed 333 minke whales in the 2015/2016 whaling season. The programme has been criticised by the Scientific Committee of the International Whaling Commission as lacking sufficient scientific justification for the use of lethal methods.

This chapter begins by outlining the context surrounding the *Whaling in Antarctic* case and identifies the problematic nature of the ICRW's exception for 'scientific whaling'. Given the incompatibility of this exception with the principles of Earth jurisprudence, this chapter employs the second method of rewriting judgments outlined by Judge Preston: redrafting the ICRW and then rewriting the ICJ whaling case based on these reforms. In particular, this chapter will draw on the Earth Charter, Earth jurisprudence literature

---

1 *Whaling in the Antarctic (Australia v Japan: New Zealand Intervening) (Judgment)* [2014] ICJ Rep 226 ('*Whaling in the Antarctic*').
2 Opened for signature 2 December 1946, 161 UNTS 72 (entered into force 10 November 1948) (ICRW).

and relevant international environmental agreements to reframe the ICRW. The reformulated ICRW will then be applied to the facts considered in the *Whaling in Antarctic* case to present a new determination which prohibits any lethal whaling activities on the basis of scientific inquiry.

II: International convention on the regulation of whaling

Whales are sentient and intelligent beings with cultural and social lives. Large whales (humpback, fin, killer and sperm whales) move in complex yet coordinated family units and have the same neurons as humans and great apes, that allow them to feel love, self-awareness and emotional suffering.[3] D'Amato and Chopra explain that '[w]hales speak to other whales in a language that appears to include abstruse mathematical poetry'.[4] Along with these intrinsic attributes, whales are considered the 'engineers of marine ecosystems' for their fundamental role in the composition and function of marine nutrient cycles.[5] Since the industrial revolution, however, whales have tended to be valued according to anthropocentric interests, including using their blubber to make whale oil and their flesh for consumption.

The ICRW was established with the goal of limiting whaling to ensure whale supplies for future populations. The preamble states that 'whale stocks are susceptible of natural increases if whaling is properly regulated, and . . . increases in the size of whale stocks will permit increases in the number of whales which may be captured without endangering these natural resources'.[6] The ICRW is supplemented by a Protocol, which made minor amendments to the Convention, and a Schedule.[7] The Schedule

3 See, e.g., E Valsecchi et al., 'Social Structure in Migrating Humpback Whales (Megaptera Novaeangliae)' (2002) 11 *Molecular Ecology* 507; L Marino et al., 'Cetaceans Have Complex Brains for Complex Cognition' (2007) 5 *PLOS Biol* e139; E Lettevall et al., 'Social Structure and Residency in Aggregations of Male Sperm Whales' (2002) 80 *Canadian Journal of Zoology* 1189; P R Hof and E Van Der Gucht, 'Structure of the Cerebral Cortex of the Humpback Whale, Megaptera Novaeangliae (Cetacea, Mysticeti, Balaenopteridae)' (2007) 290 *The Anatomical Record: Advances in Integrative Anatomy and Evolutionary Biology* 1.

4 A D'Amato and S K Chopra, 'Whales: Their Emerging Right to Life' (1991) 85 *American Journal of International Law* 21.

5 See, e.g., J Roman et al., 'Whales as Marine Ecosystem Engineers' (2014) 12 *Frontiers in Ecology and the Environment* 377. For instance, whales that feed on the bottom of the ocean release faeces as they return to the surface, which allows for an upward flow of nutrients critical in fertilising ecosystems. See also, J Roman and J J McCarthy, 'The Whale Pump: Marine Mammals Enhance Primary Productivity in a Coastal Basin' (2010) 5 *PLOS ONE* e13255.

6 ICRW preamble para 3.

7 *Protocol to the International Convention for the Regulation of Whaling*, opened for signature 19 November 1956 338 UNTS 366 (entered into force 4 May 1959).

provides catch limits for particular species of whales and defines territories where certain whaling activities are prohibited or restricted.[8] Specifically, the Schedule imposes a moratorium on all commercial whaling and prohibits commercial whaling in the Southern Ocean (termed the Southern Ocean Whale Sanctuary).[9] An exemption to the moratorium and catch limits is found in Article VIII (1) of the ICRW, which allows a State party to issue special permits that authorise whaling 'for purposes of scientific research'.

Despite the moratorium on all commercial whaling, a number of ICRW member States continue whaling. Collectively, Japan, Norway and Iceland have killed close to 30,648 whales since the moratorium came into force.[10] Stakeholders generally agree that these countries have exploited various loopholes in the ICRW, including the exception for 'scientific whaling', to continue what is, in fact, commercial whaling.[11]

In terms of governance, the Convention is administered by an intergovernmental body named the International Whaling Commission (IWC). The IWC has a Scientific Committee, which reviews all special permit whaling proposals submitted by State parties. Following acceptance of the Scientific Committee's report, the IWC passes a non-binding resolution regarding the proposal with suggested modifications.[12] The Scientific Committee also analyses the results of research programmes and reports its findings to the IWC.[13]

III: ICJ case: Whaling in the Antartic (Australia v Japan: New Zealand intervening)

This chapter is rewriting the *Whaling in the Antarctic* case heard in 2010 after Australia instituted proceedings against Japan in the International Court of Justice. The case concerned the second phase of the Japanese Whale Research Program under Special Permit in the Antarctic (JARPA II)

---

8 ICRW Schedule (effective 4 April 2015).
9 This moratorium came into full effect in 1986 and the prohibition on whaling in the Southern Ocean was introduced in 1994.
10 Whale and Dolphin Conservation, 'Stop whaling' <http://us.whales.org/wdc-in-action/whaling>.
11 Norway and Iceland argue that they are not bound by the ban on commercial whaling. Norway objected to the moratorium and Iceland filed a reservation to the commercial ban.
12 International Whaling Commission, *Rules of Procedure,* amended by the Commission at the 65th Meeting (September 2014).
13 International Whaling Commission, *Process for the Review of Special Permit Proposals and Research Results from Existing and Completed Permits* (11 October 2012).

programme. JARPA II's stated objectives were to: monitor the Antarctic ecosystem; model competition among whale species; elucidate changes in whale stock structures, and improve the management of Antarctic minke whale stocks.[14]

To achieve these objectives, the programme provided permits for the use of lethal and non-lethal methods on a number of whale species. JARPA II, in combination with its predecessor JARPA I, resulted in 20,162 whales being killed for apparent scientific purposes.[15] In particular, JARPA II led to an average of 450 minke whales being killed annually.[16]As a result of conducting such programmes, Japan is in the unenviable position of being the world's largest user of lethal whaling under the scientific exemption.[17]

Australia argued that Japan's conduct was not for scientific purposes and more specifically, Australia alleged that Japan has breached its obligation to:

- observe the zero catch limit to the killing of whales for commercial purposes (para 10(e) of the Schedule to the ICRW);
- refrain from commercial whaling in the Southern Ocean Sanctuary (para 7(b) of the Schedule to the ICRW); and
- observe the moratorium on the use of factory ships or whale catchers attached to factory ships (para 10(d) of the Schedule to the ICRW).

Furthermore, Australia alleged that Japan was in breach of two other multilateral environmental agreements. First, it was claimed that Japan breached the *Convention on International Trade in Endangered Species of Wild Fauna and Flora* (CITES),[18] which lists whales as an Appendix I specimen and bans the trade of any Appendix I specimen other than in exceptional circumstances. Second, Australia alleged that Japan was in breach of the *Convention on Biological Diversity* (CBD)[19] for failure to respect:

14 Ibid., 264 [113].
15 'Fact Check: How Does Japan Compare with Other Whaling Nations?', *ABC News*, 7 January 2015 <www.abc.net.au/news/2014–04–08/whaling-around-the-world-how-japan's-catch-compares/5361954>.
16 *Whaling in the Antarctic* case 64 [202].
17 'Fact Check: How Does Japan Compare with Other Whaling Nations?', above n 15.
18 Opened for signature 3 March 1973, 993 UNTS 243 (entered into force 1 July 1975).
19 Opened for signature 5 June 1992, 1760 UNTS 79 (entered into force 29 December 1993).

- Article 3, which requires parties to ensure that activities within their jurisdiction or control do not cause damage to the environment of other States or areas beyond limits of national jurisdiction;
- Article 5, which requires parties to co-operate with other contracting parties; and
- Article 10 (b), which requires parties to adopt measures to avoid or mitigate adverse impacts on biological diversity.

In response, Japan argued that its whaling activities were precluded from the operation of the ICRW because the whales were killed pursuant to the exception available 'for purposes of scientific research', in accordance with Article VIII (1) of the ICRW. Conversely, Australia asserted that 'scientific research' should 'aim to contribute to knowledge important to the conservation and management of stocks' and that lethal methods should only be used where there are no other means to achieve that objective.[20]

The ICJ focused on whether JARPA II was designed and implemented 'for purposes of scientific research' and did not reach a conclusion on the applicability of the CBD or CITES.[21] Thus, the majority's decision was narrowly focused on the correct statutory interpretation of the clause 'for the purposes of scientific research'. After reviewing and interpreting the provision, the majority found that the activities conducted under JARPA II could broadly be construed as 'scientific research', and that lethal methods may in some instances be necessary to advance scientific understanding.

However, the majority concluded that JARPA II was not 'for the purposes of' scientific research. This finding was based on Japan's failure to assess and use viable non-lethal alternatives, the limited research outputs from JARPA II and the number of whales killed under the programme.[22] Accordingly, the court found that the approach used by Japan was not reasonable or proportionate to meet the stated research objectives of JARPA II. As such, the research was not 'for the purposes of' scientific research, and so Japan was ordered to revoke permits and licences issued under the JARPA II research programme.

The ICJ's decision recognises that there might be instances where lethal means are appropriate in conducting scientific research on whales. This finding is not consistent with the principles of Earth jurisprudence, which

---

20 *Whaling in the Antarctic* case 33 [74].
21 The court held that it did not need to consider if JARPA II constituted 'commercial whaling', and instead the judgment focuses on the special permit exception under the ICRW. *Whaling in the Antarctic* case 31–32.
22 *Whaling in the Antarctic* case 267 [126].

as a legal philosophy establishes that animals have a right to life and eco-systems have the right to exist.[23] In addition to these main inconsistencies, the ICJ decision is limited from an Earth jurisprudence perspective as it:

- leaves open the possibility of 'scientific whaling' by Japan in the Southern Sanctuary (provided Japan can justify the use of lethal methods)[24];
- does not limit the hunting of whales in the Northern Pacific;
- cannot affect the other two nations that conduct commercial whaling (Norway and Iceland);
- does not impact upon States which engage in Aboriginal subsistence whale hunting (United States and Denmark); and
- provides no guidance as to the role of the CBD or CITES in resolving such disputes, leaving unresolved the question of how to reconcile these intersecting environmental agreements.

Following the decision, Japan discontinued JARPA II and submitted a research proposal entitled the 'New Scientific Whale Research Program in the Antarctic Ocean' ('NEWREP-A') to the Scientific Committee of the IWC for approval.[25] This programme provides for the annual killing of 333 minke whales in the Antarctic.[26] Similar to the two phases of JARPA, the new research programme's objectives are to improve the precision of biological and ecological information regarding minke whales and investigate the structures of marine ecosystems using lethal and non-lethal methods. The Scientific Committee found that Japan had not established the need for lethal methods in its new proposal because the programme's research

---

23 See e.g., *Universal Declaration on the Rights of Mother Nature*, World People's Conference on Climate Change and the Rights of Mother Earth (22 April 2010) <http://therightsofnature.org/universal-declaration/>. See also Gaia Foundation, *Earth Law Principles* (2016) <www.gaiafoundation.org/earth-law-principles>.

24 The ICJ's judgment was a catalyst for the adoption of Resolution 2014–5 by the IWC. This resolution notes that, although the Court's decision is only binding between the parties to the case, the Scientific Committee to the IWC must reform its approach to reviewing special permit research programmes in line with the judgment: International Whaling Commission, *Resolution on Whaling Under Special Permit*, Resolution 2014–5 (2014).

25 Fisheries Agency of Japan, Ministry of Foreign Affairs of Japan, *Outline of the Plan for the New Scientific Whale Research Program in the Antarctic Ocean* (2015) <www.mofa.go.jp/files/000117420.pdf>.

26 This represents 81 more minke whales than the amount actually killed in the last year of JARPA II but less than the amount of whales that were permitted to be killed under JARPA II according to 'Japan's "New" Proposal for Continued Research Whaling in the Antarctic Ocean': Greenpeace, IFAW, WWF, 15 December 2014.

objectives can be met by analysing existing data and employing new non-lethal methods.[27] Yet Japanese whaling boats left on a whaling expedition in 2015 under the new research proposal, and returned early in 2016 with 333 dead minke whales, of which 200 were pregnant females.[28] Unsurprisingly then, when the IWC conducted its biennial meeting in October 2016, the key points of contention related to the Scientific Committee's findings, Japan's resumption of whaling and the ruling in the ICJ case.[29] At this meeting, the IWC adopted a resolution proposed by Australia entitled 'Improving the Review Process for Whaling Under Special Permit'.[30] Essentially, this resolution granted the Scientific Committee more time and access to information for the review process, and established a working group to provide publically, and to the Commission, an accessible and less technical report on the Scientific Committee's findings. This Working Group is going to begin by considering Japan's current special permit scheme.

Due to the problems inherent in the ICRW, including the scientific research loophole, scholars generally agree that a renegotiation of the ICRW is required before lasting and legitimate progress on whale conservation is made.[31] Legal scholarship regarding the ICJ decision has largely focused on the Court's treatment of scientific evidence.[32] Others have used the case as an entry point to examine the geopolitical context around whaling and

---

27 International Whaling Commission, *Report of the Scientific Committee: Review report of the NEWREP-A*, Doc SC/66a/Rep06 (15 December 2015) 88 <http://www.ascobans. org/sites/default/files/document/AC22_Inf_16.1.b_IWC_SC2015.pdf>.

28 Rachael Bale, *Japan Kills 200 Pregnant Minke Whales* (25 March 2016) National Geographic<http://news.nationalgeographic.com/2016/03/160325-Japan-whaling-minke-whales-Antarctica/>.

29 International Whaling Commission, *Summary of Main Outcomes, Decisions and Required Actions from the 66th Annual Meeting*, IWC/66/Main Outcomes (28 October 2016) <https://archive.iwc.int/pages/search.php?k=&search=%21collection24471&offs et=0&order_by=relevance&sort=DESC&archive=0#>

30 International Whale Commission, *Resolution on Improving the Review Process for Whaling under Special Permit*, IWC/66/11Rev (28 October 2016). This resolution passed with 35 contracting parties for, 17 against and 10 abstentions.

31 B van Drimmelen, 'The International Mismanagement of Whaling' (1991) 10 *Pacific Basin Law Journal* 240; T Jordan, 'Revising the International Convention on the Regulation of Whaling: A Proposal to End the Stalemate within the International Whaling Commission Comment' (2011) 29 *Wisconsin International Law Journal* 833.

32 See, e.g., L C Lima, 'The Evidential Weight of Experts before the ICJ: Reflections on the Whaling in the Antarctic Case' (2015) 6 *Journal of International Dispute Settlement* 621; M Moïse Mbengue and R Das, 'The ICJ's Engagement with Science: To Interpret or Not to Interpret?' (2015) 6 *Journal of International Dispute Settlement* 568; D McCahey and S A Cole, 'Human(e) Science? Demarcation, Law, and "Scientific Whaling" in Whaling in the Antarctic' (2015) 15 *The Journal of Philosophy, Science & Law* 37.

compliance with international environmental agreements.[33] This chapter contributes a more radical perspective and examines how the case would be decided if the rights of whales and ecosystems were formally recognised in line with Earth jurisprudence.

IV: Approach to rewriting the international convention for the regulation of whaling

Our rewritten judgment is based on a reimagined ICRW. It was necessary to redraft the Convention given the fundamental incompatibility of the ICRW with the principles of Earth jurisprudence. As such, we applied Justice Preston's second approach – identifying and reforming aspects of existing law which prevent or impede earth justice.[34] To test this approach, we drafted an entirely new convention for whales, drawing on the *Universal Declaration of the Rights of Mother Earth* (the redrafted convention is included as an Annex in this chapter).[35] We then applied the new convention, which we are calling the *Whale Convention*, to the facts in the *Whaling Case*. In order to do this, we altered the claims, breaches and arguments made by the parties, while remaining as close as possible to the constructs and arguments put forward in the ICJ case.

The redrafting process involved three key changes. First, the redrafted convention moves away from the ICRW's focus on regulating whales as a resource for human benefit and instead values whales and marine life on the basis of their own intrinsic value. Second, the *Whale Convention* extends a rights-based approach, normally reserved for humans, to whales and marine life. Third, the new convention creates a Special Representative for whales that was modelled on similar offices and agencies created to represent the interests of other vulnerable, voiceless groups, such as children or future generations.[36] Our decision to incorporate this third aspect was driven by the absence of a representative for the interests of whales in the ICJ decision,

---

33 See, e.g., P J Clapham, 'Japan's Whaling Following the International Court of Justice Ruling: Brave New World – Or Business as Usual?' (2015) 51 *Marine Policy* 238; C R Payne, 'ICJ Halts Antarctic Whaling – Japan Starts Again' (2015) 4(01) *Transnational Environmental Law* 181.

34 Ibid., 7.

35 *Universal Declaration on the Rights of Mother Nature*, World People's Conference on Climate Change and the Rights of Mother Earth (22 April 2010).

36 For example, Ireland has created a position of the Ombudsman for Children: <www.oco.ie/>. Hungary has created an Ombudsman for Future Generations: <www.ajbh.hu/en/web/ajbh-en/dr.-marcel-szabo>.

even though the ICJ heard arguments from the parties and received testimony from a range of experts.

The revised judgment explores these dimensions of the re-envisaged convention further. It illustrates how recognising the right of animals to life shifts the focus from whether the ends (i.e., scientific research) justify the means (i.e., lethal methods) to an acknowledgement of the intrinsic value of non-human animals as ends in themselves. The judgment also draws on the reasoning of Judge Trindade in his separate opinion, and takes inspiration from his use of the CBD and CITES. By drawing on Trindade's reasoning, the rewritten judgment illustrates how the international rule of law can be strengthened by courts adopting an approach to legal reasoning that harmonises standards and builds consistency across the fragmented instruments that make up international environmental law.

INTERNATIONAL COURT OF JUSTICE
YEAR 2014
31 March 2014

*WHALING
IN THE ANTARCTIC*

(AUSTRALIA v JAPAN: NEW ZEALAND intervening)

Alleged violations of the Whale Convention − Origins of the Convention − Interpretation of Article XX Rights of Whales − Issuance of special permits under JARPA II − The Court not called upon to decide matters of scientific policy − The Court's only task is to determine whether Japan's issuance of special permits that allowed Whaling fall within the scope of the Whale Convention − Consideration of CBD and CITES − Conservation of Living Species

Judgment

A: Claims by Australia and the response by Japan

Australia alleges that Japan has breached various obligations under the *Whale Convention* by permitting the use of lethal methods under the research programme JARPA II.

In particular, Australia alleges that, in establishing JARPA II, Japan breached its obligations to:

a) guarantee the rights of whales individually and collectively to life and to exist (Art 1 (a)) and to be free from torture, cruel or inhumane treatment (Art 1(b));

b) protect the rights of whales to regenerate their species (Art 1(c));

c)   refrain from commercial whaling (Art 3); and

d)   refrain from using lethal methods of scientific research (Art 8).

In Australia's view, the *Whale Convention* does not allow research programmes that permit lethal processes, even if the research outcomes are expected to promote the health and wellbeing of whale species. According to Australia, it follows from this that Japan has breached, and continues to breach, the rights of whales under the *Whale Convention* as outlined in Article 1, in addition to provisions relating to scientific methods set out in Articles 5 to 8.

The Special Representative for Whales, appointed under Arts 10–11 of the *Whale Convention*, supports Australia's claims.

Japan contests all alleged breaches and argues that all whaling activities are permitted under Article 5 of the *Whale Convention,* which states:

> A contracting State may permit the capture or treatment of whales for the purposes of scientific research subject to such conditions as the State considers necessary to comply with the requirements of the Convention.

Japan further argues that the purpose of the research was to gather data about whale populations, and that such information is necessary for the long-term fulfilment of whales' rights as set out in Article 1 of the Convention.

Japan further argues that the methods employed in the JARPA II programme were necessary and proportionate to the legitimate purpose of scientific research.

Japan's continuation of whaling is well documented and not in dispute in the present proceedings.

The present proceedings concern the interpretation of the *Whale Convention* and the question of whether Japan has breached the Convention by issuing special permits allowing lethal takings for scientific research.

The task before the Court is to determine whether the research objectives of JARPA II are being carried out for the purposes of scientific research as defined by Article 5 of the *Whale Convention.*

B: History of whaling conventions

Before examining the relevant issues, the Court finds it useful to provide a general overview of the *Whale Convention* and its origins. For centuries, whales have been treated as a resource with only economic worth. With the invention of industrial whaling methods and ships, populations of whale species declined and individual whales suffered horrific emotional and physical trauma, and death.

Various international agreements have been reached since relatively early in the eighteenth century that limited whaling with the intention of helping whale populations recover so the whaling industry could continue. The current Convention (the *Whale Convention)* was preceded by the *International Convention for the Regulation of Whaling* (ICRW).

The ICRW entered into force for Australia on the 10 November 1948 and for Japan on the 21 April 1951. After initially ratifying the ICRW, New Zealand withdrew in 1968 until 15 June 1976 when it again adhered to the Convention.

When the ICRW was first negotiated, anthropocentric attitudes underpinned its enactment and primary purpose. It was recognised that some form of international cooperation and regulation was necessary to prevent whale stocks from being overexploited or species going extinct, in order to ensure their continued exploitation into the future. In reflection of this, the stated objective of the ICRW framework was to 'provide for the proper conservation of whale stocks and thus make possible the orderly development of the whaling industry'.[37] Therefore, the ICRW reaffirmed the anthropocentric reasoning that underpinned the destruction of whale populations in the first place, that is, whales continue to be constructed as 'a collection of objects' not as a 'communion of subjects'.[38]

The Preamble to the ICRW recognised 'the interest of nations of the world in safeguarding for future generations the great natural resources represented by the whale stocks'. Other references in the Preamble to 'exploitation' of a 'natural resource' and 'development' of 'whale fisheries' reinforced the Convention's overarching conceptualisation of whales as a resource to be utilised by humans.

While the ICRW acknowledged concerns about 'over-fishing' and expressed a desire to ensure 'proper and effective conservation', such references were consistently constructed as being part of an overall objective of sustainably managing whale numbers to facilitate future human exploitation. The ICRW did not contain any substantive provisions for regulating whaling or the whaling industry. These were found in the Schedule, which was subject to amendments adopted by the ICRW's governing body, the International Whaling Commission.

Similar to other multilateral environmental agreements, the Schedule to the ICRW established limits on the number of whales that could be caught. Eventually, the Schedule was amended to place a moratorium on all commercial whaling, which parties were exempt from if the whaling was for scientific research or subsistent purposes.

37 *ICRW* Preamble.
38 T Berry, 'The Spirituality of the Earth' in C Birch et al., (eds), *Liberating Life: Contemporary Approaches in Ecological Theology* (Maryknoll, 1990).

In 2006, parties to the ICRW agreed that the treaty was no longer in force. This decision was based on the lack of compatibility between the ICRW and the *Universal Declaration of the Rights of Mother Earth*, along with other international environmental treaties. The ICRW was deemed to be out-dated due to its anthropocentric nature. State parties acknowledged that the ICRW's construction of whales and whaling was antithetical and counter-productive to the conservation of whales and the protection of their intrinsic rights.

Since this time, the *Whale Convention* has been negotiated and ratified by all the previous state parties of the ICRW.

C: General overview of the Whale Convention

The *Whale Convention* re-established a system through which states can collectively regulate activities that affect whales and abide by the obligations they have assumed thereunder. The text of the Convention is annexed to this judgment for ease of reference.

Unlike its predecessors, the *Whale Convention* acknowledges that whales do not exist for human use or exploitation and that, individually and collectively, whales have particular rights, by virtue of being members of the Earth community.

The first paragraph in the Preamble recognises that 'all species are part of an indivisible, living community of interrelated and interdependent beings'. The second paragraph of the Preamble acknowledges the 'history of exploitation of whales for anthropocentric benefit', and the sixth paragraph stresses the intrinsic value of whales, which forms the basis of 'the rights of all whale species to live freely and with respect'. Under the final paragraph of the Preamble, Contracting parties agreed to 'establish a system whereby whaling is permanently prohibited and preservation of whales is assured'. In order to achieve this goal, State parties agreed to abolish and make illegal all commercial whaling under Article 3.

The *Whale Convention* also establishes a range of rights that whales are owed by humans and States. The rights operate as minimum standards for all activities that have the potential to impact on whales.

The Court acknowledges that species rights, like human rights, may at times conflict with the rights of others. A clear cut example is where the rights of invasive species are limited in order to protect the rights of native species. The *Whale Convention* addresses these potential conflicts in Article 2 by stating that:

> Where the rights of whales conflict with each other or with the rights of members of other species, such conflict should be resolved in a way which maintains the integrity, balance and health of Mother Earth, and is informed by the best available scientific information.

This provision suggests that conflicts between humans and other earth species must be resolved in a way that ensures the health of Mother Earth. Human interests are not to be given higher standing or weighting than non-human interests. Where rights conflict, Earth jurisprudence holds that laws and other socioeconomic systems must 'maintain a dynamic balance between the rights of humans and those of other members of the Earth community on the basis of what is best for Earth as a whole'.[39] Creating this balance involves weighing the rights of a specific animal with the rights of other animals and ecosystems, which courts should determine on a case-by-case basis.

D: The Convention on Biological Diversity (CBD), The Convention on International Trade in Endangered Species of Wild Fauna and Flora (CITES) and the Universal Declaration on the Rights of Mother Earth

In addition to the *Whale Convention* there are a number of multilateral environmental agreements that emphasise the conservation of nature. These documents should not be approached in isolation and should be considered from a *systemic outlook* as part of strengthening the international rule of law.[40] As such, the interpretation of the *Whale Convention* should be construed in a way to be compatible with these other instruments that promote conservation.

Lethal scientific research is not consistent with the principle of intergenerational equity, a principle endorsed in the Preambles of both the *Convention on Biological Diversity* (CBD)[41] and the *Convention on International Trade in Endangered Species of Wild Fauna and Flora* (CITES).[42] The precautionary principle found in CBD and CITES will further inform the Court's reasoning on the basis that the international community has adopted a conservation-oriented approach in environmental treaty development, which is best implemented by adopting a precautionary approach in the context of scientific whale research.[43]

---

39 Ibid., Art 1(7). See also G Wright, 'Animal Law and Earth Jurisprudence: A Comparative Analysis of the Status of Animals in Two Emerging Critical Legal Theories' (2013) 9 *Australian Animal Protection Law Journal* 5.

40 *Whaling in the Antarctic* case 135 [25] (Judge Cancado).

41 *Convention on International Trade in Endangered Species of Wild Fauna and Flora,* opened for signature 3 March 1973, 27 UST 1087; TIAS 8249; 993 UNTS 243 (entered into force 1 July 1975).

42 *Convention on Biological Diversity,* opened for signature 5 June 1992, 1760 UNTS 79; 31 ILM 818 (entered into force 29 December 1993).

43 *Whaling in the Antarctic* case 152 [70] (Judge Cancado).

While the CBD and the CITES do not adopt an ecocentric approach to the regulation of non-human species, they do seek to prioritise conservation over human exploitation of flora and fauna. In this way, they are not counter to the re-alignment of governance required for Earth jurisprudence.

The principle of *lex specialis derogate legi generalis* applies in instances where the more specific rule (here the *Whale Convention*) applies over more general rules (here the CBD and CITES). As such, any conflicts between the *Whale Convention* and the CBD/ CITES should be resolved by prioritising the *Whale Convention*, as it is the more specific instrument on whales.

Equally, the development of Earth jurisprudence under the *Universal Declaration on the Rights of Mother Earth* should be borne in mind in the interpretation of international instruments including the CBD, CITES and the *Whale Convention*. This is founded in Article 3(c) of the *Universal Declaration on the Rights of Mother Earth,* which requires courts and other institutions to 'promote and participate in . . . analysis, interpretation and communication about how to live in harmony with Mother Earth in accordance with this Declaration'. Likewise, courts must contribute to the establishment of 'precautionary and restrictive measures to prevent human activities from causing species extinction, the destruction of ecosystems or the disruption of ecological cycles' in accordance with article 3(i).

E: JARPA II and scientific purposes under the Whale Convention

The JARPA II Research Plan has four research objectives: (1) Monitoring of the Antarctic ecosystem; (2) Modelling competition among whale species and future management objectives; (3) Elucidation of temporal and spatial changes in stock structure; and (4) Improving the management procedure for Antarctic minke whale stocks.

In order to realise these objectives, JARPA II allows the use of lethal methods. In each season, JARPA II provides that 50 fin and humpback whales and 850 minke whales can be taken to satisfy objectives (1) and (2). This plan has resulted in approximately 3,891 whales being killed. The question before the Court is whether JARPA II is 'scientific research' under the *Whale Convention*.

Japan relies on Article 5 of the *Whale Convention* to justify JARPA II. As noted above, Article 5 allows for scientific research projects focused on whales 'subject to such conditions as the State considers necessary to comply with the requirements of the Convention'.

According to Article 6 of the *Whale Convention*, 'scientific research' means 'activities carried out for the predominant purpose of gathering information for legitimate scientific objectives, and with the ultimate objective of progressing the rights of whales recognised in Article 1'.

Article 7 supplements Article 6 by stating that 'The means and methods employed for scientific research must be humane, necessary and proportionate to the scientific objectives, and must result in minimum possible impact on whales and marine ecosystems'. Article 8 further qualifies Articles 5 to 7 by placing a prohibition on the use of lethal research methods. Lethal methods are, in accordance with Article 8, 'not necessary or proportionate approaches for the purposes of achieving scientific objectives and cannot be considered as "scientific research"'.

The Court observes the quantity of whale meat generated in the course of JARPA II and the commercial sale of that meat. It is considered that these factors cast significant doubt on whether the killing, taking and treating of whales was indeed for the purposes of scientific research and for the progressive realisation of the rights of whales.

The Court further notes the limited amount of scientific output from JARPA II, which is essentially based on the production of two peer-reviewed papers. This output is considered especially minimal given the number of whales killed in the process.

Japan's response was that the sale of meat funds the research and that research could be slow to produce significant outputs. Thus, Japan claimed these aspects should not delegitimise the activities carried out under JARPA II. The Court is not persuaded by this argument, by reason of the combination of the following factors: the scale of the takings, the lack of scientific outputs and the commercial sale of the meat. The Court is therefore of the view that JARPA II is not being carried out for 'legitimate scientific objectives' in line with Article 6. It follows then that Article 5 of the *Whale Convention* does not condone the permits provided under JARPA II.

The Court sees no basis for Japan's assertion that lethal methods are allowed under the 'scientific research' provision within the *Whale Convention*. Article 8 of the *Whale Convention* explicitly prohibits the use of lethal methods, stating that they do not fall within the concept of 'scientific research'.

In any event, such methods cannot be considered 'humane', the standard required by Article 7. The Special Representative for Whales and other scientific advisors and experts who presented evidence in this court tendered convincing evidence that the use of lethal methods causes the whale pain and distress, resulting in the loss of life, and affects other whales with familial and social bonds.

The Court observes that, as a matter of scientific opinion, the removal of whales from their habitat at large scales causes significant harm to various marine ecosystems. The Court concludes, therefore, that, even ignoring the effect of Article 8, the use of lethal methods is not 'humane' and does not result in the 'minimum possible impact on whales and marine ecosystems' in line with Article 7.

In addition, the Court observes that the conduct of JARPA II, even if carried out for the purpose of scientific research, amounts *prima facie* to a breach of the rights to life and the right to be free from cruel and inhumane treatment to which whales are individually and collectively entitled (Art 1 (a) and (b)). The evidence before the Court illustrates that a number of pregnant whales have been killed as a result of the use of lethal sampling, impacting upon future generations of whale species. As a result, Japan has breached the collective right of whales to regenerate their species (Art 1(c) ICRW).

Given the methods employed and output produced by the JARPA II programme, its activities cannot be considered to fall within the scope of 'scientific research'. The Court finds it more appropriate to classify the programme as commercial whaling, under Article 4 of the *Whale Convention*. This makes it then necessary to consider the position on commercial whaling under international law, specifically whether Japan is in breach of Article 3 of the *Whale Convention*.

F: JARPA II and the interpretation of 'commercial whaling'

Article 3 of the ICRW prohibits 'commercial whaling', which is defined in Article 4 as 'any capturing, killing or other treatment of whales which has, among its purposes, the attainment of commercial benefit'.

Australia asserts that because JARPA II resulted in the generation and sale of whale meat it should be considered as a programme that permits 'commercial whaling'. Australia relies on press reports containing statements by Japanese officials which indicate that JARPA II has commercial objectives.

Japan argues that the sale of whale meat is an incidental outcome and not among the purposes of JARPA II. As noted above, Japan argues that the sale of whale meat is not for commercial reasons but to fund the ongoing research undertaken as part of JARPA II.

The Court observes that States often seek to achieve more than one goal when formulating and implementing a particular policy.

It is significant that the definition of 'commercial whaling' in the *Whale Convention* is drafted broadly to include situations where the attainment of commercial benefit is only one purpose of a programme.

Taken together, JARPA II has as one of its purposes the commercial sale of whale meat and *prima facie* the programme is not 'for the purposes of scientific research'. The Court finds, therefore, that Japan permitted the undertaking of 'commercial whaling' and in doing so breached Article 3 of the *Whale Convention*.

G: Remedies

For these reasons,
    THE COURT,

(1) Unanimously
    Finds that the special permits granted by Japan in connection with
    JARPA II do not fall within the definition of 'scientific research' for the
    purposes of Article 5 of the *Whale Convention.*
(2) Unanimously
    Finds that Japan violated the rights of whales in Article 1(a)–(c) by
    allowing for the use of lethal methods.
(3) Unanimously
    Finds that Japan has breached Article 3 (the prohibition on commer-
    cial whaling).
(4) Unanimously
    Decides that Japan will revoke any permits or licences granted in
    relation to JARPA II, and refrain from granting any further permits in
    pursuance of that programme.
(5) Unanimously
    Decides that Japan must refrain from authorising or implementing
    any special permit for whaling that employs lethal sampling methods, or
    otherwise interferes with the rights of whales as provided in the ICRW.

Whale Convention

Preamble

We the peoples and nations of the world,

> *Considering* that all species are part of an indivisible, living community
> of interrelated and interdependent beings;
> *Affirming* the intrinsic value of whales and wanting to recognise the
> rights of all whale species to live freely and with respect;
> *Recognising* that the history of exploitation of whales for anthropocen-
> tric benefit has caused significant destruction of whale numbers and
> damage to the Earth community;
> *Recognising* also that previous regulation of whaling has had a positive
> effect on whale conservation, but that nonetheless activities damaging
> to whales continue to take place;
> *Desiring* to establish a system whereby whaling is permanently prohib-
> ited and preservation of whales is assured;
> *Having decided* to conclude an agreement for the protection of all whale
> species, agree as follows:

Rights of whales

1   Contracting States guarantee that all whales (both as individuals and as collective species) have the following inherent rights:

(a) The right to life and to exist;

(b) The right to be respected and to be free from torture, cruel or inhumane treatment;

(c) The right to regenerate their species' bio-capacity and continue vital cycles and processes free from human disruptions;

(d) The right to health;

(e) The right to clean air;

(f) The right to clean water;

(e) The right not to have their genetic structure modified or disrupted in a manner which threatens its integrity or vital and healthy functioning.

2   Where the rights of whales conflict with each other or with the rights of members of other species, such conflict should be resolved in a way which maintains the integrity, balance and health of Mother Earth, and is informed by the best available scientific information.

Commercial whaling

3   All commercial whaling is prohibited.

4   'Commercial whaling' is defined as any capturing, killing or other treatment of whales which has, among its purposes, the attainment of commercial benefit.

Scientific research

5   A contracting State may permit the capture or treatment of whales for the purposes of scientific research, subject to such conditions as the State considers necessary to comply with the requirements of the Convention;

6   'Scientific research' means activities carried out for the predominant purpose of gathering information for legitimate scientific objectives and with the ultimate objective of progressing the rights of whales recognised in Article 1;

7   The means and methods employed for scientific research must be humane, necessary and proportionate to the scientific objectives, and must result in minimum possible impact on whales and marine ecosystems;

8   Lethal methods are not necessary or proportionate approaches for the purposes of achieving scientific objectives and cannot be considered as 'scientific research' defined in Article 6.

9 Contracting States, individually and through international cooperation, must develop and employ non-lethal methods for scientific research in the context of whales that are consistent with the rights of whales established in Article 1.

Special Representative for Whales

10 The office of 'Special Representative for Whales' is hereby created.
11 The Special Representative shall be empowered to represent the interests of all whale species before international courts, tribunals, meetings and other fora in which matters relevant to the wellbeing of whales are being discussed or determined.

Chapter 18

Restoring the transboundary harm principle in international environmental law

Rewriting the judgment in the San Juan River case

Afshin Akhtar-Khavari

Commentary

One of the oldest and central principles of international environmental law is the transboundary harm principle. This sets out the requirement that States should prevent activities on their territory causing significant harm or damage to the environment of another State or 'areas beyond national control'.[1] In selecting a very old and established principle to analyse in my judgment, I aim to show how international law as machinery for global governance has been and continues to be complicit in our excessive consumption of nature in the Anthropocene.[2] The transboundary harm principle continues to support the idea that States in international law are its only subjects and

---

1 This environmental principle, which will be referred to as the transboundary harm principle, has its origins in customary international law in the case of *Corfu Channel* (*United Kingdom v Albania*) (*Merits*) [1949] ICJ Rep 22. However, the idea of avoiding extraterritorial harm to the environment of another State has its origins in an arbitration, which is known as the *Trail Smelter Case* (*United States v Canada*) (1941) 3 RIAA 1905. For an initial expression of this principle of prevention in relation to the environment, see the work of the International Court of Justice in the *Advisory Opinion on the Legality of the Threat or Use of Nuclear Weapons* [1996] ICJ Rep 226 [29].
2 The Anthropocene epoch is a technical term in geology referring to the stratigraphic footprint of human beings similar to those of meteorites or other significant geological events in the past. The term Anthropocene has been increasingly used to more generally describe the significance of human beings for the natural world. In Earth systems science, for instance, it has been used to explore the idea that human beings are influencing and even determining the functions of Earth systems, such as the climate system. For more, see A Zlasiewic, M Williams, W Steffen and P Crutzen, 'The New World of the Anthropocene' (2010) 44 *Environmental Science Technology* 2228; and F Biermann et al., 'Down to Earth: Contextualizing The Anthropocene Epoch' (2016) 39 *Global Environmental Change* 341.

ignores the wider and broader interests that Planet Earth itself would have in protecting itself against the narrow interests of States.[3]

The transboundary harm principle was most recently discussed in the context of the *San Juan River* case.[4] In this case, the Court did not really develop the principle and related rules supporting it, because it had already done so in the *Pulp Mills* decision.[5] However, in the *San Juan River* case, the transboundary harm principle and norms central to its implementation were put to the test when Costa Rica and Nicaragua took actions against each other claiming, amongst other things, that they had each breached the requirement in international law to carry out an environment impact assessment (EIA) and to consult with the other as a result of their activities along the San Juan river. Nicaragua argued that it had rights in relations to a channel from the San Juan river which flowed through the territory of Costa Rica. It also argued that it was under no obligation to carry out an EIA because the dredging along the channel did not risk causing significant environmental harm to Costa Rica generally and also the wetlands of international significance that were within the reach of the channel. The Court found that the channel flowing from the San Juan river belonged to Costa Rica and not Nicaragua as had been argued, but that the dredging work did not risk significant harm to Costa Rica and its wetlands.

In a separate action, which the International Court of Justice combined with the case that Costa Rica had brought against Nicaragua, the Court was also asked to consider whether the road construction along the San Juan river obligated Costa Rica to carry out an EIA. The road construction, which had already begun, has a planned length of 159.7 kilometres, and 108.2 kilometres of it will be along the San Juan river.[6] The Court accepted evidence of the fact that a significant portion of the road is very close to the river and in some areas is only 'five metres on the right bank of the

---

3   The term Planet Earth is used as a noun to emphasize not just our relationship with each other and nature in an abstract sense, but the entirety of the planet as a subject that can influence our consciousness. Michel Serres in *The Natural Contract* (Elizabeth MacArthur and William Paulsen trans, University of Michigan Press, 1995) refers to the Earth to suggest that our consciousness about our relationship to nature needs to be elevated from the local to the global.

4   The term *San Juan River* case is used to refer to the combined International Court of Justice (ICJ) cases of: *Certain Activities Carried out by Nicaragua in the Border Area (Costa Rica v Nicaragua)* and also the *Construction of a Road in Costa Rica Along the San Juan River (Nicaragua v Costa Rica)*, ICJ 16 December 2015 (the case is as yet unreported in the ICJ Reports but is otherwise available at <www.icj-cij.org/docket/files/150/18848.pdf>). In this case the ICJ combined the case initially brought by Costa Rica against Nicaragua with that which Nicaragua had also launched against Costa Rica.

5   *Pulp Mills on the River Uruguay (Argentina v Uruguay)* ICJ Judgment of 20 April 2010 (I).

6   *San Juan River* case [155].

river'.[7] This proximity is seen as making it easier for the sediment to 'discharge . . . into the river'.[8] Also, around a quarter of the road is being 'built in areas that were previously forested'.[9] Lastly, the road, once constructed, would have impacts on two wetlands of international significance, one of which is located in Costa Rica, the other in Nicaragua.

The Court found that the construction of the road posed a risk of 'significant transboundary harm' and as such Costa Rica was under an obligation to carry out an EIA or to consult with Nicaragua about the project.[10] Although Costa Rica had studied the impact of the construction on the environment, this had been done after it had begun building the road, and so the Court found that Cost Rica had not satisfied its procedural obligations towards Nicaragua. Despite the failure to carry out an EIA before the construction had begun, the critical problem for the Court was whether in constructing the road Cost Rica had caused 'significant harm' to Nicaragua. The Court found that the road would contribute at most an extra two per cent sediment to the San Juan 'river's total load'.[11] From this conclusion the Court went on to examine whether the extra two per cent would cause 'significant harm' to the morphology of the river and its navigability, which it took as the most serious repercussions of the road construction. It concluded that all of Nicaragua's 'modelling and estimates' of the impacts of the sediment deposits had 'not been substantiated by empirical data'.[12] The Court came to similar conclusions when asked to rule on whether the sediment deposits had any impact on the quality of the water or the aquatic ecosystem of the San Juan river.[13] Given these findings, the Court did not rule in favour of Nicaragua's requests for reparations, which amongst others, included an argument about having Costa Rica restore the area back to its preconstruction state.[14]

The separate opinion written below is a dissent from the Court's decision, particularly in relation to its finding that there was no 'significant' harm to Nicaragua and the San Juan river, and that Costa Rica had not breached its obligations under the transboundary harm principle. The separate opinion shows how the Court should not have ignored Planet Earth as a potential subject of the harm caused by Costa Rica's urbanisation of the riparian areas of the San Juan river. It argues that the Court and the parties before

7 Ibid.
8 Ibid.
9 Ibid.
10 Ibid., [146–162].
11 Ibid., [194].
12 Ibid., [203].
13 Ibid., [208–213].
14 Ibid., [224–228].

it limited themselves by looking just at the scientific information available on the immediate impacts of the road construction and the depositing of the sediments in the San Juan river. In narrowing their views of what they recognised as 'harm' to the 'local' environment, the Court limited itself in terms of what it considered as 'significant'. In that process it left out of its consideration the real global dimensions of the harm and instead only considered immediate effects that science could measure in the local and the present context. As a 'geological force' human beings are influencing the way that Earth systems function.[15] This separate opinion seeks to counter the force and power of international environment law's central principle, which remains complicit in embracing the idea of the Anthropocene. Critically, one real consequence of the *San Juan River* decision is that the Court decided against asking Costa Rica to restore the riparian areas of the river back to its original historical trajectory.[16] Had the Court gone further in its analysis of Planet Earth's interests as an interested subject in this suite, it may have found that the inability of important ecosystems to restore themselves was 'significant harm' which should have affected Costa Rica's ability to construct the road.

This separate opinion is inspired by the remarkable and influential separate opinion that the former Judge Weeramantry of the International Court of Justice wrote in what is now known popularly as the *Danube Dam Case*.[17] In that separate opinion he drew from historical, cultural and religious sources to argue how the principle of sustainable development had universal normative appeal and that it should be recognised as a principle of international environmental law.[18] It was the first sustained argument in the work of any judge of the International Court of Justice in developing the jurisprudential foundations for the principle of sustainable development.

---

15 There is much written on the Anthropocene but the following explores the governance implications of the concept: Biermann et al., above n 2.

16 For more on restoration, see E Higgs, *Nature by Design: People, Natural Process, and Ecological Restoration* (Massachusetts Institute of Technology, 2003); and W R Jordan III, *The Sunflower Forest: Ecological Restoration and the New Communion with Nature* (University of California Press, 2012).

17 *Case Concerning the Gabcikovo-Nagymaros Project (Hungary v Slovakia)* [1997] ICJ Rep 7.

18 See for instance, A Akhtar-Khavari and D Rothwell, 'The ICJ and the Danube Dam Case: A Missed Opportunity for International Environmental Law' (1998) 22 *Melbourne University Law Review* 507.

Separate opinion delivered in response to the *San Juan River* Case

One of the oldest and central principles of international environmental law is the requirement that States must prevent activities on their territory causing significant harm or damage to the environment of another State or 'areas beyond national control'.[19] The transboundary harm principle has more recently been extended in international law by requiring that States should act with due diligence towards other States. In the *Case Concerning Pulp Mills on the River Uruguay* this Court identified that due diligence required that States should undertake an environmental impact assessment (EIA) when activities within their territory carry a risk of 'significant transboundary harm'.[20] In the *Pulp Mills* case this Court decided that the content and procedures for an EIA had to be determined by States themselves, bearing in mind that each situation would present a unique set of circumstances based on the size and significance of the activities being undertaken within that State.[21] If the EIA reveals that activities within the State are likely to cause significant transboundary harm then that State will be under an obligation to consult affected parties to 'prevent or mitigate' against the likely harm. The upshot of these procedural requirements is that they aim to support States to take measures to avoid causing harm to the territory of another State, which remains a significant substantive norm of international environmental law.

The Court found that Costa Rica had breached its procedural obligations because it had not carried out the appropriate EIA, which was needed because of the risk of 'significant harm' to the environment. It went on, however, to find that the construction of the road itself did not breach the substantive obligations which Costa Rica owed to Nicaragua, despite the impact that the road construction has on the environment of the San Juan river and its riparian areas which I will discuss below in more detail. In particular, in this separate opinion I want to point out my disagreement with the Court in the way that it only justifies certain things as 'harm' or 'damage', as well as the narrow construction of what it assumes is the 'environment' that the road construction will affect. Given that the principle of transboundary harm was used to limit the level of care for Planet Earth, it should no longer have support from this Court in this limited sense.

19 *Corfu Channel (United Kingdom* v *Albania) (Merits)* [1949] ICJ Rep 22; and *Trail Smelter Case (United States* v *Canada)* (1941) 3 RIAA 1905.
20 *Pulp Mills on the River Uruguay (Argentina* v *Uruguay)* [2010] ICJ Rep 14 [204].
21 Ibid., [205].

A: Harmful activities in the Anthropocene

In the past decade, the use of the term Anthropocene has become popular and describes the general impact of human beings as a geological force on Planet Earth. In the geological sciences, the Anthropocene describes the variety of stratigraphical signals that human beings leave on Planet Earth. For Earth system scientists the Anthropocene points to the significance of the influence that human beings have on the planetary level systems like that of the climate system. Some Earth system scientists have suggested that human beings in the Anthropocene contribute to a 'radical anthropogenic alteration of the planet's natural cycles and systems'.[22] The Anthropocene Working Group of the International Commission of Stratigraphy voted in 2016 to accept the idea that Planet Earth has officially entered into the Anthropocene Epoch. However, the idea has yet to go through several stages of bureaucracy before the International Union of Geological Sciences officially approves a change to the International Chronostratigraphic Chart which will mark the official change in the geological evolution of epochs for Planet Earth. This decision will mark an end to the Holocene, which was a very stable and mild stage in climatic conditions allowing the human population to prosper and grow in numbers. The Anthropocene potentially started in the industrial age or even as far back as the time when humans cleared trees and cultivated land for the purposes of agriculture or domesticated animals. Some scientists have argued that, from the 1850s onwards, the increased levels of carbon dioxide in the atmosphere as a result of increased fossil fuel usage in urbanized areas have contributed significantly to changing the human impact on Earth systems.

The concept of the Anthropocene itself makes no demands, but it does enable us to ask important questions, such as whether Planet Earth has a potential carrying capacity which could be breached. Earth system scientists have already responded to the idea that human beings in the Anthropocene can potentially overwhelm the normal functioning of Earth system. They use the term planetary boundaries to identify a scientific threshold which, if breached, will have severe consequences for the proper functioning of the Earth system in question. The Anthropocene, however, more than anything, creates the potential for uncertainty in terms of how Earth systems and the natural world generally will respond to the dominance of human beings as geological agents. Given the potential implications for Planet Earth because of the Anthropocene, institutions and international environmental law need to develop to cope with the new reality that we will increasingly have to face.

22 Biermann, above n 16.

It is with this in mind that I want to dissent from the Court and argue that international law and its principles have to evolve to account for Planet Earth in an entirely new way, and recognise it as a potential subject in conversations, agreements, disputes and other social and physical relationships that exist around us. There is nothing natural about the traditional institution of international law itself, which is driven by States and is based on having States as its only subjects. Every other living or corporeal entity is traditionally seen as simply being the object that international law can say something about. The international law system has helped to create competition only amongst entities that it calls and legally recognises as States and in that process it has determined who can ultimately compete for everything on Planet Earth. International law has for too long existed and allowed only States to feature in battles and duals at the global level, without having to measure their performance against the harm that is being done to Planet Earth itself. This is not to suggest that international law has not done anything to protect the environment, but it has been defined more by compromises and the self-interest of States than by what Planet Earth itself may have agreed to had it been at the drafting table of major international agreements. In this dissent I want to suggest that the Anthropocene requires that Planet Earth to be seen as an important subject and character in terms of how we think about and respond to conflicts and battles when they relate to the use and consumption of the environment.

Returning to the facts of the case, this Court carefully assessed the implications of the constructions of the 159.7 kilometre road and concluded that the project would have no impact on the river or the river's morphology as well as navigational rights on the San Juan river.[23] In part, the Court appears to have been influenced by the *Trail Smelter Arbitration Case* where, in relation to the transboundary harm principle, the Tribunal made the following comment:

> Under the principles of international law, no State has the right to use or permit the use of its territory in such a manner as to cause injury by fumes in or to the territory of another or the properties or persons therein, when the case is of serious consequence and the injury is established by clear and convincing evidence.[24]

The requirement that the parties must present 'clear and convincing evidence' sets a clear threshold for what is acceptable, but it can only be satisfied by having to look at very specific local and small scale changes to

23 *San Juan River* case [187–216].
24 *Trail Smelter Case (United States v Canada)* (1941) 3 RIAA 1905, 1907.

the environment being considered. The Court in this case was concerned to ensure that the additional two per cent sediment that had already been deposited in the San Juan river would have an insignificant impact on navigation, water quality, and the environment of the riparian areas.

The Court's approach to both the existing law and its assessment of facts is surprising. The Court ignored its own judgment in the 1996 *Advisory Opinion on the Legality of the Threat or Use of Nuclear Weapons*, where it indicated that 'activities within' the jurisdiction of a State had to 'respect the environment of other States or of areas beyond national control'.[25] The term 'respect' as used by the Court suggests that in this case it should have looked more carefully at the impact that the road construction would have on the two wetlands of international significance, only one of which is in Costa Rica whereas the other is in Nicaragua. The Court also did not take into account the potential impact that the future users of the road would have on the wetlands. It was presented with, but seemed to have also ignored the evidence of the large areas of forests that were cut down to accommodate the road construction. The Court also failed to concern itself with the impact of deforestation on the ecology of the riparian areas to the San Juan river and was more concerned with sediment deposits instead. The Court should have gone further in asking whether Costa Rica showed the requisite level of 'respect' towards the 'environment of' Nicaragua. Notwithstanding, my separate opinion will continue to discuss a more fundamental failure by the Court.

This Court cannot continue to narrowly interpret the transboundary harm principle to ignore larger, more global and complex implications of activities for States as a whole, but Planet Earth as well. In the 1996 *Advisory Opinion on the Legality of the Threat or Use of Nuclear Weapons* this Court included 'areas beyond national control' as a zone that States should 'respect' with their activities.[26] This position also limits the harm done to areas of the world that are not owned by States but it does target very specific areas such as the high seas. The protection of limited legal geographies in the form of State territories or those beyond national control ignores that the narrow interpretation of the transboundary harm principle enabled States to harm the health and integrity of Planet Earth itself.

States, through their activities, can influence and determine the functioning of Earth systems that span across different legal jurisdictions. As an example, desertification is a significant problem for many States because they lose out on a range of ecosystem services. Desertification can also have

---

25 *Advisory Opinion on the Legality of the Threat or Use of Nuclear Weapons* [1996] ICJ Rep. 226 [29].
26 Ibid.

global impacts because of the pressures it puts on other areas of the world for the services that have been lost. Local factors – such as deforestation – lead to desertification but it would be wrong to suggest that the problems do not have their origin in global contexts as well. Climate variations, for instance, have great impacts on the speed of desertification in some areas and the Earth system driving weather patterns is influenced by activities within all States around the world. In interpreting the transboundary harm principle we can no longer presume, as the Court does in this case, that in the Anthropocene the activities of States are not influenced by global changes and significantly, that they do not have global implications.

The problem with the narrow interpretation of the transboundary harm principle is best illustrated by this case. Urbanisation, which is traditionally associated with turning land surfaces into roads, buildings, and other structures, has been significant in transforming Planet Earth. It is in all possibility one of the most significant hallmarks of the Anthropocene and there is no evidence that urbanisation is slowing down. One research has argued that by 2030 we are likely to add another 1,527,000 square kilometres in 'urban land cover'.[27] Urbanisation is sometimes necessary because of increasing populations that need to be accommodated and fed. However, a significant amount of urbanisation related activities are also linked simply to increasing the economic growth potentials of a State.

Not all measures are bad in that certain kinds of urban developments allow populations to live closely together, thereby reducing a range of costs on the environment. However, the general challenge with urbanisation is not just that we lose access to productive land, which may have once been used to provide human beings with ecosystem services. It has irreversible impacts on the biosphere and can lead to the loss of important habitats and species. Its impacts on the biogeochemical and climate systems are being increasingly recognised, and given the interconnected nature of these systems we may not always be able to measure the exact nature of the harm that is taking place. A significant amount of research has been done in the past decade assessing the impacts of urbanisation on local, regional and global environmental and ecological change. There seems to be little doubt that urbanisation is an important feature of global environmental change and that not all cities or countries contribute equally to it. It is conceivable therefore that urbanisation itself, depending on its reach and location, will act as a kind of pollution with implications for Planet Earth. This is different to the way that the Court interpreted the transboundary harm principle which focuses on a specific, tangible and measurable harm that a State causes and then only to the local

---

27 K C Seto et al., 'A Meta-Analysis of Global Urban Land Expansion' (2011) 6(8) PLoS One e23777 1–9.

environment of the neighbouring State. This narrow approach ignores the range and cumulative impact of harms that are being done to Planet Earth and which get conceptually silenced by the Court.

Given the quintessentially local benefits of urbanisation and the decision-making processes that are involved, it is not surprising that it is difficult to embed concerns for it in international environmental law. However, assessing the effects of urbanisation from the viewpoint of Planet Earth and not from the perspective of States – who will want to avoid blaming each other for harming the environment through urban developments – will enable us to more clearly appreciate why it has transboundary effects. The impacts of urbanisation on the environment go deeper than described above because of its long-term and cultural influences, which can at times encourage us to treat nature simply as an inert and usable object rather than as an important subject with whom we negotiate before we destroy it in the name of consumption. The irreversible effects of urbanisation are therefore not just material but cultural. International environmental law has to come to terms with the effects of urbanisation on Planet Earth.

In this case, Nicaragua initiated some arguments that resemble claims associated with the urbanisation of the San Juan river by Costa Rica. It was suggested by Nicaragua that the road construction could increase agriculture and commercial activities along the right bank of the San Juan river which belongs to Costa Rica.[28] Nicaragua also indicated that the use of the road could increase the risks of 'spills of toxic materials into the river whenever hazardous substances are transported on the road'.[29] In response, however, Costa Rica pointed to the speculative nature of Nicaragua's claims and also that it 'did not explain the legal basis of its claims'.[30] The Court dismissed Nicaragua's claims by simply suggesting that they were speculative.[31]

Given that the length of the road is estimated as being 159.7 kilometres, the construction can be viewed as a large urban development project. The impact of the construction on the riparian areas needed to be assessed not just for sediment deposits in the San Juan river but for the significant forest coverage that was cleared, and also its short and long-term impacts on the wetlands of international significance. Further, the Court – and I should say that the submissions by Nicaragua did not add hope to my desire to see more rigorous arguments put forward about the harm done to the environment – does not seem concerned with measuring the impacts of the urban development project on the actual or potential loss of ecosystem services for Nicaragua.

28 *San Juan River* case [214].
29 Ibid.
30 Ibid.
31 Ibid., [216].

This could range from the air pollution potentially caused by the transport vehicles using the road, to the visual and noise pollution from having the road span along the San Juan river. Although the Court suggested that some of these losses that I have just described are speculative, it is more the case that Nicaragua did not gather the kind of evidence needed to convince us of these material and emotional losses and connections to the ecosystems. This is not entirely Nicaragua's fault, in that the conceptual language inherent in the transboundary harm principle and the precedent set by the Court more generally would not have encouraged their legal counsel to examine the effects of Costa Rica's urbanisation on Nicaragua.

It is also my claim in this separate opinion that the transboundary harm principle has to help answer the complex and important question, which is whether the urbanisation of the San Juan river is itself pollution that causes harm to Planet Earth? This suggestion goes further than what is discussed in the previous paragraph in that the rise of overall activity – commercial and otherwise – from the use of the road and potential urban development projects alongside it in the future will lead to pressures being placed on a range of Earth systems. The Court cannot escape dealing with this issue by referring to it as being 'speculative'. True, it is speculative to indicate exactly *how* significant the urbanisation process is likely to be, but the process has already begun with the construction of such a large and major road. The construction of the road itself and the significant harm being done to the natural environment will remove established capacity from Planet Earth to cope with a range of human induced pollution from other sources from around the globe, including those in Costa Rica. This argument does not in any way suggest that all urban developments put all States in potential breach of the transboundary harm principle. Rather, it is the location of this project – alongside a river and traversing through wetlands and forests – and the length of the road being constructed that makes it important to Planet Earth as an interested silent participant in this litigation. It is also that the project reduces the available amount of land area in the world that is not developed. Significantly, it also opens up an area in Costa Rica allowing the potential for possible future urban development projects in that region. Creating the potential for future urban sprawls by a significant road construction has to be seen as part of the irreversible impact that human beings have on Planet Earth in the Anthropocene. It is beyond speculation to suggest that major road construction encourages and supports urban sprawls and developments.

B: Restoration and the nature of harm

The premise behind the transboundary harm principle is that States must 'protect' the territory of neighbouring States from harm resulting from activities that they control. The Court in this case – as I have already indicated – narrows how they conceive of the notion of harm and also 'protection' that is embedded in the transboundary harm principle. Much of environmental law is built on the assumption that our involvement with nature is to protect it from our harmful activities. There is nothing wrong with this but environmental law in this sense continues to be built on the assumption that passive restoration is sufficient to help with the recovery of ecosystems. For instance, an important normative development in the past decade, being the principle of sustainable development – which this Court recognised in the *Danube Dam case*[32] – assumes that certain levels and kinds of consumption patterns will be supported by the ability of nature to restore itself without human intervention. This assumption is no longer viable given the significance and nature of the geological footprint of human beings on earth. Changes in climate systems and the impact of invasive species on the ongoing health and integrity of ecosystems around the world are examples of important changes that affect the ability of ecosystems to restore themselves without human intervention. In light of the Anthropocene, the idea that we simply need to protect ecosystems from overconsumption or immediate harm is simplistic and we have to rethink how we approach the subject matter of restoration given that environmental law rules and principles are mostly built on protecting rather than restoring nature.

The term restoration can mean different things to different groups of people but one useful definition is that which the International Society for Ecological Restoration (SER) has come up with. SER has described restoration as the 'process of assisting with the recovery of an ecosystem that has been degraded, damaged, or destroyed'.[33] This definition indicates that the important point for human intervention is when an ecosystem is degraded, damaged or destroyed. Until that happens we can measure the health and integrity of an ecosystem by its capacity to restore itself. Significantly, the focus of the definition is on an ecosystem rather than a single harmful event, which may or may not have implications for the functioning of an ecosystem. This leads me to two different conclusions about the relevance of restoration for the way in which we think about the transboundary harm principle. The first of which is that we have to 'protect' from harm the capacity of an ecosystem to restore itself without human intervention. The second is the requirement to restore an ecosystem when it has been degraded, damaged,

---

32 *Case Concerning the Gabcikovo-Nagymaros Project (Hungary v Slovakia)* [1997] ICJ Rep 7.

33 Society for Ecological Restoration, Science and Policy Working Group, SER *International Primer on Ecological Restoration* (Version 2, October, 2004) <www.ser.org/resources/resources-detail-view/ser-international-primer-on-ecological-restoration>.

or destroyed. In the latter case, we are talking about restoration as part of the remedies that need to be available, should a State fail to protect an eco-system. However, in the second case we have to stop the activities that are having effects on the capacity of an ecosystem to restore itself. The remedy in the first instance is to acknowledge that the cause of the anthropogenic activities will have to stop to give the ecosystem an opportunity to restore itself.

In this case, the Court focused heavily on sediment deposits as a possible cause for harm but seems to have neglected examining whether the capacity of ecosystems in the area to naturally restore were affected or not by the overall project. There is little doubt that riparian areas of rivers influence the ecology of the river as a whole and nearby ecosystems in which a river is located. In this case the urbanisation process – which as discussed above is more than just depositing sediments in the San Juan river – in Costa Rica could have long-term effects on the capacity of the ecosystem – within which the road is located – to recover on its own. This is due to the impact of the road construction process but also because of the land clearing that was required to build the road. These activities, including the potential urban-isation of the region in the future, and the use of the road will no doubt influence the capacity of the ecosystems spanning the length of the road to restore themselves. Taking away the capacity to restore from certain parts of the environment in order to build the road will have implications not just for Costa Rica but also ecosystems in Nicaragua. This is because of the con-nectivity that exists between ecosystems and which no doubt is a feature of the landscape through which the road was built.

The narrow reading of the transboundary harm principle appears to have precluded the parties to the litigation from really addressing these points by providing the Court with scientific evidence or information on the func-tioning of ecosystems within the area of the construction. Given the law as it currently stands, they were unable to ask and to answer two important questions: whether active intervention is required to restore any of the eco-systems in the region; and whether the capacity of the ecosystems to restore themselves was lost in important parts of the region under consideration. The Court was asked by Nicaragua to order the restoration of the entire area within which the road was constructed.[34] The Court responded by pointing out that they had not found any wrongdoing and therefore did not impose an obligation on Costa Rica to restore the area back to its original state. However, this approach to restoration is more akin to the traditional man-ner in which Courts compensate parties by requiring that those in the wrong put the other one in the same position as if the harm had not been done in

---

34 *San Juan River* case [226].

the first place. The suggestion being made in this separate opinion is that the capacity of an ecosystem to restore itself has to be seen as an integral dimension of what it means to protect nature and assist States to avoid harm being done to territories outside of their jurisdiction. The Court's discussion of restoration ignores how it was further narrowing the environmental protection measures otherwise available to Planet Earth.

The Court erred in continuing to rely on a narrow reading of the transboundary harm principle. The facts of this case presented it with an opportunity to interpret the principle to reflect the priorities of the geological epoch that Planet Earth is in. I have not disagreed with the Court and the precedent it has already set, which is to look for 'significant harm' before a State is liable for the transboundary effects of its activities. However, I have argued that the Court needs to account for the interests of other subjects, such as Planet Earth, in identifying what is relevant for its consideration of 'harm'. Urbanisation is a cultural phenomenon that has critical implications for Planet Earth. Certain kinds of urban sprawls and development activities can have implications for Earth systems and it is the indiscriminate support for such initiatives that can 'significantly harm' Planet Earth. I did not, however, limit myself to critiquing the Court by focusing just on the activities of States but have also argued that the nature of 'harm' itself has to be interpreted outside of the traditional protection paradigm that has restricted environmental law's developments. The ability of ecosystems to recover and restore has to be seen as central to how we protect nature in the Anthropocene. Taking away the capacity of ecosystems to be resilient has to be seen as significant to Planet Earth. As such I completely disagree with the Court that Costa Rica has not significantly harmed the territory of Nicaragua – and Planet Earth – in constructing the road along the San Juan river.

Criminal law and environmental activism

Part V

Criminal law and
environmental activism

Chapter 19

Stand with Jono

Culture-jamming, civil disobedience and corporate regulation in an age of climate change

Matthew Rimmer

Introduction

The Whitehaven Coal hoax was a striking, wild event, which has been the subject of legal debate, political discussion and media controversy.

On the 7th January 2013, Jonathan Moylan issued a hoax press release on an ANZ letterhead saying the bank had withdrawn its $1.2 billion loan facility from Whitehaven's Maules Creek Coal Project on the grounds that the project would harm the environment and the climate. Whitehaven's share price fell before recovering. In an interview, Moylan sought to explain his action:

> Well, I certainly didn't intend any harm to shareholders in Whitehaven and, you know, for the record, I do apologise. Though I won't apologise for exposing ANZ's dirty investments in Whitehaven Coal and the process where the local community has been totally. . . . We're up against, you know, a big company here and change doesn't happen without people taking risks and I think that, you know, this kind of thing is likely to happen in the future, perhaps not me, but people are going to be taking more and more risks to ensure that our rights and our environmental rights and the rights of landholders are respected and our children and grandchildren have a future.[1]

Moylan was charged under section 1041E of the *Corporations Act 2001* (Cth) for making false or misleading statements, and faced up to ten years in prison and a $765,000 fine. After initially contesting the matter, Jonathan

---

1 C Duffy and B Knight, 'Environmental activist draws anger, support and possible charges', *ABC 7:30 Report*, 9 January 2013 <www.abc.net.au/7.30/content/2013/s3667080.htm>.

Moylan decided to plead guilty to the offence under section 1041E of the *Corporations Act 2001* (Cth) in May 2014.[2]

On the 25th July 2014, Justice David Davies sentenced Jonathan Moylan at the Supreme Court of New South Wales for a breach of section 1041E (1) of the *Corporations Act 2001* (Cth).[3] The ruling is a careful and deliberate decision, showing equipoise. Justice Davies has a reputation for being a thoughtful and philosophical adjudicator.[4] The judge convicted and sentenced Moylan to imprisonment for one year and eight months. The judge ordered that Moylan be 'immediately released upon giving security by way of recognisance in the sum of $1,000 to be of good behaviour for a period of two years commencing today'.[5]

This chapter provides a 'wild law' reading of the 'Stand with Jono' controversy.[6] Nonetheless, it adopts a somewhat different approach from a number of other chapters in this collection. This chapter does not engage in writing a shadow text of an existing judgment. The work is not written in the style of an antinomian, dissenting judgment. As a matter of style and substance, it seemed to be inappropriate to engage in judicial impersonation in respect of a case of impersonation. Instead, a somewhat more narrative voice seemed appropriate in covering the various strands of the Whitehaven Coal hoax. Rather than adopt a judicial voice, this chapter seeks to chart the polyvocal debate of the 'Stand with Jono' controversy. In particular, the piece highlights the clash between the stentorian legal system and the wild response of civil disobedience. Hopefully, the effect will be to encourage a 'wild law' interpretation of the 'Stand with Jono' controversy, much like the other chapters in this collection.

This chapter has several parts. First, it explores the question of hoaxes, culture-jamming, and subvertising – drawing comparisons with The Yes Men and Bidder 70. Second, this chapter considers the role of the law in dealing with civil disobedience. In particular, this analysis draws upon the work of Bill McKibben and Naomi Klein. Third, this chapter examines the

---

2 G Wilkins, 'Email hoaxster Jonathan Moylan pleads guilty', *The Sydney Morning Herald* (online), 24 May 2014 <www.smh.com.au/business/email-hoaxster-jonathon-moylan-pleads-guilty-20140523–38uce.html>.

3 *R v Moylan* [2014] NSWSC 944.

4 Spiegelman CJ and Judges of the Supreme Court of New South Wales, Speech delivered at the swearing in ceremony of the Honourable David Lloyd Davies SC as a Judge of the Supreme Court of New South Wales, Supreme Court of New South Wales Banco Court, 29 June 2009 <www.supremecourt.justice.nsw.gov.au/Documents/davies290609.pdf>.

5 *R v Moylan* [2014] NSWSC 944 [105].

6 C Cullinan, *Wild Law: A Manifesto for Earth Justice* (Green Books, 2003); P Burdon (ed.), *Exploring Wild Law: The Philosophy of Earth Jurisprudence* (Wakefield Press, 2011); M Maloney and P Burdon (eds), *Wild Law in Practice* (Routledge, 2014); and P Burdon, *Earth Jurisprudence: Private Property and the Environment* (Routledge, 2015).

question of the role of the regulator ASIC, and corporate regulation. It also considers the role of the media in reporting upon such matters. Finally, this chapter considers the Whitehaven Coal dispute as part of a larger push for fossil fuel divestment, both in Australia and elsewhere.

A: Culture-jamming

The Whitehaven dispute raised legal issues in respect of hoaxes, resistance and culture-jamming. There has been a long history of controversies over hoaxes in Australia – from the Ern Malley/Angry Penguins scandal[7] to the Demidenko affair.[8] The legal system has often been confused and confounded by the complications of hoaxes.

The media were upset at being duped by the hoax.[9] Moylan issued a fake press release from ANZ. Journalist Ben Cubby followed up the release with Jonathan Moylan: 'It was reasonably convincing. I thought that it looked like an interesting story there'. He published the story on Twitter, but it raised alarms with Cubby's editor who suspected that there was an impersonation of the ANZ employee. After the hoax was discovered, Cubby deleted the original tweet and issued a correction. *The Australian Financial Review*, *Business Spectator*, and *AAP* all ran the story without contacting Moylan. Ben Cubby remained aggrieved about the hoax: 'There will be people out there saying that media are much more responsible for people losing money than Mr Moylan, but at the end of the day if you make a conscious, premeditated decision to lie to a journalist and then try and conceal those lies, you should take some responsibility for that as well'.[10] In his view, the hoax and the impersonation were problematic in ethical and legal terms. By contrast, Jonathan Moylan's mother, Marion, suggested that the media should share some of the blame for what happened: 'It's the job of a journalist to check their sources, and they didn't. They just wanted to be first with the news, and they never got in trouble for anything'.[11]

Jonathan Moylan likened his actions to those of the satirical group, The Chaser. The Chaser have been infamous for testing APEC security in 2007, during a visit of United States President Barack Obama, with a fake

7 M Heyward, *The Ern Malley Affair* (The University of Queensland Press, 1993); and John Tranter (ed.), 'Special hoax issue' (2002) 17 *Jacket* <http://jacketmagazine.com/17/index.shtml>.
8 M Rimmer, 'The Demidenko affair: Copyright law, plagiarism and ridicule' (2000) 5 (3) *Media And Arts Law Review* 159.
9 SBS, 'Jonathan Moylan and the 300 million dollar hoax', *SBS Two*, 4 March 2015 <www.sbs.com.au/news/thefeed/story/jonathon-moylan-and-300-million-dollar-hoax>.
10 Ibid.
11 Ibid.

motorcade, and one of the comedians dressed as Osama bin Laden.[12] After the members of The Chaser were arrested and questioned by police, all charges were eventually dropped by the New South Wales Director of Public Prosecutions in 2008.[13] Nicole Rogers has explored how The Chaser engaged in culture-jamming of APEC.[14] She noted in this analysis that 'the rational play of law is ill-suited to controlling the arbitrary and the frivolous, the satirical and parodic, the carnivalesque'.[15]

Human rights lawyer Professor Sarah Joseph provides a thoughtful analysis of the controversy in respect of the Whitehaven hoax.[16] She noted that there was a long tradition of hoaxes in Australian media and politics. Joseph noted that hoaxes 'were a staple on Australian FM radio until a 2Day FM's prank call to the London hospital treating a pregnant Kate Middleton went horribly wrong late last year, when one of the nurses who was "pranked" committed suicide'.[17] She observed: 'FM radio style hoaxes are meant to amuse listeners and aren't political'.[18] Joseph considers a number of satirical hoaxes, which are designed to engage in political speech: 'Those of the Chaser are also meant to make people laugh, but have a political element in embarrassing political figures like arch climate sceptic Lord Monckton or the essential apparatus of the State, as in its infamous APEC stunt which thoroughly embarrassed NSW police'.[19] Joseph also considers the case of The Yes Men who engage in identity correction of corporations, industry associations, governments and international organisations, like the World Trade Organization.

The philosopher Edward Spence from Charles Sturt University argued that the Whitehaven hoax was an unethical act, which was harmful to all.[20] He maintained that this was a case of 'identity theft':

12 ABC News, 'Chaser stunt raises questions about APEC security', *ABC News*, 7 September 2007 <www.abc.net.au/news/2007–09–07/chaser-stunt-raises-questions-about-apec-security/662730>.
13 B Miller, 'APEC charges dropped against Chaser satirists', *ABC PM*, 28 April 2008 <www.abc.net.au/pm/content/2008/s2229598.htm>.
14 N Rogers, 'Law and the fool', (2010) 14 *Law Text Culture* 286.
15 Ibid., 305.
16 Sarah Joseph, 'The Whitehaven hoax: ratbag act or legitimate protest?', *The Conversation* (online), 15 January 2013 <https://theconversation.com/the-whitehaven-hoax-ratbag-act-or-legitimate-protest-11604>.
17 Ibid.
18 Ibid.
19 Ibid.
20 E Spence, 'Whitehaven hoax was an unethical act that was harmful to all', *The Conversation* (online), 11 January 2013 <https://theconversation.com/whitehaven-hoax-was-an-unethical-act-that-was-harmful-to-all-11571>.

Moylan engaged in identity theft by creating a false identity for ANZ Bank and then falsely impersonating one of the ANZ employees. Stealing someone else's identity for whatever misguided reason for whatever ends is ethically wrong. One can reasonably presume that neither Ms Milne nor Mr Brown would appreciate having their own virtual identities stolen and misused to embarrass them or damage their personal or political reputations, for that would also be ethically wrong.[21]

Moreover, Spence maintained that 'Moylan engaged in information corruption'. He objected that Moylan 'purposely used false information (disinformation) to corrupt the integrity of the digital informational environment'.[22] Such accusations of unethical conduct seem somewhat unstable. It is not clear in strict legal terms that this was indeed a case of 'identity theft'. It is not necessarily easy to fit the actions of Jonathan Moylan into a category commonly associated with criminal law.[23]

In his ruling, Justice Davies explained the nature of the offence in section 1041E of the *Corporations Act 2001* (Cth):

Section 1041E, by its broad terms, captures a wide range of prohibited conduct. Persons ordinarily charged under this section have tended to disseminate false information for the purpose of receiving some gain for themselves or for some company with which they are involved. However, the section extends beyond those types of cases. At various times during the sentencing hearing reference was made to 'white collar crime' with the Crown, in particular, drawing attention to analogous cases such as cases involving insider trading, tipping or making false statements for the purpose of personal gain. Those types of cases have some elements in common including the loss of control over the false information, the involvement of and damage to the market, and the difficulties of detection, investigation and proof beyond reasonable doubt.[24]

The judge commented: 'Whilst it is true that the present offence did not involve personal financial gain, and that is relevant in the sentencing process, the Memorandum refers in a number of places to the need to increase

---

21 Ibid.
22 Ibid.
23 Australian Federal Police, 'Identity Crime' <http://afp.gov.au/policing/fraud/identity-crime>.
24 *R v Moylan* [2014] NSWSC 944 [55].

penalties for "market manipulation" and "market misconduct", which was precisely what happened here'.[25]

Robert Sutherland – the lawyer for Jonathan Moylan – previously argued that the media should have sought to verify the provenance and accuracy of the hoax email. In sentencing, Justice Davies noted that the *Newcastle Herald*, AAP, the *Australian Financial Review* and *Bloomberg* had published the hoax.[26] The judge noted 'the Offender's Senior Counsel went so far as to say that the journalists more than the Offender ought to be held to account for the ultimate effect on the market'.[27] His Honour rejected that argument:

> I completely reject that submission. Whatever responsibility the journalists might have had to check the story, it is quite hypocritical of the Offender to point the finger at them when he set up the false media release intending (as he accepted) that at least some of them would accept it as genuine. The whole point of the exercise was that the recipients of the media release should, at least for a period of time, accept the genuineness of what was given to them in order to embarrass and encourage the ANZ to say publicly that it was supporting the project or, what would be far less likely, to say that it would withdraw the funding.[28]

The judge commented: 'If the Offender had really expected the journalists not to publish until an official statement from the ASX's announcement platform, the hoax would not have got off the ground'.[29] Justice Davies observed: 'If the Offender knew anything about the announcement platform and how it worked, he must have expected that at least some journalists would not wait for that sort of confirmation or the hoax would have had no effect'.[30]

Sutherland observed that Moylan had intended to carry out a culture-jamming performance like *The Chaser*: '[Moylan] has found himself in much deeper water than he ever anticipated'.[31] The judge noted that Moylan had been inspired by The Yes Men, amongst others. He cited a conversation between Rob Harrison, an investment broker, and Jonathan Moylan:

25 Ibid.
26 Ibid.
27 Ibid.
28 Ibid.
29 Ibid.
30 Ibid.
31 M Davey, 'Jonathan Moylan: Whitehaven hoax doesn't fit usual mould, his lawyer says', *The Guardian* (online), 11 July 2014 <www.theguardian.com/world/2014/jul/11/jonathon-moylan-whitehaven-hoax-doesnt-fit-usual-mould-his-lawyer-says>.

Mr Harrison: Do you understand the impact of putting a false and fraudulent media release out to the public has on wealth and shareholdings in stock? Did you consider that before you did this?

The Offender: Well we see this as similar to The Yes Men's announcement about Union Carbide compensating victims of the Dow Chemical spill in India.[32]

Justice Davies observed that Jonathan Moylan had considered the case of The Yes Men before taking action. While recognising that Moylan was not motivated by 'personal profit', Davies observed that his actions did have significant impact in that: 'Some investors lost money or their investment in Whitehaven completely'.[33]

The judge considered the question of contrition and Moylan's guilty plea. Moylan offered an apology to the court, observing:

The fact that the release, which was intended to generate publicity about the ANZ loan, led to trading on the stock exchange came as a complete surprise to me. It was my failure to consider the market-sensitivity of the statements that led to trading that would not have otherwise occurred. Wantonly causing harm or loss to others in no way forms part of my commitment to non-violence. Those who traded on that day have every right to feel deceived and angry about the consequences of my actions. Additionally, I am well aware that many people have their retirement savings managed by a broker and invested in Whitehaven, and such people may have lost money that they have worked hard to earn. Those people in particular deserve an apology and explanation for their unjust and undeserved loss.[34]

The judge observed: 'The statements made to Mr Harrison of BBY, Jamie Freed of AFR and Conor Duffy from the ABC seem to me to amount to qualified apologies to those who may have lost money'.[35] In sentencing, the judge read and noted a number of decisions concerning market misconduct offences. His Honour noted that personal profit was the dominating motive in the cases: 'In all cases sentences of imprisonment were imposed, although in some cases these were suspended in whole or in part'.[36] He observed: 'The lack of such motive in the present case is what distinguishes it from those

32 *R v Moylan* [2014] NSWSC 944 [47].
33 Ibid. [103].
34 Ibid. [94].
35 Ibid. [95].
36 Ibid. [101].

cases'.[37] Nonetheless, his Honour stressed 'this was much more than some sort of public mischief offence where, for example a false report is made of a crime for some private purpose'.[38]

The case of Jonathan Moylan could be compared to a number of disputes in the United States in respect of climate change, culture-jamming and civil disobedience. The Yes Men are culture jamming activists based in the United States.[39] The Yes Men are particularly fond of the tactic of 'identity correction' – impersonating representatives of companies, governments, and international institutions to criticise the absurdity of their discourse.[40] In October 2009, The Yes Men staged a press conference, pretending to be the United States Chamber of Commerce. The group announced the Chamber had decided to support substantive legislative action on climate change. The Yes Men also published a press release, and established a website.[41]

In response, the United States Chamber of Commerce sought 'redress for Defendants' fraudulent acts and misappropriation of its valuable intellectual property'.[42] The Chamber complained: 'The acts are nothing less than commercial identity theft masquerading as social activism . . . such conduct is destructive of public discourse, and cannot be tolerated under the law'.[43] The Chamber alleged that the Yes Men had engaged in copyright infringement, trademark infringement, trademark dilution, unfair competition, false advertising, and cyber-squatting, amongst other things. Such arguments seemed to run against the Chamber's past stance in respect of the First Amendment and the freedom of speech.

Calling for the case to be dismissed, the Electronic Frontier Foundation commented that 'The Chamber took a controversial position on a vital political matter, climate change' and the 'Defendants engaged in a parody to criticise that position. . . . Trademark rights do not encompass the right to silence criticism'.[44] The Electronic Frontier Foundation highlighted the

37 Ibid. [101].
38 Ibid. [102].
39 The Yes Men <http://theyesmen.org> and Yes Lab <www.yeslab.org>.
40 A Boyd and D O Mitchell (eds), *Beautiful Trouble: a Toolbox for Revolution* (Or Books, 2012).
41 K Sheppard, 'The Yes Men punk the Chamber', *Mother Jones* (online), 19 October 2009 <www.motherjones.com/mojo/2009/10/yes-men-punk-chamber>.
42 'Complaint – jury trial demanded' in *United States Chamber of Commerce v Jacques Servin and others* 1:2009cv02014 District of Columbia District Court; see <www.eff. org/cases/chamber-commerce-v-servin> and <www.eff.org/document/complaint-42>.
43 Ibid.
44 'Reply in support of Defendants' Motion to Dismiss Plaintiff's First Amended Complaint' in *United States Chamber of Commerce v Jacques Servin and others* 1:2009 cv02014 District of Columbia District Court; see <www.eff.org/cases/chamber-commerce-v-servin> and <www.eff.org/files/filenode/yesmen/yesmen-mtd-reply.pdf>.

importance of the First Amendment in protecting and safeguarding the freedom of speech in the United States. In June 2013, four years after the complaint, the Chamber withdrew the action against The Yes Men.[45] The Yes Men included the prank against the Chamber in their 2015 documentary, *The Yes Men Are Revolting*.[46]

The conflict between The Yes Men and the United States Chamber raises a number of policy concerns about the legal status of culture-jamming. This highlights the dangers of authoritarian efforts to demand identity registration, and moves to outlaw impersonation. The Yes Men were bemused by the action in Australia against Jonathan Moylan.

The Youth for Climate Truth were also the subject of litigation because of a climate spoof.[47] The Koch Industries brought an action against the anonymous members of this group after they released a fake press release, which said that the Koch Brothers had changed from their position of climate denial. The Koch Industries argued that the anonymous 'John Does' had engaged in 'trademark infringement, cybersquatting, and unfair competition'.[48] In its complaint, the Koch Industries emphasized: 'Koch brings this action to redress Defendants' misappropriation of Koch's intellectual property and impersonation of Koch for the purpose of deceiving the public and promoting Defendants' agenda'.[49] The Koch Industries stressed: 'Koch supports open and honest public discourse'.[50] The Koch Industries argued that the 'Defendants' impersonation of Koch and misappropriation of its intellectual property are antithetical to honest discourse because they deceive the public respecting Koch's true position on Issues and respecting Defendant's motives'.[51] Justice Dale Kimball for the United States District Court for the District of Utah Central Division granted the motion to dismiss by the Defendants and determined that Koch's complaint did not state any claims upon which relief may be granted. Moreover, the judge granted a protective order preventing the disclosure of any individual's identity that

---

45 C McSherry, 'Chamber of Commerce cries uncle, abandons spurious trademark lawsuit against the Yes Men', *The Electronic Frontier Foundation*, 13 June 2013 <www.eff.org/deeplinks/2013/06/chamber-commerce-abandon-spurious-trademark-lawsuit-against-yes-men>.

46 The Yes Men, *The Yes Men are Revolting*, Felt Films, 2014 <www.imdb.com/title/tt2531282>.

47 T Zeller Jr, 'Koch Industries unamused by climate spoof', *The New York Times*, 27 January 2011 <http://green.blogs.nytimes.com/2011/01/27/koch-industries-unamused-by-climate-spoof>.

48 Ibid.

49 *Koch Industries v John Does (Youth for Climate Truth)* (2011) Case No. 2:10CV1275DAK <www.citizen.org/documents/Koch-v-Does-District-Court-Opinion.pdf>.

50 Ibid.

51 Ibid.

may have been disclosed through the third parties' compliance with the sub-poenas. Public Citizen hailed the decision as a victory for freedom of speech: 'We are gratified that the court affirmed our clients' First Amendment right to engage in anonymous political speech and rejected Koch's baseless legal theories'.[52]

The victory of The Yes Men and The Youth for Climate Truth could be contrasted with the case of Tim DeChristopher.[53] DeChristopher – known as Bidder 70 – protested a Bureau of Land Management oil and gas lease auction by successfully bidding on 14 parcels of land, without any intention of paying for the purchases. Tim DeChristopher argued: 'I want you to join me in valuing this country's rich history of nonviolent civil disobedience'.[54] He commented:

> When a corrupted government is no longer willing to uphold the rule of law, I advocate that citizens step up to that responsibility. This is really the heart of what this case is about. The rule of law is dependent upon a government that is willing to abide by the law. Disrespect for the rule of law begins when the government believes itself and its corporate sponsors to be above the law.[55]

DeChristopher stressed that the United States of America is 'a place where the rule of law was created through acts of civil disobedience'.[56] He commented: 'Since those bedrock acts of civil disobedience by our founding fathers, the rule of law in this country has continued to grow closer to our shared higher moral code through the civil disobedience that drew attention to legalized injustice'.[57] DeChristopher argued: 'The authority of the government exists to the degree that the rule of law reflects the higher moral code of the citizens, and throughout American history, it has been civil disobedience that has bound them together'.[58]

DeChristopher was convicted of an indictment, and sentenced for two years. On appeal, the judge noted that 'mixed in with his argument about selective

52 Public Citizen, 'Federal judge favors free speech, rules that political spoof against Koch Industries did not break law', *Public Citizen*, 9 May 2011 <www.citizen.org/pressroom/pressroomredirect.cfm?ID=3336>.
53 B Gage and G Gage, *Bidder 70*, 2012 <www.imdb.com/title/tt2085759>.
54 T DeChristopher, 'Official Statement at sentencing hearing', 26 July 2011 <www.peace fuluprising.org/tims-official-statement-at-his-sentencing-hearing-20110726>.
55 Ibid.
56 Ibid.
57 Ibid.
58 Ibid.

prosecution, the Defendant raises the spectre of retaliatory sentencing'.[59] The judge observed that the 'Defendant's statements that he would "continue to fight" and his view that it was "fine to break the law" were highly relevant to these sentencing factors'.[60] The climate activist was released from Federal Prison after 21 months of imprisonment.[61]

It should be noted that Chief Justice Gleeson of the High Court of Australia considered the question of impersonation in a 2001 defamation matter.[62] He referred to two famous Australian and New Zealand satirists, Max Gillies and John Clarke. Chief Justice Gleeson observed:

> There is a man called John Clarke who is very entertaining on television, and he attributes outrageous statements to public figures and it is amusing to watch him. It is not unthinkable that somebody might attempt to sue him for defamation, but I would be surprised to see him sued for injurious falsehood, so obviously made up are the statements.[63]

Such comments show, perhaps, a certain latitude for impersonation under Australian law. The suggestion from the judge would seem to be that there needs to be some greater harm established for an action to be successful in respect of a hoax.

B: Civil disobedience

The case of Jonathan Moylan led to a wider discussion about freedom of speech, civil disobedience and the rule of law.

The climate activist Bill McKibben has discussed the rise of the 'fossil fuel resistance'.[64] He observed: 'After decades of scant organized response to climate change, a powerful movement is quickly emerging around the country and around the world, building on the work of scattered front-line organizers who've been fighting the fossil-fuel industry for decades'.[65]

---

59 *United States v Tim DeChristopher* (Ct App 10th Cir, Utah, No 2:09-CR-00183-DB-1, 14 September 2012) slip op <www.ca10.uscourts.gov/opinions/11/11–4151.pdf>.
60 Ibid.
61 The Huffington Post, 'Tim DeChristopher released: climate activist leaves federal prison after serving 21 months', *The Huffington Post*, 23 April 2013 <www.huffingtonpost. com/2013/04/22/tim-dechristopher-release_n_3133026.html?ir=Australia>.
62 *Palmer-Bruyn and Parker Pty Ltd v Parsons* S8/2001 [2001] HCATrans 259 (21 June 2001).
63 *Palmer-Bruyn and Parker Pty Ltd v Parsons* S8/2001 [2001] HCATrans 259 (21 June 2001).
64 B McKibben, 'Fossil fuel resistance', *Rolling Stone* (online), 11 April 2013 <www.rolling stone.com/politics/news/the-fossil-fuel-resistance-20130411>.
65 Ibid.

McKibben commented that the climate movement was a decentralised network of activists: 'It has no great charismatic leader and no central organization; it battles on a thousand fronts'.[66]

In his work, *Oil and Honey*, Bill McKibben discusses the rise of civil disobedience as a means of addressing climate change.[67] He draws inspiration from the work of Henry David Thoreau on civil disobedience. McKibben discussed how he drew upon inspiration from teaching on the civil rights movement and Martin Luther King Jr. in opposing the Keystone XL Pipeline:

> By the time I was done with the semester, I'd decided that 350.org should organize the first major civil disobedience action for the climate movement. . . . [I]t was time to stop changing lightbulbs and start changing systems. If we were going to shake things up, we'd need to use the power King had tapped: the power of direct action and unearned suffering. We'd need to go to jail.[68]

McKibben was conscious that youths like Tim DeChristopher had borne the brunt of civil disobedience actions. He called for veterans and elders to take the lead in civil disobedience: 'Now it's time for people who've spent their lives pouring carbon into the atmosphere to step up, too, just as many of us did in earlier battles for civil rights or for peace'.[69] Speaking on the case of *R v Moylan*, Bill McKibben said that 'Harshly punishing nonviolent action against the coal barons is not just spiteful and mean, it's stupid'.[70] He observed: 'Mr Moylan clearly speaks for the future'.[71]

In her book, *This Changes Everything*, Naomi Klein considers the rise of community opposition to extractive industries.[72] She envisages a new zone of conflict, which she labels 'Blockadia': 'Blockadia is not a specific location on a map but rather a roving transnational conflict zone that is cropping up with increasing frequency and intensity wherever extractive projects are attempting to dig and drill, whether for open-pit mines, or gas fracking, or

66 Ibid.
67 B McKibben, *Oil and Honey: The Education of an Unlikely Activist* (Times Books 2013).
68 Ibid., 17.
69 Ibid., 22.
70 Stand with Jono, 'Jonathan Moylan verdict: 20 months suspended sentence, on condition of 2 years good behaviour', 25 July 2014 <www.standwithjono.org/jonathan-moylan-pleads-guilty>.
71 Ibid.
72 N Klein, *This Changes Everything: Capitalism vs The Climate* (Simon and Schuster, 2014).

tar sands oil pipelines'.[73] In Australia, the Green-Brown alliance of Lock the Gate has been particularly prominent in opposing coal seam gas and mining in lands of great agricultural and environmental value. The various actions against Whitehaven Coal could be seen as part of the Blockadia movement. Naomi Klein observed:

> In New South Wales, Australia, opposition to new coal mining operations grows more serious and sustained by the month. Beginning in August 2012, a coalition of groups established what they call the 'first blockade camp of a coal mine in Australia's history,' where . . . activists have chained themselves to various entrances of the Maules Creek project – the largest mine under construction in the country, which along with others in the area is set to decimate up to half of the 7,500 hectare (18,500 acre) Leard State Forest and to wield a greenhouse gas footprint, representing more than 5 per cent of Australia's annual emissions according to one estimate.[74]

In North America, the battle against the Keystone XL Pipeline has united a number of disparate groups in collective civil disobedience. The disaster of the BP oil spill in the Gulf of Mexico has also galvanised community action. Naomi Klein sees a shift from away from the language of risk assessment to a renewed focus upon the precautionary principle: 'Blockadia is turning the tables, insisting that it is up to industry to prove that its methods are safe – and in the era of extreme energy that is something that simply cannot be done'.[75]

Bob Brown – the former leader of the Australian Greens – defended the act of civil disobedience by Jonathan Moylan.[76] He highlighted the long tradition of civil disobedience, as practised by Gandhi, Mandela, and Martin Luther King. Bob Brown considered the history of environmental activism in Australia. He noted that the battles to save the Gordon and Franklin rivers involved civil disobedience: 'Through Australia's long, proud history of environmental activism runs a common thread – it has been led by women and men committed to non-violence'.[77] Bob Brown cited the views of climate scientist James Hansen: 'CEOs of fossil energy companies know

73 Ibid., 296.
74 Ibid., 301.
75 Ibid., 335.
76 B Brown, 'It's coalminers, not Moylan, who are costing us the earth', *The Age* (online), 11 January 2013 <www.theage.com.au/federal-politics/political-opinion/its-coalminers-not-moylan-who-are-costing-us-the-earth-20130110–2cix6>. See also B Brown, *Optimism: Reflections on a Life of Action* (Hardie Grant Books, 2014).
77 Ibid.

what they are doing and are aware of long-term consequences of continued business as usual'.[78] He noted the view of the climate scientist that 'these CEOs should be tried for high crimes against humanity and nature'.[79] Bob Brown paraphrased the climate scientist: 'James Hansen's reasonable scientific assessment is that business-as-usual in coal mining is a high crime against humanity'.[80] He maintained: 'But here in Australia in 2013, while Whitehaven's mine will help cost us the Earth, it is Moylan's actions that have excited outrage and may cost him his freedom'.[81]

Clive Hamilton, a Professor of Public Ethics at Charles Sturt University, wrote a couple of pieces in defence of Jonathan Moylan.[82] He has been more generally concerned about climate politics in Australia.[83] Hamilton observed: 'Recognising this new reality, perhaps Jonathan Moylan and Tim DeChristopher are pioneering a new phase of climate campaigning aimed at making it more difficult for coal and oil companies to do business'. He suggested: 'What might be dubbed "virtuous malfeasance" – hostile actions motivated by the public good aimed at damaging a company's interests – may be a new form of civil disobedience practiced by a market-savvy generation of young activists'.[84] Hamilton noted that 'Often those who engage in civil disobedience are otherwise the most law-abiding citizens' and 'have most regard for the social interest and the keenest understanding of the democratic process, including its failures'.[85] Hamilton contended: 'With runaway climate change now jeopardising the stable, prosperous and civilised community that our laws are designed to protect, some are now asking whether the time has arrived when their obligations to their fellow humans and the wider natural world entitle them to break laws that protect those who continue to pollute the atmosphere in a way that threatens our survival'.[86] He concluded: 'When

78　Ibid.
79　Ibid.
80　Ibid.
81　Ibid.
82　C Hamilton, 'ANZ imposter takes up new climate tactic', *The Conversation* (online), 8 January 2013 <https://theconversation.com/anz-imposter-takes-up-new-climate-tactic-11482> and C Hamilton, 'ASIC and the great coal hoax', *The Conversation* (online), 18 January 2013 <https://theconversation.com/asic-and-the-great-coal-hoax-11669>.
83　C Hamilton, *Scorcher: The Dirty Politics of Climate Change* (Black Inc, 2007); C Hamilton, *Requiem for a Species: Why We Resist the Truth about Climate Change* (Earthscan, 2010); and C Hamilton, *Earthmasters: The Dawn of the Age of Climate Engineering* (Yale University Press, 2013).
84　C Hamilton, 'ANZ imposter takes up new climate tactic', *The Conversation* (online), 8 January 2013 <https://theconversation.com/anz-imposter-takes-up-new-climate-tactic-11482>.
85　Ibid.
86　Ibid.

talking to young climate activists it soon becomes apparent that they feel they have been abandoned by their elders, whom they see as bequeathing them a world no-one would want to live in'.[87]

Amanda Kennedy from the University of New England suggested that civil disobedience was the result of the lack of democratic processes to object to mining developments.[88] She noted: 'Activists, and in many instances, communities, feel increasingly compelled to engage in extreme actions to have their voices heard'.[89] Kennedy suggested: 'Within a regulatory regime that prioritises development interests, and provides limited opportunity for independent arbitration of development decisions, this is hardly surprising'.[90] Kennedy highlighted that 'The mining boom has created unprecedented land use conflicts, particularly in rural and regional Australia'.[91] She observed that the legal system was not necessarily very good at resolving such conflicts: 'Agricultural communities and environmentalists are engaged in disputes with mining and gas companies over the use and management of natural resources. In many cases, these conflicts have eluded resolution'.[92] Kennedy suggested: 'The regulatory framework for mining development fails to meet the expectations of the public, increasingly driving some to take alternative direct actions'.[93] She maintained that the 'hoax email case demonstrates that when we rely on legal arrangements that simplistically consider the role of communities and the nature of land use conflict, we create the potential for adverse consequences'. In her view, 'We need to think more creatively about how communities and resource users resolve disputes, as land use demands will only increase in intensity and complexity'.[94]

Human rights lawyer Professor Sarah Joseph considered the tradition of non-violent action.[95] She observed: 'There is indeed a long and rich tradition of non-violent actions which highlight injustices (or perceived injustices) and sometimes play a role in redressing them'.[96] Joseph highlighted peaceful protests by Mahatma Ganhi and Rosa Parks. She also compared the case of Jonathan Moylan to that of Pussy Riot – the Russian culture-jammers who

87 Ibid.
88 A Kennedy, 'Whitehaven hoax shows NSW planning system can't cope with community concern', *The Conversation* (online), 11 January 2013 <https://theconversation.com/whitehaven-hoax-shows-nsw-planning-system-cant-cope-with-community-concern-11537>.
89 Ibid.
90 Ibid.
91 Ibid.
92 Ibid.
93 Ibid.
94 Ibid.
95 Joseph, above n 16.
96 Ibid.

have protested against Vladimir Putin and the Russian Orthodox Church.[97] Joseph also highlights the generation of non-violent online internet activism. She highlights the work of Wikileaks,[98] Anonymous,[99] and Aaron Swartz.[100] Moreover, since the case of Jonathan Moylan, there has been much debate and controversy over Edward Snowden who revealed mass dragnet surveillance by the National Security Agency and the Five Eyes.[101]

Professor Michael Adams – the Dean of the Western Sydney University School of Law – was sceptical whether the actions of Jonathan Moylan could be classified as an act of 'civil disobedience'.[102] He noted: 'It has been suggested a hoax by anti-coal activist Jonathan Moylan wiping millions of dollars from Whitehaven Coal's share price was an act of "civil disobedience", akin to chaining a person to a tree, a public protest or even a prank call on the radio'.[103] Professor Michael Adams noted that 'the difference between a public nuisance and civil disobedience is the impact on the general community'.[104] Adams said that: 'A protest normally provides publicity for a cause and brings the matter to the general public's attention, but causes little harm to the community'.[105] He was worried: 'A fraud – and in particular one that impacts on the share market – has huge consequences'.[106] Adams

97 S Joseph, 'A closer look at the Pussy Riot phenomenon', Castan Centre for Human Rights Law, 24 August 2012 <http://castancentre.com/2012/08/24/a-closer-look-at-the-pussy-riot-phenomenon>. See also Pussy Riot, *Pussy Riot!: a punk prayer for freedom* (The Feminist Press at CUNY, 2012); Pussy Riot, *Pussy Riot: a punk prayer*, Docurama, 2014 <www.amazon.com/Pussy-Riot-A-Punk-Prayer/dp/B00GHH9HAO> and M Gessen, *Words will Break Cement: The Passion of Pussy Riot* (Riverhead Books, 2014).

98 WikiLeaks and J Assange, *The WikiLeaks Files: The World According to US Empire* (Verso, 2015).

99 B Knappenberger, *We are Legion: the story of the hactivists*, Luminant Media, 2012 <www.imdb.com/title/tt2177843/?ref_=nm_knf_t2>t and G Coleman, *Hacker, Hoaxer, Whistleblower, Spy: The Many Faces of Anonymous* (Verso, 2014).

100 B Knappenberger, *The Internet's Own Boy: the Story of Aaron Swartz*, Alive Mind, 2015 <www.imdb.com/title/tt3268458>; A Swartz and L Lessig, *The Boy who Could Change the World: the Writings of Aaron Swartz* (The New Press, 2016); and J Peters, *The Idealist: Aaron Swartz and the Rise of Free Culture on the Internet* (Slate Magazine, 2016).

101 L Poitras, *Citizenfour*, Praxis Films, 2014 <www.imdb.com/title/tt4044364/?ref_=fn_al_tt_1> and G Greenwald, *No Place to Hide: Edward Snowden, the NSA, and the US Surveillance State* (Metropolitan Books, 2014).

102 M Adams, 'Public nuisance – or fraud? Whitehaven hoax puts market credibility at risk', *The Conversation* (online), 10 January 2013 <https://theconversation.com/public-nuisance-or-fraud-whitehaven-hoax-puts-market-creditability-at-risk-11544>.

103 Ibid.
104 Ibid.
105 Ibid.
106 Ibid.

acknowledged: 'It is important that activists should use every method and in particular the power of the media and social networking to get the relevant message out to the public or target audience, but to stay within the boundaries of the law'.[107] He warned: 'Having any criminal conviction against your name does not help the environment nor your personal standing'.[108] This is quite a measured analysis by Professor Michael Adams. Adams has elsewhere explored the problem of greenwashing – so he has also been concerned about misleading and deceptive representations.[109]

In the case of *R v Moylan*, Justice Davies distinguished the case of Jonathan Moylan from the United Kingdom case about protesters against the Iraq war in *R v Jones*.[110] In this case, Lord Hoffmann discussed the role of civil disobedience:

> Civil disobedience on conscientious grounds has a long and honourable history in this country. People who break the law to affirm their belief in the injustice of a law or government action are sometimes vindicated by history. The suffragettes are an example which comes immediately to mind. It is the mark of a civilised community that it can accommodate protests and demonstrations of this kind. But there are conventions which are generally accepted by the law-breakers on one side and the law-enforcers on the other. The protesters behave with a sense of proportion and do not cause excessive damage or inconvenience. And they vouch the sincerity of their beliefs by accepting the penalties imposed by the law. The police and prosecutors, on the other hand, behave with restraint and the magistrates impose sentences which take the conscientious motives of the protesters into account. The conditional discharges ordered by the magistrates in the cases which came before them exemplifies their sensitivity to these conventions. These appeals and similar cases concerned with controversial activities such as animal experiments, fox hunting, genetically modified crops, nuclear weapons and the like, suggest the emergence of a new phenomenon, namely litigation as the continuation of protest by other means.[111]

In this famous case, Hoffmann called for respect for the tradition of civil disobedience: 'If there is an issue as to whether the defendants were justified

---

107 Ibid.
108 Ibid.
109 M Nehme and M Adams 'Section 18 of the Australian consumer law and environmental issues' (2012) 24 *Bond Law Review* 30.
110 *R v Jones* [2007] 1 AC 136.
111 Ibid.

in doing acts which would otherwise be criminal, the burden is upon the prosecution to negative that defence'.[112]

By contrast, Justice Davies was disinclined to accept the argument that Jonathan Moylan was engaged in civil disobedience.[113] Justice Davies observed: 'As far as the Offender's prospects of rehabilitation are concerned, I am inclined to accept the assessment of Mr Johnston, who assessed the Offender as a low risk of re-offending'.[114] The judge commented: 'He is not a criminal in the classic sense of one who needs rehabilitation, although I consider that there is some likelihood that he may continue to engage in what might be regarded as minor breaches of the law as acts of disobedience to further his beliefs and purposes'.[115] In sentencing, Justice Davies told Jonathan Moylan:

> I have considered all the available sentences and I am satisfied that no other sentence than imprisonment is appropriate in all the circumstances. But for the guilty plea, I would have imposed a sentence of 2 years' imprisonment. With a 15 per cent discount for the plea the appropriate sentence is a period of 1 year and 8 months (rounded down). However, taking into account the guilty plea, the fact that the hoax was readily admitted within a short period of time, the fact that the Offender has not previously been convicted of a serious offence, and the fact that the offence was not committed for the purpose of personal financial gain, nor was any obtained, I consider that the Offender should be immediately released upon giving security of $1000 upon the condition to be of good behaviour for 2 years with such sentence commencing today.[116]

The judge warned Moylan:

> If you are not of good behaviour during the 2-year recognisance, which at a minimum means that you do not commit any further offences, or if you fail without reasonable excuse to fulfil or comply with the conditions of your recognisance, your recognisance may be forfeited and you will be brought back before the Court and the orders I have made may be revoked or cancelled.[117]

---

112 Ibid.
113 *R v Moylan* [2014] NSWSC 944.
114 Ibid. [98].
115 Ibid. [100].
116 Ibid. [104–5].
117 Ibid. [109].

For his part, Stephen Galilee, the CEO of the NSW Mineral Council, was disappointed in the lightness of the sentence. 'It was disappointing for us to see a relatively light sentence handed down because a strong signal should have been sent that this was a serious act'.[118]

It is notable that the ruling in *R v Moylan* does not seem to have stopped community action against Whitehaven Coal. There was a concerted Leard blockade against the Maules Creek Coal Mine.[119] This protest involved 350.org, Front Line Action on Coal, Greenpeace, the Nature Conservation Council, and Lock the Gate. There was a concern about the various impacts of the Maules Creek Coal Mine – particularly in terms of the destruction of the Leard State Forest, the cost to farmers, the effect upon Indigenous cultural sites and climate impacts. Notably, Australian Wallabies rugby union star David Pocock was arrested at a Maules Creek Coal Mine protest on the 30th November 2014.[120] The charges against the rugby union celebrity were dismissed the following year.[121] Subsequently, David Pocock called for a moratorium on new coal mines at the Paris climate talks.[122]

In the Australian Senate in 2015, Senator Christine Milne, then the leader of the Australian Greens said: 'It is hard to believe the stupidity, the backwardness, the ignorance of knocking down the Leard State Forest to build a coalmine in an age of global warming, when the coal should be staying in the ground'.[123]

Senator Lee Rhiannon has offered support for Jonathan Moylan.[124] In a 2015 speech, she said that 'the world is turning its back on coal'.[125] She said: 'In Australia, because of weak planning laws at a state level and weak environment laws at a federal level, it is so hard for communities to be heard'.[126] Senator Rhiannon praised community action: 'I congratulate the

118 SBS, above n 9.
119 Leard Blockade <http://350.org.au/maules-creek/why-maules-creek-must-be-stopped>.
120 Staff Writer, 'Wallaby David Pocock arrested in Maules Creek coalmine protest', *The Sydney Morning Herald* (online), 30 November 2014 <www.smh.com.au/nsw/wallaby-david-pocock-arrested-in-maules-creek-coalmine-protest-20141130–11x5w3.html>.
121 M Lowe, 'David Pocock Maules Creek mine protest case dismissed, no conviction recorded', *The Sydney Morning Herald* (online), 4 February 2015 <www.smh.com.au/nsw/david-pocock-maules-creek-mine-protest-case-dismissed-no-conviction-recorded-20150204–135mhx.html>.
122 L Taylor, 'Prominent Australians ask world leaders to consider ban on new coalmines', *The Guardian* (online), 27 October 2015 <www.theguardian.com/environment/2015/oct/27/prominent-australians-ask-world-leaders-to-consider-ban-on-new-coalmines>.
123 Commonwealth, *Parliamentary Debates*, Senate, 10 February 2015, 300 (Christine Milne).
124 Commonwealth, *Parliamentary Debates*, Senate, 10 February 2015, 312 (Lee Rhiannon).
125 Ibid.
126 Ibid.

many farmers and supporters across the country and, indeed, across the world who are working so hard to ensure the right thing is done here and that protection for the Leard forest is put in place, that Aboriginal sacred sites and special places are not destroyed and that the climate action that we so urgently need is taken seriously and becomes a reality'.[127] She noted: 'More than 300 have now been arrested in direct action'.[128] Rhiannon praised the actions of climate activists: 'Jonathan Moylan, a Second World War veteran of the Kokoda Trail, and Bill Ryan are some of those people who are regularly there adding their voices and alerting the world to the crimes of Whitehaven'.[129]

Her colleague Senator Larissa Waters also objected to the 'approval of a 13-megaton coalmine in the middle of a critically endangered forest'.[130] She was concerned whether Whitehaven Coal had used false and misleading information in respect of offsets in order to gain an approval from the Federal Department of the Environment. Waters observed:

> Of course, we saw one of the activists involved in trying to protect this forest, Jonathan Moylan, issue a fake press release which caused a temporary dip in the share price of the bank that was proposing to fund this mine. He had the book thrown at him and was subjected to charges which would have borne a 10-year penalty. Thankfully, he got off, but it highlights the difference of approach between a company which has provided false information to get an approval and then has the investigation dropped and an activist who was faced with a potential 10 years in jail.[131]

Waters objected to the project: 'What a mockery it is for this mega-coalmine to be approved right when the cost of thermal coal has halved in the last three years'.[132] She noted that 'This mine is not financial' and 'it is going to have an absolutely atrocious effect on the climate'.[133] Waters concluded that 'It would pump out 30 million tonnes of CO_2, which is equivalent to the annual emissions of New Zealand's entire energy sector – massive climate disaster, massive biodiversity disaster'.[134]

127 Ibid.
128 Ibid.
129 Ibid.
130 Commonwealth, *Parliamentary Debates*, Senate, 10 February 2015, 307 (Larissa Waters).
131 Ibid.
132 Ibid.
133 Ibid.
134 Ibid.

In the wake of the Jonathon Moylan case, there have been a number of controversies in respect of climate change, civil disobedience, and necessity. In the United Kingdom, thirteen Heathrow airport climate protestors were found guilty of aggravated trespass in July 2015.[135] At the Willesden Magistrates' Court, District Judge Deborah Wright expressed concern about the cost of the disruption. She dismissed the defence of the protestors that their actions were necessary. In her view, the action of the protestors was 'symbolic and designed to make a point, not to save lives'.[136] Justice Wright told the protestors that they could face jail. In response, the members of the Plane Stupid campaign released a statement, discussing their civil disobedience:

> Today's judgment demonstrates that the legal system does not yet recognise that climate defence is not an offence. We took action because we saw that it was sorely needed. When the democratic, legislative and processes have failed, it takes the actions of ordinary people to change them.[137]

Tim Sanderson offered support for his daughter and the other Heathrow climate protestors.[138] He observed: 'Politicians should be brave enough to expose the real choices, which are quite simple: either continue with our destructive habit of dependency on air travel (and ignore climate change) or make radical changes to our frequent-flying lifestyle, rethink our strategies for economic growth, and – hopefully – preserve the planet in a habitable state for future generations'.[139] The environmentalist George Monbiot called the climate protestors 'freedom fighters'.[140] He observed that there would be two verdicts: one will relate to the legal status of what the protesters did, and the other will concern the moral status. Monbiot maintained that the protestors could be popular heroes, even if they were locked up: 'Vilified,

---

135 Press Association, 'Heathrow climate protesters found guilty of aggravated trespass', *The Guardian* (online), 26 January 2016 <www.theguardian.com/environment/2016/jan/25/heathrow-climate-protesters-found-guilty-of-aggravated-trespass?CMP=share_btn_tw>.

136 Ibid.

137 Ibid.

138 T Sanderson, 'I'm proud of my daughter and the other brave Heathrow protestors. Here's why', *The Guardian* (online), 26 January 2016 <www.theguardian.com/commentisfree/2016/jan/25/heathrow-protesters-plane-stupid?CMP=soc_3156>.

139 Ibid.

140 G Monbiot, 'The Heathrow "hooligans" are our modern day freedom fighters', *The Guardian* (online), 20 January 2016 <www.theguardian.com/commentisfree/2016/jan/20/heathrow-third-runway-protesters-trial-freedom-fighters>.

prosecuted, but – in the court of public opinion – ultimately vindicated: this is what happens to the heroes of democracy'.[141]

In the United States, there has been a similar battle over climate change, civil disobedience and necessity in the Delta Five case in Washington State.[142] In September 2015, five climate activists from Rising Tide Seattle halted the passage of a crude oil train. They were arrested and were charged with criminal trespass and blocking a train. A spokesperson for Rising Tide said the activists 'will be the first ever to argue that their actions were justified because of the threat of climate change, using the "necessity defense"'.[143] They said: 'The outcome of [the] trial could set national precedent for climate related civil disobedience and is being carefully watched'.[144] Discussing the case, Tim DeChristopher observed: 'Like all civil disobedience, this new wave of climate disobedience is an inherent critique of the moral authority of government'.[145] He commented: 'The necessity defense is an opportunity to elaborate that implicit critique into a fully developed legal argument for the responsibility of citizen action in the face of governmental failure'.[146]

In terms of wild law, there will be much further discussion about climate change, civil disobedience, and necessity.

C: Corporations law

The decision of Justice Davies will receive a great deal of attention because it raises larger questions about corporations law, the role of ASIC, culture-jamming, climate change, and civil disobedience.[147]

In a statement to the broadcaster SBS, ASIC defended its action against Jonathan Moylan: 'ASIC is all about ensuring investors – the millions of

141 Ibid.
142 J Mikulka, 'Delta 5 trial set to make history with "necessity defense" for climate action', *Desmog* (online), 10 January 2016 <www.desmogblog.com/2016/01/10/delta-5-trial-set-make-history-necessity-defense-climate-action> and K Herzog, 'Activists on trial for blocking oil train will argue it was justified by climate change', *Mother Jones*, 11 January 2016 <www.motherjones.com/environment/2016/01/oil-train-activists-trial-climate-change>.
143 Herzog, above n 142.
144 Ibid.
145 T DeChristopher, 'Civil disobedience often leads to jail. But now, protesters can explain themselves', *The Guardian* (online), 13 January 2016 <www.theguardian.com/commentisfree/2016/jan/13/civil-disobedience-often-leads-to-jail-but-now-protestors-can-explain-themselves?CMP=edit_2221>.
146 Ibid.
147 M Rimmer, 'Coal in court: Whitehaven, climate change and civil disobedience', *The Conversation*, 19 July 2013 <https://theconversation.com/coal-in-court-whitehaven-climate-change-and-civil-disobedience-15991>.

Mums and Dads with superannuation – can have trust and confidence in our financial markets and that is why we took action against Mr Moylan'.[148] ASIC denied that it had ignored more serious types of insider trading: 'Our actions are about upholding investor trust and confidence in the integrity of our markets'.[149] ASIC maintained it was 'satisfied with the sentence'.[150] ASIC refused to comment upon why it did not pursue a case against media organisations, which published Moylan's press release without checking the information with official bodies such as the ASX: 'We have no comment on the standards of journalism in Australia'.[151] SBS asked: 'How concerned is ASIC about copycat hoaxes in the future and what has been put in place to ensure it doesn't occur again?'[152] ASIC maintained: 'Our action in the Moylan case was certainly about sending a message that we take the integrity of Australia's financial markets very very seriously'.[153]

In the wake of the ruling, Jonathan Moylan published an opinion in *The Guardian* – 'My hoax wasn't meant to crash the market but save the environment'.[154] He commented: 'When our politicians continue to break their election promises and cosy up to coal interests, our eclectic mix of citizens – men and women willing to engage in civil disobedience – are the best hope we have for the future'.[155] Moylan noted: 'Hoax press releases are nothing new, and taking the piss with a purpose is part of a long tradition of creative mischief'.[156] He observed: 'For the record, I was not trying to crash the market. I was trying to highlight the fact that ANZ are funding the biggest open cut coal mine currently being constructed in Australia'. Moylan acknowledged: 'I bear the consequences for my decision to bluff ANZ bank with a fake press release'.[157] He observed that 'the day will come when banks will have to choose between coal and their reputations'.[158] He hoped that investors 'should also be fully informed about the environmental impact of their investments in order for the planet to work, because it's the only home we have'.[159]

---

148 SBS, above n 9.
149 Ibid.
150 Ibid.
151 Ibid.
152 Ibid.
153 Ibid.
154 J Moylan, 'My hoax wasn't meant to crash the market but save the environment', *The Guardian* (online), 25 July 2014 <www.theguardian.com/commentisfree/2014/jul/25/my-hoax-wasnt-meant-to-crash-the-market-but-save-the-environment>.
155 Ibid.
156 Ibid.
157 Ibid.
158 Ibid.
159 Ibid.

Jonathan Moylan has received vocal support from farmers, environmentalists, and climate activists. Rick Laird, a farmer from Maules Creek, travelled to Sydney for the sentencing hearing. He observed: 'To most people ANZ is just a bank, but to our community at Maules Creek their loan to Whitehaven Coal threatens to put an end to 150 years of farming in the region'.[160] Laird stressed: 'We've been fighting this mine for years but what Jono did means the world knows what is happening to Maules Creek farms and the Leard State Forest'.[161] He commented that the community has been showing solidarity with Jonathan Moylan: 'Jono stood with us, so we're here to stand with Jono'.[162]

In a campaign video, barrister Brian Walters SC stated, 'Jonathan served ANZ shareholders by making them aware their resources are being used unethically'.[163]

However, a number of commentators have lamented the failure of ASIC to pursue other cases involving false information. Paddy Manning – for *Crikey News* – noted the lack of charges by ASIC for Richard Macphillamy after an investigation into spreading of false rumours to affect share prices surrounding Macquarie Bank during the Global Financial Crisis.[164] He noted:

> The defence team for Whitehaven Coal hoaxer Jonathan Moylan, facing a possible jail sentence for disseminating false information to the market in 2013, has used a sentencing hearing today to raise the case of Richard Macphillamy. Macphillamy was banned from providing financial services for 18 months after an investigation into rumourtrage (the spreading of false rumours to affect share prices) surrounding Macquarie Bank at the height of the financial crisis.[165]

He sought to contrast this case with the controversy around Jonathan Moylan.

160 Stand with Jono, above n 70.
161 Ibid.
162 Ibid.
163 Stand with Jono, 'The facts: Jonathan Moylan, ASIC, and Whitehaven', *YouTube*, 22 September 2013 <https://youtu.be/ACikpCkwRyY>.
164 P Manning, 'Will Whitehaven coal hoaxer Jonathan Moylan go to jail?', *Crikey News*, 11 July 2014 <www.crikey.com.au/2014/07/11/will-whitehaven-coal-hoaxer-jonathan-moylan-go-to-jail>.
165 Ibid.

Michael West – of the Fairfax newspaper, *The Sydney Morning Herald*, has referred to the failure of ASIC to pursue action against the Commonwealth Bank.[166] He commented:

> The problem with corporate regulation in this country is essentially cultural, however. It is about people cowed by the big end of town, spooked into inaction. It is about regulation by press release. It is about the appearance of regulation. Now the people running the large institutions have been sent an unequivocal message that they will not be prosecuted come hell or high water, it is worth contemplating once again what the most concerted enforcement action of the past couple of years has been – the only raid. This was the raid on the campsite of a lone green activist, Jonathan Moylan, in Singleton, NSW. Moylan's crime had been to issue a fake press release to rabble a coal company and its bankers. Ironically, this was an action taken from principle, not from personal gain. Yet the message from our government is: act from principle and you will be dragged through the courts, act for a big bank, churn people's life savings, and the worst fate to befall you might be the embarrassment of a Senate inquiry and a few 'enforceable undertakings'.[167]

In his view, it was striking that Jonathan Moylan was the subject of legal action by ASIC – but no action was taken in respect of the Commonwealth Bank financial planning scandal.

Bernard Keane in *Crikey News* accused ASIC of double standards: 'The corporate regulator is too scared and incompetent to take on large companies, but has come down hard on the man who sent out a fake press release to expose the stupidity of finance journalists'.[168] He lamented:

> If you're online and ASIC knows you're just an individual, without corporate resources to fight it, then it will come down very hard on you indeed. That's why the Whitehaven Coal hoaxer Jonathan Moylan has this week been charged by ASIC and now faces 10 years' jail for raising awareness of the laziness of investors and journalists and Whitehaven's

---

166 M West, 'The response to the Commonwealth financial planning scandal shows banks are really above the law', *The Sydney Morning Herald* (online), 26 October 2014 <www.smh.com.au/business/comment-and-analysis/the-response-to-the-commonwealth-financial-planning-scandal-shows-banks-really-are-above-the-law-201410 26–11c13d.html>.

167 Ibid.

168 B Keane, 'Double standards as gutless ASIC targets the little guy', *Crikey News* (online), 5 July 2013 <www.crikey.com.au/2013/07/05/double-standards-as-gutless-asic-targets-the-little-guy>.

potentially highly damaging CSG projects. Moylan, who was on the receiving end of a rare raid by ASIC officers after the hoax, had issued a fake media release about the company from his laptop. Moylan has been charged with breaching the *Corporations Act*.[169]

Keane wondered: 'If Moylan was backed by a large company, you can bet ASIC would at worst have demanded an enforceable undertaking not to do it again, or maybe instigated a multi-year investigation that quietly ends without charges'.[170]

In 2015, there was a striking sequel to the litigation in respect of Jonathan Moylan. Greenpeace and Environmental Justice Australia lodged a complaint with ASIC, alleging that Whitehaven Coal had made misleading representations in respect of future coal demand.

The inspiration for the litigation came in part from the United States. In 2015, Peabody Energy, a major US coal company, was found by the New York attorney-general to be in violation of New York securities law because of the failure to disclose climate risks.[171]

In November 2015, Greenpeace called for an investigation into Australian coal companies over climate risk.[172] Marina Lou, Legal Advisor for Greenpeace International, commented that research into the Australian coal companies showed that four of them, including Rio, Whitehaven, AGL and Wesfarmers, have not disclosed risks to their business based on a 2°C warming scenario. She was concerned about this failure:

> The finding of the New York attorney-general was precedent-setting. Climate change, and the regulatory risk it poses to fossil fuel companies, can have a material impact. If known, it should be disclosed to investors. That's why we are urging Australian regulators to conduct investigations into companies listed on the ASX, for potentially similar misconduct. Greenpeace International has been systematically scrutinising disclosure of fossil fuel-exposed companies like Exxon and Peabody for a long time. With other major coal mining countries like Indonesia cutting their production and export of coal, Australia could soon become the world's biggest supplier of seaborne coal. The

169 Ibid.
170 Ibid.
171 Attorney-General of New York, 'Peabody Energy Assurance', Assurance No 15–242 <http://ag.ny.gov/pdfs/Peabody-Energy-Assurance-signed.pdf>.
172 Greenpeace, 'Greenpeace calls for investigation into top Australian coal companies over climate risk', Press Release, 19 November 2015 <www.greenpeace.org/australia/en/mediacentre/media-releases/climate/Greenpeace-calls-for-investigation-into-top-Australian-coal-companies-over-climate-risk>.

exposure this poses to the Australian securities market could have marked impact on all Australian investors.[173]

Greenpeace and Environmental Justice Australia wrote to ASIC, asking for an investigation into the state of climate disclosure for these companies.

In December 2015, Environment Justice Australia and Greenpeace complained to ASIC that Whitehaven Coal may have provided investors with misleading forecasts of future coal demand.[174] The focus of the complaint was a slide presentation made by the company to the Australian Stock Exchange on the 16 November 2015. Whitehaven Coal appeared to rely upon the 'new policies scenario' of the International Energy Agency.

The lawyer for Environment Justice Australia, David Barnden, commented: 'For Whitehaven to concentrate on that scenario, taking three pages of its 23-page presentation to illustrate how strong demand for coal will be, but not disclosing the risks, it appears to be misleading to us'.[175] He contended: 'On the face of it, the information and the weight that is given to it in that investor presentation would suggest it could be misleading and might induce investors to trade shares or hold on to them'.[176] He observed: 'Coal companies or energy companies with high exposures to carbon need to appropriately disclose the risk of what demand for those products could look like in a world with strict regulation around the consumption of coal or carbon'.[177]

In response, Whitehaven Coal was indignant at the criticism that it received. A spokesperson said:

> The IEA report is written on the basis of the New Policies Scenario, which is the central scenario used in the World Energy Outlook series of reports. Whitehaven's charts reference this central scenario. In contrast, the bridge scenario amounts to 341 words in the last pages of the report, which is otherwise 48,000 words long. We reject any suggestion we've used the information out of context. Any reading of the 48,000-word IEA report clearly illustrates that future demand for coal in SE Asia is real and growing. Activists need to accept this fact. Whitehaven is extremely positive about the outlook for its high-quality coal, as

---

173 Ibid.
174 P Ker, 'Whitehaven Coal Complaint with ASIC Over Use of IEA Coal Demand Forecasts', *The Sydney Morning Herald* (online), 18 December 2015 <www.smh.com.au/business/mining-and-resources/asic-asked-to-investigate-whitehaven-coals-use-of-iea-coal-demand-forecasts-20151217-glpvdm.html>.
175 Ibid.
176 Ibid.
177 Ibid.

reflected in the updated profit guidance the company issued this week. To the extent that we have used IEA data to support this outlook, we have appropriately referenced the full source document.[178]

It will be interesting to see whether this issue will be reviewed and tested by ASIC.

There is a larger discussion going on about rewilding corporations law. Former Liberal leader John Hewson has expressed concern about the failure of corporations to address climate change risk and management.[179] There has been a push to hold companies – such as ExxonMobil and Volkswagen – accountable for misleading and deceptive statements about climate change and carbon emissions. Such a discussion highlights the ways in which corporations law could be read to promote a vision of environmental protection and climate justice.

D: Fossil fuel divestment

The case of Whitehaven Coal has also been the subject of a larger debate about fossil fuel divestment. There has been a push for Australian universities to be divested of their fossil fuel stocks in respect of coal, oil, and gas. Investments in Whitehaven Coal have come under particular scrutiny from environmental and climate activists.

After much criticism of its investment policies, the Australian National University engaged in a partial divestment of its holdings in respect of fossil fuel companies. The Council of the Australian National University agreed to a proposal from the Vice Chancellor Professor Ian Young to engage in the divestment of stocks in seven companies following an independent review of the University's domestic equities.[180] In particular, the University decided to divest its holdings in Iluka Resources, Independence Group, Newcrest Mining, Sandfire Resources, Oil Search, Santos and Sirius Resources. The Abbott government and the mining industry were outraged by the decision.[181] Nonetheless, the partial divestment has had

---

178 Ibid.
179 S Phillips, 'World headed to recession if climate change ignored, John Hewson warns', *ABC News*, 5 April 2016 <www.abc.net.au/news/2016–04–05/world-to-face-recession-if-climate-change-ignored:-john-hewson/7298770>.
180 Australian National University, 'University to divest holdings in seven companies', Press Release, 3 October 2014 <www.anu.edu.au/news/all-news/university-to-divest-holdings-in-seven-companies>.
181 F Jotzo, 'Outrage at ANU divestment shows the power of its idea', *The Conversation* (online), 13 October 2014 <https://theconversation.com/outrage-at-anu-divestment-shows-the-power-of-its-idea-32736>.

a positive financial impact on the Australian National University. Fossil Free ANU has pushed for a broader fossil fuel divestment by the Australian National University.[182]

There has been much controversy over the University of Sydney's investments in Whitehaven Coal.[183] Professor of Politics at the University of Sydney, John Keane, has considered the controversy at his educational institution.[184] He noted: 'Pushed and pulled in different directions by government policies and market forces, modern universities try hard to be public institutions for the public good'.[185] Keane commented:

> A recent investigation by Greenpeace suggests that the University of Sydney, at least for the moment, falls short of its own principles. The carefully-researched report reveals that the University of Sydney has around $1 million of its endowment invested in Whitehaven Coal. The actions of this company place it in direct conflict with the University's public commitments to the environment, indigenous Australia, and the public welfare of our society more broadly.
>
> Whitehaven Coal is developing the largest coal mine currently under construction in Australia, at Maules Creek in northern New South Wales, in the face of staunch opposition from a diverse network of committed Australian citizens. Local farmers fear for the impact on their livelihoods of the mine's colossal water use, with the water table in nearby areas set to drop by several metres. The mine has damaged or destroyed dozens of cultural heritage sites of the indigenous Gomeroi traditional owners, who have been denied access to sacred sites to perform ceremony.[186]

Keane emphasized that the University of Sydney 'must decide whether it is an institution which abides by its important commitments to the environment and to Indigenous Australians, or an organisation which chooses to side with entrenched political power at the expense of a planet already in poor health'.[187] In his view, 'The decision requires more than an investment

182 Fossil Free ANU <https://fossilfreeanu.wordpress.com>.
183 M Green, 'Climate activism's new frontier is targeting fossil fuel investors', *The Sydney Morning Herald* (online), 15 September 2014 <www.smh.com.au/national/climate-activisms-new-frontier-is-targeting-fossil-fuel-investors-20140912–10fxoc.html>.
184 J Keane, 'Coal, divestment and democracy', *The Conversation* (online), 17 September 2014 <https://theconversation.com/coal-divestment-and-democracy-31764>.
185 Ibid.
186 Ibid.
187 Ibid.

review, or another proclamation'.[188] Keane called for 'determined action – for the sake of the good reputation of the University, and for the public good'.[189]

Moreover, there have been various actions by climate activists in respect of financial institutions providing support for fossil fuel projects.[190]

In response, the Whitehaven Coal chief Paul Flynn called fossil fuel divestment 'green imperialism at its worst'.[191] He complained: 'I wouldn't tell people how to invest their money and they shouldn't purport to be doing that either'.[192] Flynn maintained: 'Obviously that side is very selective in the facts that they use, it doesn't suit them to have a broader discussion about all the facts'.[193]

The Minerals Council of Australia has wanted to use the case of Jonathan Moylan as a precedent by which to pursue other environmentalists and climate activists. In a piece commissioned by the Minerals Council of Australia, RMIT economist and Institute of Public Affairs fellow, Sinclair Davidson, argued that section 1041E of the *Corporations Act 2001* (Cth) should also be applied to fossil fuel divestment activists.[194] He argues: 'To the extent that stigmatisation deliberately causes investors to make valuation errors and consequently rebalance their portfolios away from fossil fuel stocks, a violation of the *2001* (Cth) has occurred'.[195] This would appear to be a weak, tendentious, and meritless argument, to anyone with legal training. While the offence is a broad one, it is not so open-ended, as Sinclair Davidson would maintain.

Fossil fuel divestment has been advocated by a wide range of policy-makers, intellectuals and activists – including former United States President Barack Obama; climate leader and former Vice President of the United States, Al Gore; Christiana Figueres, the Executive Secretary of the *United Nations Framework Convention on Climate Change* 1992; United Nations special envoy on climate change, Mary Robinson; religious leader, Desmond Tutu;

188 Ibid.
189 Ibid.
190 350.Org, Market Forces, and Greenpeace, *Disconnected: ANZ's fossil fuel financing*, December 2013 <http://350.org.au/wp-content/uploads/2014/06/Annual_Report_ANZ_Final.pdf>.
191 A Saunders, 'Whitehaven coal boss lashes "green imperialism"', *The Sydney Morning Herald* (online), 29 August 2014 <www.smh.com.au/business/mining-and-resources/whitehaven-coal-boss-lashes-green-imperialism-20140828–109kln.html>.
192 Ibid.
193 Ibid.
194 S Lewis, 'A critique of the coal divestment campaign', Minerals Council of Australia, June 2014 <www.minerals.org.au/file_upload/files/reports/A_critique_of_the_coal_divestment_campaign_Sinclair_Davidson_Jun_2014.pdf>.
195 Ibid.

the Carbon Tracker; and Bill McKibben. It would indeed be absurd and ridiculous to enforce section 1041E of the *Corporations Act 2001* (Cth) against those engaged in political communication about environmental protection, climate action, and fossil fuel divestment.

In the future, there is a need to ensure that the *Corporations Act 2001* (Cth) is not used as a crude means of quelling dissent in respect of matters of public importance. Moreover, fossil fuel divestment seems to be an effective means of rewriting financial investment norms and rules.

Conclusion

The dispute in *R v Moylan* is an unruly case about the Whitehaven Coal hoax, which defies easy classification or categorisation. The judgment could be rewritten in a number of ways under a 'wild law' methodology. The litigation in *R v Moylan* could be considered in light of cultural debates over culture-jamming, identity theft, and impersonation. The dispute could also be seen in light of legal, ethical, and philosophical debates in respect of freedom of speech, non-violent action, necessity, and civil disobedience. The dispute in *R v Moylan* raises larger questions about corporations law and the role of the regulator of ASIC. There is further scope for revising and reforming corporations law, so that it takes concerns about environmental damage and climate risk seriously. In addition to the analysis of the Whitehaven Coal hoax, it is notable that Whitehaven Coal's statements have been the subject of much debate and analysis. The case of *R v Moylan* raises larger questions about climate change, fossil fuel divestment, and the adoption of renewable energy. The conflict in *R v Moylan* will no doubt be an important precursor to further battles over climate change between governments, fossil fuel corporations, and civil society activists.

Chapter 20

Magee v Wallace

Susan Bird[1]

Commentary

Magee v Wallace [2014] VSC 643 may appear, *prima facie*, as an insignificant case with little relevance for wild law. It is an appeal from a Magistrates' Court decision, *Wallace v Magee*, about minor criminal damage under section 10 of the *Summary Offences Act 1966* (Vic). Kyle Magee pleaded guilty to charges of causing damage to an advertising hoarding. Magee pasted over, with flour and water, an advertisement by a multinational corporation with bill posters containing his philosophy. While Magee's act could be viewed as a simple case of criminal damage, it becomes more complicated when explored in context. Magee is an activist who is acutely aware of the impact many corporations have on the environment. Magee's painting and posting over corporate advertising is a form of direct action aimed at raising awareness, and sparking a conversation about how we can save the planet. This can be achieved, Kyle believes, through a free and independent media, where ordinary citizens have a right to be heard. His court actions rely on the Victorian *Charter of Rights and Responsibilities Act 2006*.

The appeal in *Magee v Wallace* raises a defence of consent contained in the *Summary Offences Act*. Magee argues that consent to his actions may be implied via an application of the *Charter of Human Rights and Responsibilities Act*. Kyle Magee, representing himself, mounts the argument that section 10 of the *Summary Offences Act* can be interpreted to allow a body, other than the owner of property, to consent to his bill posting. Section 10(1) states:

> Any person who posts any placard bill sticker or other document on or writes or paints on or otherwise defaces any road bridge or footpath or any house building hoarding wall fence gate tree tree-guard post pillar hydrant fire-alarm petrol pump or other structure whatsoever without

1 For Hallie.

the consent of the occupier or owner of the premises concerned *or of any person or body having authority to give such consent* shall be guilty of an offence.[2] [Author emphasis].

Magee argues that the Victorian government is a body which has the authority to give consent where his bill posting occurs in public spaces, such as train stations or bus stops. Further, he argues the government does give implied consent to acts which constitute freedom of expression under section 15 of the *Charter of Human Rights and Responsibilities Act.* Justice Ginanne disputed this interpretation of the Charter and held that Magee did not establish the defence of consent:

> Mr Magee did not refer to authority for that proposition and I do not accept it. The consent required to establish a defence to actions which would otherwise contravene s 10 of the Summary Offences Act is consent *by the owner of the property* or someone with the authority to give consent. The rights recognised by the Charter do not confer that consent.[3]

The decision relies quite heavily on the outcome in *Magee v Delaney* [2012] VSC. In *Magee v Delaney,* Magee argued that freedom of expression was a 'lawful excuse' for criminal damage under sections 197(1) and 199(a)(i) of the *Crimes Act 1958* (Vic).

Magee v Delaney came within the first five years of operation of the *Charter of Human Rights and Responsibilities Act.* Before the passing of the *Human Rights Act 2004* (ACT), there were no comprehensive legislative provisions in place for protecting human rights in Australia.[4] In 2004, the then Attorney General Robert Hulls created a Human Rights Consultation Committee to explore whether Victoria would benefit from similar legislation to the ACT.[5] After receiving over 2,000 written submissions, the committee concluded that a 'substantial majority' of Victorians desired greater protection of their human rights enshrined within the law.[6] This led to the tabling of a 190 page report which included a Draft Charter, on which the current legislation is based.[7] At the time of Magee's first hearing, the scope

---

2 *Summary Offences Act 1966* (Vic) s 10(1).
3 *Magee v Wallace* [2014] VSC 643 [14] (Ginnane J).
4 Human Rights Law Centre, 'Updated Chapter Five: Victorian Charter of Human Rights and Responsibilities' *Human Rights Law Centre Manual* (2009) <http://hrlc.org.au/hrlrc-guide-victorian-charter>. See also G Williams, 'The Victorian Charter of Human Rights and Responsibilities' in H Sykes (ed.), *Future Justice* (Future Leaders, 2010) 136.
5 Human Rights Law Centre, above n 4, 5.
6 Ibid.
7 Ibid.

of the Act and its application were still in the process of being tested in how it limited the application of existing legislation such as the *Summary Offences Act*. In this case, counsel for Magee, and later Magee representing himself, ran an argument based on section 15 of the Act which protects freedom of expression. A question for the court was whether Kyle's acts in refacing billboards could be considered 'expression' and therefore be protected under this section, and whether these rights should be upheld where they encroach on the 'rights' of property owners – in this case, multinational corporations – to display their advertising.

While the presiding Judge, Justice Kyrou, held that Magee's acts in painting over advertising hoardings could constitute a form of expression, he held that freedom of expression 'was subject to lawful restrictions reasonably necessary to respect the property rights of other persons, irrespective of whether those persons are human beings, companies, government bodies or other types of legal entities'. Here the judge does not distinguish the rights of multinational corporations from the citizens of the state of Victoria, whom the Charter was designed to protect. In the wild law judgement, my reasoning seeks to restore the true purpose of the Charter in protecting human citizens engaging in peaceful protest over the rights of corporate entities. Wild law seeks to redress the imbalance and allow freedom of expression for those raising awareness of environmental destruction.

Kyle Magee is a repeat offender who has spent much of his adult life in protest against corporate advertising. Magee began by painting over billboards at the age of 21,[8] and has continued to do so, even after being repeatedly caught and charged by authorities. His *philosophical* stance on protecting democracy via the refacing of billboards so motivates him that he states in his blog, 'global liberal media please', that he is compelled to act. He describes doing nothing as akin to watching a rape without lifting a finger to help.[9]

Kyle Magee argues that if we are to have a true democratic system, one which is in a position to respond to global issues such as the destruction of the environment, then we must have a free media which is not under the control of corporations.[10] Corporate power, Magee asserts, controls what is reported in the media, and what is not. Public spaces are of particular interest to Magee, as it is here where he believes that free speech can truly flourish if liberated from the shackles of corporate advertising, and opened up for debate between citizens. In the city of Melbourne, it is impossible to

8   S Bird and D Vakalis, 'Kyle Magee: adbusting, exclusion and the urban environment' (2011) 14 *Southern Cross University Law Review* 163.
9   <https://globalliberalmediaplease.net>.
10 Ibid.

walk more than a few metres without being exposed to myriad messages from multinational corporations. By comparison, messages from members of the public are not visible in the city. Graffiti is highly regulated via the *Graffiti Prevention Act 2007* (Vic), to the extent where the burden of proof is reversed where persons are found with implements that could potentially 'mark graffiti' on public/private property.[11]

This debate feeds into a larger argument around the privatisation of public spaces, which has gained momentum worldwide.[12] It was only just over two hundred years ago, in the same year that Australia was 'discovered' by the British, that there were major uprisings in Britain caused by the *Enclosure Acts*[13], which sought to remove land from public ownership and vest the interests into the hands of wealthy farm owners. This issue has reignited around the world as neoliberal governments from both the right and left of politics[14] subscribe to the idea that services are more efficiently run by private companies.[15] Nowhere has this experience been more rapid and marked than Victoria during the 'Kennett era'. Jeff Kennett, a neoliberal politician, won the Victorian election in 1992, and succeeded in dismantling and selling off more public resources than any other government in Australia's history. As Mark Lowes writes:

> For several years following its electoral victory in 1992, the Kennett government proved to be the most active, controversial and ideological administration in twentieth century Victoria. In its energetic implementation of an ideologically charged program of downsizing, deregulation and privatization – the essential tenets of neoliberalism – the Kennett regime 'changed the face of Victoria'.[16]

---

11 *Graffiti Prevention Act 2007* (Vic) s 7. For commentary on this issue, see also: S Bird (2009) 'The aesthetic of authority and the outlaw of the street' (2009) 3 *Public Space: The Journal of Law and Social Justice.*

12 A Minton, *What Kind of World are we Building? The Privatisation of Public Space* (Royal Institute of Chartered Surveyors, 2006).

13 The *Enclosure Acts* were a series of Acts passed in England from the sixteenth century onwards.

14 T Ali, *The Extreme Centre: A Warning* (Verso, 2015).

15 M Lowes, 'Neoliberal power politics and the controversial siting of the Australian Grand Prix motorsport event in an urban park' (2004) 27 *Society and Leisure* 70. Garrett Hardin's thesis in 'The Tragedy of the Commons' (1968) 162 *Science* 1243 underpins the argument of many neoliberal thinkers who refer to this article as 'proof' that communal ownership leads to 'tragedy', or violent conflict (for example, M O'Keefe 'The tragedy of anarchy' (2009)11 *Journal of Military studies* 4).

16 M Lowes, 'Neoliberal power politics and the controversial siting of the Australian Grand Prix motorsport event in an urban park' (2004) 27 *Society and Leisure* 70.

These changes did not go without protest. Kennett became a very unpopular leader, and was voted out in 1999, but the damage was already done.[17]

The issue again came to the fore during the Occupy protests in the city square of Melbourne, where it was argued by the then Mayor Robert Doyle that the protesters were: 'a self-righteous, narcissistic, self-indulgent rabble . . . denying use of popular and busy public space to anyone except themselves'.[18] However, as asserted in *New Matilda*, this argument cannot hold weight as the city square has been sold to private interests:

> The fact that [Doyle] can say that in a context in which the streets in question self-evidently don't belong to everyone — Jeff Kennett sold them to a hotel chain — speaks volumes. Neoliberalism introduces the market into every aspect of social life, reshaping citizenship as consumption. The notion of a collective public disappears, as the paradigm for every interaction becomes the market exchange. In the neoliberal mind, corporations become better citizens than people; they more perfectly manifest a market logic.[19]

This is a prevailing attitude in the case of Kyle Magee, where the right of the private property owner, in the form of a multinational corporation, is given a greater right to freedom of speech than the citizen. It appears that in the neoliberal era, property rights trump human and environmental rights at every turn.

There is a growing concern about the lack of political alternatives given the race to the 'extreme centre' by both major parties in Australia.[20] This has resulted in movements that reject traditional democracy as a solution to the world's problems. The Occupy movement is one of these, as is the drive to reinstall the commons as a form of organisation for managing

---

17 That said, there have been no attempts by labour governments to buy back the assets sold by Kennett. Indeed the swathe of privatisation continues with the recent negotiations by the left wing Andrews government to sell the Port of Melbourne to the highest bidder. The hold up in this deal has been the Coalition (see J Edwards, 'Victorian Parliament unlikely to approve Port of Melbourne privatisation without significant changes' *ABC News*, 8 December 2015 <www.abc.net.au/news/2015–12–08/port-of-melbourne-privatisation-likely-to-be-rejected/7011650). This confirms Tariq Ali's argument in *The Extreme Centre* that 'left wing' governments are now just as likely to support neoliberal reforms as right wing ones; Ali, above n 14.

18 Robert Doyle quoted in 'Public space for sale' *New Matilda* (online), 26 July 2012 <https://newmatilda.com/2012/07/26/public-space-sale>.

19 Ibid.

20 This is also a global phenomenon; see Ali, above n 14.

shared resources.[21] Those in the 'Commons Strategies Group' argue for a reconsideration of rights discourse that takes into account environmental concerns. They argue for new ways of governing resources based on an ethic of sharing rather than one of greed.[22]

While Magee still holds out hope for the traditional democratic process to protect the environment, he cannot see it effectively operating without a loosening of the corporate stranglehold on the media. Should the media operate in a free and democratic manner, allowing the voice of the people to be heard, Magee has trust that humanity will respond accordingly, and face up to its responsibilities in preventing environmental destruction.

In this wild law judgment, Justice Bird will build on the notion of Green Governance to assert the extension of human rights to the environment, and allow Magee's appeal under the *Charter of Human Rights and Responsibilities Act 2006* (Vic).

IN THE WILD LAW COURT OF VICTORIA   Not Restricted
AT MELBOURNE
COMMON LAW DIVISION
JUDICIAL REVIEW AND APPEALS LIST
S CI 2013 05722

KYLE MAGEE   Appellant

v

SHAYNE WALLACE   Respondent

—

21 See e.g., B Weston and D Bollier, *Green Governance: Ecological Survival, Human Rights, and the Law of the Commons* (Cambridge University Press, 2014). See also Elinor Ostrom's comprehensive work on the commons as a form of governance that won her a Nobel Memorial Prize in Economic Sciences: E Ostrom, *Governing the Commons: The Evolution of Institutions for Collective Action* (Cambridge University Press, 1990).

22 Overconsumption contributes to many global problems, including catastrophic climate change. Scientists argue that even looking at the essentials to survival, such as food production, the world is tipped to exceed its limits if agricultural practices, in particular meat production, do not become environmentally conscious. See for example D Tilman, 'Global environmental impacts of agricultural expansion: the need for sustainable and efficient practices' *Proceedings of the National Academy of Scientists of the United States of America* vol 96 no 11,5995. This fits with Garrett Hardin's 'Tragedy of the Commons' thesis. However, there are alternatives.

JUDGE:	Bird J
WHERE HELD:	Melbourne
DATE OF HEARING:	7 August 2014
DATE OF JUDGMENT:	9 January 2016
CASE MAY BE CITED AS:	Magee v Wallace
MEDIUM NEUTRAL CITATION:	[2016] WLC 123

CRIMINAL LAW – Appeal – Summary offences – Bill posting – Whether defence of consent established – *Summary Offences Act 1966* s 10.

———

HUMAN RIGHTS – Freedom of expression – Lawful restrictions reasonably necessary – Protection of public order – Rights of other persons including non-human persons – Reasonable limits on rights – *Charter of Human Rights and Responsibilities Act 2006* ss 3(1), 7(2), 15, 32(1), 36.

HER HONOUR:

The appellant, Mr Kyle Magee, is opposed to commercial advertising in public places. His actions have brought him before this Court previously in the case of *Magee v Delaney*.[23] At the Southern Cross railway station on 14 February 2013, he posted or attached pieces of black paper and documents covering over an advertising board displaying an advertisement for iced tea. The documents contained statements of his opposition to commercial advertising. He was charged with the summary offence of bill posting. He claimed that he was exercising his human right of freedom of expression which is recognised by the *Charter of Human Rights and Responsibilities Act 2006* ('the Charter') and thereby had a defence to the charge. The Magistrate did not agree and fined him $400. Mr Magee appeals against that decision.[24]

For the reasons that follow, I have concluded that Mr Magee has established his grounds of appeal. The appeal is allowed.

Magee has been charged under section 10 of the *Summary Offences Act 1966* (Vic) ('the Act'). Whilst not disputing the facts as presented by the attending officer, Magee has disagreed on the application of the law, arguing that his actions are protected by section 15 of the Charter, which allows freedom of expression.

One could well ask why the State is using the full force of the criminal law against Mr Magee. The State is here acting as an agent of a multinational

———

23 [2012] VSC 407.
24 *Magee v Wallace* [2016] VSC 123.

corporation to protect that corporation's commercial interests. If Mr Magee were to be called before a court, it ought, in my judgment, to be as a party to a civil action brought by the corporation, for the $40.17 damage caused by Magee's actions.

However, as this case is now before me, as an appeal from a conviction in the Magistrates' Court which resulted in a sentence of imprisonment, it is necessary to now determine the issues before me. I will apply the Charter, in particular section 15, to the undisputed facts of the case. I will assess whether the disruption caused by Mr Magee's actions were great enough to override the freedoms prescribed in the Charter. In interpreting the Act, a member of the judiciary must preserve, to the greatest extent possible, the human rights which are enshrined in the Act. The Court may also refer to cases outside of the Victorian jurisdiction relevant to the interpretation of the Act. This is set out within the Act at section 32:

(1) So far as it is possible to do so consistently with their purpose, all statutory provisions must be interpreted in a way which is compatible with human rights
(2) International law and the judgements of domestic and foreign and international courts and tribunals relevant to a human right may be considered in the interpretation of statutory provisions

When interpreting the Act, the courts need to look to the purpose of the Act, and can use extrinsic materials to do so.[25] The Victorian Charter was designed to protect the rights of Victorian citizens and to ensure their freedoms in our liberal democratic society. These rights were drawn from those enshrined in the *International Covenant on Civil and Political Rights*, upheld in international law, and applied in Australia through its being a signatory.

These rights were not enacted to provide protection to the commercial interests of multinational corporations, operating largely outside of Australia, and often violating the rights of human and non-human populations, and the environment, in their pursuit of wealth and power. In fact it is explicitly set out in the Act that the rights and responsibilities contained therein pertain to human persons.[26] It is not inconsequential that the billboard posted over by Mr Magee belonged to Nestlé, which has a highly controversial history. The company was famously boycotted in the 1970s due to its promotion of unhealthy infant formula to mothers in less economically

---

25 *Interpretation of Legislation Act 1984* (Vic) s 35.
26 *Charter of Human Rights and Responsibilities Act 2006* (Vic) s 3 states that '*person* means a human being' – not a non-human person such as a corporation.

developed nations.[27] Nestlé has more recently come under legal scrutiny for its environmental destruction in using palm oil in Kit Kat chocolate bars,[28] child slavery on its cocoa supplier's plantations,[29] and for tapping into the United States water table in various drought stricken districts, including in Michigan where a court held that it must curb its 400 gallon per minute depletion of the supply.[30]

Magee has spoken out in favour of environmental concerns in an atmosphere where governments have been extremely slow to respond to environmental destruction, even when faced with the looming global threat of climate change. The impact of unbridled capitalism on the environment cannot be ignored by this Court, and I take judicial notice of it. As stated by Weston and Bollier, 'The consequences visited upon our natural environment [by humans] . . . have been ruinous'.[31] There is overwhelming evidence from the scientific community that supports this statement.[32] Given the degree of evidence supplied to the general public on this issue, it is surprising how little has been done to avert disaster. As was stated by Elizabeth Kolbert, 'It may seem impossible to imagine that a technologically advanced society could choose, in essence, to destroy itself, but that is what we are now in the process of doing'.[33] Mr Magee has sought to draw attention to this matter using a form of non-violent protest, and must be congratulated for his efforts in doing so.

Under section 15 of the *Human Rights and Responsibilities Act 2006* (Vic), there is a right to freedom of expression which may include any form of expression chosen. Indeed, in upholding a similar section in the New Zealand equivalent human rights legislation, Justice Heath of the High Court

27 F D Miller Jr., *Out of the Mouths of Babes: the Infant Formula Controversy* (Social Philosophical and Policy Centre Bowling Green State University, 1983).
28 See <https://corporatewatch.org/company-profiles/nestl%C3%A9-sa-corporate-crimes>.
29 'Nestle suffers loss in child slavery case', *Sky News* (online), 12 January 2016 <www.skynews.com.au/business/business/world/2016/01/12/nestle-suffers-loss-in-child-slavery-case.html>.
30 *Michigan Citizens for Water Conservation v Nestle Waters* 269 Mich App 25; 709 NW2d 174 (2005); aff'd in part, rev'd in part 479 Mich 280; 737 NW2d 447 (2007). The Michigan Supreme Court, while agreeing with the previous judgement, held that citizens had standing to bring actions under the Michigan *Environmental Protection Act*.
31 B H Weston and D Bollier, 'Regenerating the human right to a clean and healthy environment in the commons renaissance' *Creative Commons* (online) 2011 <http://commonslawproject.org/sites/default/files/Regenerating%20Essay,%20Part%20I.pdf>.
32 For example, the Reports of the Intergovernmental Panel on Climate Change (IPCC) <www.ipcc.ch/publications_and_data/publications_and_data_reports.shtml>.
33 E Kolbert, *Field notes from a Catastrophe: Man, Nature and Climate Change* (Bloomsbury, 2006) 189.

held that jogging in the park naked could constitute freedom of expression for a 'genuine naturist'.[34] In *Pointon v Police* [2012] NZHC 3208, it was found that freedom of expression could include 'the irritating' and 'the unwelcome'.[35] Magistrate Mealy was correct in his original decision that Magee's actions did in fact constitute a form of expression. This is not in dispute here so I need not explore this matter further.

In this case, the appellant has chosen to post bills using non-toxic paste, which is deliberately easy to remove when posted onto the glass of a transport shelter hoarding. This action causes minimal damage. In fact, it causes no damage to property at all, apart from the inconvenience of a person having to be employed to clean the glass, should the corporate advertiser choose to reveal the billboard, still intact, beneath it. As argued by the Appellant no more damage is caused than when a person chooses to stand in front of the billboard, causing some visibility of its message to be lost.[36] The cost of removal of Magee's refacing was estimated by one court as standing at $40.17. This action resulted in three months jail time, costing the tax payer an average of $328.10 per day in the state of Victoria,[37] not to mention the untold harms of jail time on many young men.[38]

The hoardings containing the advertising messages are placed in public spaces, where in a free and liberal democracy all individuals should have the same rights to freedom of expression. These rights should not be limited to those who have the financial resources to pay for them. Further, the rights of expression of a multinational corporation should not be privileged over the rights of a citizen of the state of Victoria, the protection of whose rights should be paramount under the Act. Indeed the Act states that the definition of persons is 'human beings' and therefore freedom of expression is not protected for corporations at all under this Act. It does not follow that a mere inconvenience, caused by damage 'particularly low on the scale of criminal activity'[39] should result in the conviction of Mr Magee resulting in a jail term.

Mr Magee was merely displaying a community announcement; he is not a commercial enterprise, and it would therefore be impossible for the court

---

34 *Pointon v Police* [2012] NZHC 3208.
35 Ibid., citing Sedley LJ in *Redmond-Bate v Director of Public Prosecutions* [1999] EWHC Admin 733.
36 *Magee v Wallace* [2016] VSC 123.
37 J Thomas, 'How much does it cost to keep people in Australian jails?' *SBS News* (online), 4 February 2015 <www.sbs.com.au/news/article/2015/02/02/how-much-does-it-cost-keep-people-australian-jails>.
38 D Heilpern, *Fear or Favour: Sexual Assault of Young Prisoners* (Southern Cross University Press, 1998).
39 *Re Kyle Magee* [2009] VSC 384 [27] (Forrest J).

to begin to argue that Magee is guilty of the more serious offence contained in section 10(3) of the Act. In terms of the consent which must be obtained as a defence to the offence of bill posting – the bill posting was conducted within a public space, and therefore it could be considered that the consent of the public must be garnered before such bill posting can occur. However, bill posting has already occurred at this site without public consent – in the form of a corporate billboard. These advertisements from such multinational corporations are a blight on our public spaces, creating an eye-sore, and inflicting a commercial aesthetic onto the beautiful streets of Melbourne.[40] In the interests of protecting the environment from the damage caused by over-consumption, I, for one, would rather see these offensive displays removed altogether.

While it was stated in *Magee v Wallace* that 'the capacity of individuals or businesses to obtain access to property to advertise their products or services is a feature of Australian society',[41] it is not an intrinsic feature. The fact that there are advertising hoardings is not in itself a strong enough argument for their continued display. There are certainly parts of the world, even those of a strong capitalist ethos, that have chosen to restrict billboard advertising. For example, São Paulo long ago passed legislation banning billboards.[42] It is reported that this ban is still in place and has been successful in:

> [R]ecover[ing] certain fundamental rights of citizenship that had been lost over time. The right to live in a city that respects the urban space, the heritage and the integrity of the . . . architecture. The right to a freer and more secure relationship with the public areas. The Clean City Law means the supremacy of the common good [over] any corporate interest.[43]

In the US, there have been many 'chapters' in the fight against billboards. In the 9th Circuit Court of Appeal, Justice Wardlaw held that the question was no longer whether billboards could be restricted, but how much they could be restricted. In the judgment, Justice Wardlaw refers to precedent

---

40 See the case of *DPP v Shoan* [2007] VSCA 220, where Justice Buchannan held that Noam Shoan, a graffiti artist had 'defaced and rendered ugly a great deal of the scenery that people pass by. At the very least, he unilaterally imposed his notions of art and decoration on the rest of the world'. For a further discussion of the power relationships behind the aesthetics of the city, see Bird, above n 11.

41 *Magee v Wallace* [2016] VSC 123 [44].

42 Lei Cidade Limpa or 'Clean City laws' (Act Municipal no. 14,223 / 06).

43 Clean City Law <ww2.prefeitura.sp.gov.br/cidadelimpa/conheca_lei/conheca_lei.html> (trans., Google).

which holds that non-commercial speech has a greater right to be displayed in public spaces that commercial messages:

> In accord with decades of judicial guidance, the City's Sign Ordinance, Article 4.4 of Chapter 1 of the LAMC, *regulates commercial speech far more extensively than it does non-commercial speech.*[44]

First Amendment rights in the US protect the voices of human persons more vigorously than those of corporate persons. While Australia does not have a Bill of Rights at federal level, the Victorian *Charter* was enacted to have much the same effect within the state of Victoria. In following this international precedent, Mr Magee's expression should be allowed a greater freedom than any commercial message.

Further, in *Magee v Wallace*, it was stated that Mr Magee's actions 'have the potential to interfere with public order'.[45] The interference with public order is perceived to stem from his actions leading to 'some form of public disturbance involving persons seeking to stop these actions'.[46] Kyle Magee has been engaging in activities involving the refacing of billboards since he was 21.[47] He has therefore been engaging in his protest for a period of close to 12 years. In this time, he has found quite a large amount of support from onlookers.[48] The only people who have sought to stop him have been officers of the law. These members of the constabulary have found Mr Magee to be polite and courteous. He has never resisted arrest or caused any interruption approximating a 'public disturbance' through his actions.

In the Canadian case *Peterborough (city) v Ramsden*,[49] Justice Iacobucci dismissed an appeal by the Peterborough City Council which argued that the restrictions on bill posting should be enforced even though bill posting had been found to be a form of expression protected under the *Canadian Charter of Rights and Freedoms*. It was held that although some inconvenience was caused by the bill posting, this was not great enough to override an important right under the *Charter*. In the case, Mr Ramsden was displaying posters advertising a performance by his band. It is argued here that messages of a political nature would be more capable of establishing a balance in their favour over any inconvenience caused, in particular in Australia,

---

44 See *Metro Lights* 551 F 3d 906 n 9 ('[W]e have held that the Los Angeles Sign Ordinance only prohibits commercial offsite signs'.); *Outdoor Sys, Inc v City of Mesa* 997 F 2d 604, 613 (9th Cir 1993), Bird J emphasis.

45 *Magee v Wallace* [2016] VSC 123 [44].

46 Ibid.

47 Bird and Vakalis, above n 8, 163.

48 <www.globalliberalmediaplease.org>.

49 *Peterborough (City) v Ramsden* [1993] 2 SCR.

where the right to freedom of political speech has been found to exist as an implied right in the *Australian Constitution Act 1900* (UK). Moreover, Mr Magee's posters are displayed in locations that already contain messages, and therefore cannot be a distraction or safety hazard to workers as was argued in *Peterborough*.

Indeed it could be contended that Mr Magee is increasing public safety through his refacing of billboards, given the known impacts that overconsumption have on the environment, and hence, the public. He is engaging in a public service in his mission to raise awareness of environmental concerns. He does not seek any reward from this project, but does it completely out of an ethic of altruism.

In this day and age it is becoming increasingly difficult for the courts not to recognise the rights of the environment, either as an extension of human rights, or as an autonomous right. Movements to achieve environmental rights as an autonomous set of rights are gaining momentum worldwide. Ecuador has, for example, enshrined the rights of nature in its Constitution. A Universal Declaration of the Rights of Mother Earth, drafted in 2008 at the People's Conference in Bolivia, has been submitted to the UN General Assembly for consideration. Burns Weston and David Bollier are amongst a growing community of scholars who are looking to expand human rights to include a right to the environment. For this to be effective, the law needs to move away from individualistic rights discourse, and toward a communitarian ecological governance:[50]

> Increasingly international human rights and environmental law scholars and practitioners are calling for or seriously entertaining an 'expansive' right to environment as a means to enhance environmental protection. They do so, understandably, out of concern over current scientific forecasts, but also out of dissatisfaction with 'traditional' international legal process which, they persuasively argue, is not up to the ecological challenges now facing the planet.[51]

Mr Magee is not the first person to appear before an Australian court for refacing a billboard. In the 1980s, a campaign of refacing billboards was conducted in protest against tobacco advertising.[52] Numerous members

50 Weston and Bollier, above n 31.
51 Ibid., 18.
52 Simon Chapman quotes the classic BUGA-UP argument for refacing billboards in 'Civil disobedience and tobacco control: the case of BUGA-UP' (1996) 5 *Tobacco Control* 179. Chapman wrote: 'What is the worse crime? To vandalise the paper sheeting on an advertising hoarding, or to meekly accept the right of wealthy corporations to promote carcinogenic products to children?'

appeared before the courts, before this direct action led to a change in the law. In 1993, three women refaced a billboard in Sydney, and coming before Magistrate O'Shane, had their charges dismissed. In the reasoning for the case, O'Shane argued that:

> The real crime in this matter was the erection of these extremely offensive advertisements. . . . And what redress does . . . the population have? Absolutely none . . . because of the massive power that is exercised through huge financial resources. It is an absolute outrage.[53]

In this case, Magistrate O'Shane acknowledged that the billboards are erected by those with the financial resources to do so, with the general public being given few options of redress. She was able to weigh up the level of offending with the offence caused by the billboard and find that the women involved should not be charged under section 195 of the *Crimes Act 1900* (NSW) with destroying or damaging property.

It was held in the lower courts by Magistrate Mealy that Mr Magee has several options available to him in protesting against capitalism and for an end to environmental destruction. I would beg to differ. As Mr Magee rightly asserts, capitalist interests have control over the media, and do not give airtime to stories that critique their sponsors. Overwhelming concern for environmental destruction in the scientific community has thus far led to little change from law makers who generally share a neoliberal, economic rationalist orientation. Even where there have been attempts at reaching agreement, this has not led to any great change in human activities. There have been over 300 multilateral treaties and 900 bilateral treaties on the biosphere alone.[54] During the time that these treaties were signed, the environmental situation has deteriorated. The *Environment Protection Act 1970* (Vic) has been in operation for nearly 50 years but has done little to alleviate environmental destruction. Indeed, in section 1B, this Act states that 'the measures adopted should be cost-effective and in proportion to the significance of the environmental problems being addressed'.

It seems that, as asserted by Mr Magee, we need to get to the heart of the problem, which is at the very core of our political system:

> [I]t all comes down to a simple but profound truth: that as long as ecological governance remains in the grip of essentially unregulated (liberal or neoliberal) capitalism—a regime responsible for much if not most of the plunder and theft of our ecological wealth in the last, roughly

53 Ibid.
54 Weston and Bollier, above n 31.

338 Criminal law and environmental activism

150 years—there never will be a human right to environment widely recognized and honored across the globe in any formal/official sense, least of all an autonomous one.[55]

If we are to preserve the planet for future generations, a new political system is needed. As is written by James Gustave Speth in *The Bridge at the End of the World*, what is needed is 'a revitalisation of politics through direct citizen participation'.[56] Mr Magee is making an attempt here at direct participation, through his peaceful protest in public space.

In allowing this appeal, my conclusions arise from the principle that environmental rights should be given precedence over the rights of the corporation. While Australia does not recognise environmental rights as an extension of human rights, I think that it would be timely to do so. In order to continue living on this planet, humans will need to take quick and decisive action to protect it. I would like to see the rights of the environment enshrined in the Victorian Charter. Until that time, we rely on the work of individuals such as Kyle Magee in raising vital awareness of these issues.

The Appeal is allowed with costs.

55 Ibid., 25
56 J G Speth, *The Bridge at the End of the World: Capitalism, the Environment and Crossing from Crisis to Sustainability* (Caravan Books, 2008) 86, 225.

Chapter 21

Duck rescuers and the freedom to protest

Levy v Victoria

Nicole Rogers

Commentary

It was the same every time: a wobbly formation of creatures in slow flight would appear and wend its way down a corridor of death. Shotguns blasted from every direction. One or two victims would fall like wet rags. If the first series of shots missed the birds, they would be blasted out of the air further down the line.[1]

Lawrence Levy has actively campaigned for a Victorian ban[2] on the shooting of native birds for the last thirty years. He and other activists have been present at the opening weekend of the Victorian annual duck shooting season since 1986,[3] in order to rescue wounded birds, retrieve dead birds for display outside politicians' offices and the Victorian Parliament House and attract media and in particular television coverage of the plight of native water birds. In that thirty year period, the number of licenced shooters in Victoria has dropped from 100,000 to 26,000. There has also, however, been a decline in the number of native water birds and their wetland habitat is now seriously compromised.[4]

In 1994, Levy was charged with three offences under Regulation 5 of the *Wildlife (Game) (Hunting Season) Regulations 1994* (Vic), on the basis that he had entered a designated hunting area at Lake Buloke during the opening weekend of the season without an authority to do so. Under Regulation 5, only licenced shooters were permitted to enter hunting areas during this

---

1 I Cohen, *Green Fire* (Angus and Robertson, 1997) 216.
2 New South Wales, the Australian Capital Territory, Western Australia and Queensland have all banned duck shooting.
3 S Carbone, 'Banned: for the first time in 30 years "waterbirds hero" is silenced' *The Age* (online), 16 December 2015 <www.theage.com.au/victoria/banned-for-the-first-time-in-30-years-waterbirds-hero-is-silenced-20151216-glow0b.html>.
4 Ibid.

period.[5] Levy brought an action in the High Court, seeking a declaration that the Regulations infringed the constitutional implied freedom of political communication as found in both the Commonwealth Constitution and the Victorian Constitution.[6] He argued that the restrictive provisions in the Regulations impeded and reduced the effectiveness of his protest activities. In his amended Statement of Claim, he focused on the validity of Regulation 5, and its blanket ban on entry to hunting areas during the critical period. The defendants demurred to the whole of Levy's amended Statement of Claim and consequently the case was heard as a demurrer, which restricted the factual evidence and background material which could be raised in argument.

The case of *Levy v Victoria*[7] needs to be considered in context as one of many court battles fought by the seemingly indefatigable activist. In 2015, after being banned from the wetlands for six months, Levy commented that 'we've fought so many court cases over the years, I've lost track of them all. It's another battleground. We fight on the wetlands and we fight in the court'.[8] At the same time, it is viewed as a highly significant Australian constitutional law case, part of the implied freedom of political communication jurisprudence which has been evolving since 1992.

A: *Levy v Victoria* and the implied freedom of political communication jurisprudence

Levy was heard only five years after the High Court had uncovered an implied freedom of political communication in the Australian Constitution, in 1992. The timing of the *Levy* case, as one of the two 'fourth generation' cases[9] in the implied freedom of political communication jurisprudence which followed the expansive decisions handed down in the first[10] and second

---

5 Under the current regulatory framework, duck rescuers are excluded from specific hunting areas for the entire open season; *Wildlife (Game) Regulations 2012* (Vic) Reg 70.
6 *Constitution Act 1975* (Vic).
7 *Levy v Victoria* (1997) 189 CLR 579.
8 Carbone, above n 3.
9 *Lange v Australian Broadcasting Corporation* (1997) 189 CLR 520; *Levy v Victoria* (1997) 189 CLR 579. I have adopted the terminology of Anne Twomey in describing these as fourth generation cases; A Twomey, 'Dead ducks and endangered political communication – *Levy v State of Victoria* and *Lange v Australian Broadcasting Corporation*' (1997) 19 *Sydney Law Review* 76. She describes the 'third generation' cases in 1996, *Langer v Commonwealth* (1996) 186 CLR 302 and *Muldowney v South Australia* (1996) 186 CLR 352, as representing an initial scaling back of the implied freedom.
10 *Australian Capital Television v Commonwealth* (1992) 177 CLR 106; *Nationwide News v Wills* (1992) 177 CLR 1.

generation cases,[11] was perhaps unfortunate for Levy given the diminished scope of the implied freedom in those later decisions.[12] It was also unfortunate from a practical perspective for an indigent activist relying on pro bono representation. The application by the Solicitor-General of Victoria to reopen the decisions in the 'second generation' cases was heard concurrently with the *Lange* case[13] and involved the intervention of the Attorney-Generals of the Commonwealth, States, Territories and a number of media organisations, thus requiring both Levy and his barrister to participate in a protracted hearing involving much broader issues than originally envisaged in the demurrer.[14]

The unanimous decision in the *Lange* case[15] was handed down some three weeks before the decision in *Levy* and in the earlier decision, the two step test for determining whether a particular piece of legislation was invalidated by virtue of the implied freedom was conclusively established. This was, therefore, the test applied by all the judges who chose, in contrast to their unanimity in *Lange,* to write separate judgments[16] but reached the same conclusion: the Victorian Regulations were valid despite the burden imposed upon the freedom of political communication and, according to Justice Gaudron, upon an implied freedom of movement.

As two writers commented in their review of 20 years of implied freedom jurisprudence, 'any fears that an activist High Court would wield the implied freedom as a sword with which to cut swathes through laws and regulations deemed undesirable have not materialised'.[17] The Court's interpretation of the second limb of the so-called *Lange* test ensured that challenges on the basis of the implied freedom failed in the *Levy* case and in the cases which followed until, in 2013, the Court invalidated legislation[18] for the first time since 1992.

---

11 *Theophanous v Herald and Weekly Times Ltd* (1994) 182 CLR 104; *Stephens v West Australian Newspapers* (1994) 182 CLR 211.

12 The two cases have been described as 'marking the end of an activist period in the Court's history'; G W Anderson, 'Corporations, democracy and the implied freedom of political communication: towards a pluralistic analysis of constitutional law' (1998) 22 *Melbourne University Law Review* 1, 2.

13 *Lange v Australian Broadcasting Corporation* (1997) 189 CLR 520.

14 See transcript *Levy v State of Victoria & Ors* M42/1995 [1996] HCATrans 393 (1 October 1996).

15 *Lange v Australian Broadcasting Corporation* (1997) 189 CLR 520.

16 With the exception of Justices Toohey and Gummow, who wrote a joint judgment.

17 T Campbell and S Crilly, 'The implied freedom of political communication, twenty years on' (2011) 30 *University of Queensland Law Journal* 59, 66.

18 In *Unions NSW v New South Wales* (2013) 252 CLR 530.

B: The Levy case and environmental protest

In 1996, the year before the *Levy* decision was handed down, political activist Albert Langer had failed to convince the Court to invalidate section 329A of the *Commonwealth Electoral Act* 1918 on the basis of the implied freedom of political communication.[19] However, the *Levy* case was the first attempt by an environmental protester to use the implied freedom of political communication to challenge legislation which appeared to be, at least in part, designed to silence environmental protest.

In the case, the High Court judges held that political protest of all forms, including non-verbal protest which relied on spectacle, imagery and performance rather than rhetoric but also including 'false, unreasoned and emotional' forms of communication,[20] was encapsulated within the term political communication. *Levy* is, in this latter sense, the precursor to subsequent cases[21] in which, as Adrienne Stone has observed,[22] the Court has adopted what she terms an 'anti-civility' stance in interpreting political communication such that it can include 'insult and emotion, calumny and invective'.[23] Stone speculatively attributes this jurisprudential development to the idiosyncratic quality of Australian political culture and, in particular, to our cultural tradition of 'irreverence' in the face of authority.[24] As Iain McIntyre has documented, there is a great Australian tradition of protest[25] which is, arguably, acknowledged in this expansive reading of political communication.

Images and spectacle create the most potent forms of political communication for protesters. In acknowledging the central importance of spectacle and imagery in political communication by protesters, the Court in *Levy* also recognised the pre-eminent role of television as a mechanism to transmit such messages. Justice McHugh, in particular, described television as 'the most effective medium in the modern world for communicating with

19 *Langer v Commonwealth* (1996) 186 CLR 302.
20 *Levy v Victoria* (1997) 189 CLR 579, 623 (McHugh J). This interesting juxtaposition of adjectives suggests that emotional and unreasoned messages are false, by way of contrast with rational and reasoned forms of communication.
21 In particular *Coleman v Power* (2004) 220 CLR 1 and *Roberts v Bass* (2002) 212 CLR 1.
22 A Stone, 'Insult and emotion, calumny and invective: twenty years of freedom of political communication' (2011) 30(1) *University of Queensland Law Journal* 79.
23 *Coleman v Power* (2004) 220 CLR 1, 91 (Kirby J).
24 Stone, above n 22, 97.
25 I McIntyre, *How to Make Trouble and Influence People: Pranks, Hoaxes, Graffiti and Political Mischief Making from Across Australia* (Breakdown Press, 2009).

large masses of people'.[26] Today the proliferation of social media options such as Instagram, Facebook and Twitter have rendered television less significant as a vehicle for political communication but reinforced the ongoing importance of spectacle and imagery for protesters and activists.

The Court's recognition of protest as a form of political communication was an important and promising development. Nevertheless, by 1997 commentators had already identified a corporatist bias in the implied freedom of political jurisprudence,[27] and in subsequent decisions[28] protesters and dissenters would find that the implied freedom offered them little protection. Michael Head has argued that three cases,[29] in particular, 'reveal the enormous scope for governments to gag political opponents and block political protests in the name of "legitimate" official objectives'.[30] The outcome of the *Levy* case was disappointing for animal rights activists and activists generally, but representative of the ongoing conservatism in the implied freedom of political communication jurisprudence.

In the *Levy* case, it was the judges' application of the second step of the *Lange* test and their reading of the term 'legitimate end' which ensured that legislation which criminalised protest events would nevertheless be valid, provided that it could be justified as protecting public safety or even, more generally, upholding public order. In fact, according to Chief Justice Brennnan, non-verbal conduct may by its very nature require more legislative regulation than the speaking of words, which is not 'inherently dangerous'.[31]

Legislation designed to prevent and criminalise protest is increasingly a feature of the Australian political landscape.[32] It is a sobering thought

---

26 *Levy v Victoria* (1997) 189 CLR 579, 623 (McHugh J).
27 See, for instance, A Fraser, 'False hopes, implied rights and popular sovereignty in the Australian Constitution' (1994) 16 *Sydney Law Review* 213. For a slightly later critique, see G W Anderson, 'Corporations, democracy and the implied freedom of political communication: towards a pluralistic analysis of constitutional law' (1998) 22 *Melbourne University Law Review* 1.
28 See *Brown v Classification Review Board* (1998) 82 FCR 225 and *Muldoon v Melbourne City Council* (2013) 217 FCR 450.
29 *Attorney-General (SA) v Corporation of the City of Adelaide* (2013) 249 CLR 1, *Monis v the Queen* (2013) 249 CLR 92 and *Wotton v Queensland* (2012) 246 CLR 1.
30 M Head, 'High Court further erodes free speech' (2013) 38(3) *Alternative Law Journal* 147, 151.
31 *Levy v Victoria* (1997) 189 CLR 579, 595 (Brennan CJ).
32 Examples include the *Summary Offences and Sentencing Amendment Act 2014* (Vic), the *Workplaces (Protection from Protesters) Act 2014* (Tas), the *Inclosed Lands, Crimes and Law Enforcement Amendment (Interference) Act 2016* (NSW) and the *Criminal Code Amendment (Prevention of Lawful Activity) Bill 2015* (WA); these recent legislative initiatives are discussed in A Ricketts, 'Freedom from political communication: the rhetoric behind anti-protest laws' (2015) 40(4) *Alternative Law Journal* 234.

that a different outcome in the *Levy* case could have prevented such legislative developments.

C: *The rewriting*

I deliberately made use of emotive terms in rewriting the judgment. For instance, in the opening paragraph, I describe the practice of duck shooting as barbaric, even though judges rarely adopt such value-laden terminology. The deployment of abstract principle as a distancing and neutralising technique in High Court constitutional law judgments has been remarked upon by a number of commentators, but its use and deficiencies in the context of genocide are, perhaps, best summed up by Margaret Thornton in her reference to the constitutional law decision on the Stolen Generation, *Kruger v Commonwealth*:[33]

> The sorrow of the Aboriginal 'Stolen Children' evaporates in the face of a legalistic excursus on the legislative scope of the Territories power.[34]

A judge unswayed by anthropocentric assumptions would view and judge the mass murder of animals with the same abhorrence with which she would view and judge the mass murder of human beings.

In the *Levy* case, much initial argument in the proceedings centred on the question of whether the implied freedom of political communication fettered State legislatures.[35] The judges did not find it necessary to decide this question, although Justice Kirby did comment that it was 'neither fanciful nor unreasonable' to assume that State legislatures were subject to the same limitations in relation to freedom of political communication as the Federal Parliament.[36] In my rewriting, given that I reached a different conclusion about the invalidity of the legislation, it was necessary to address this issue. I address it as expeditiously as possible; in later cases,[37] the Court would in fact establish that State legislation was subject to this constraint.

My reasoning in relation to the inclusion of protest activities within the ambit of political communication is consistent with the conclusions of the High Court judges. In the original judgments, however, there is no reference to the synergy which I have identified between protest activities and

---

33 *Kruger v Commonwealth* (1997) 190 CLR 1.
34 M Thornton, 'Towards embodied justice: wrestling with legal ethics in the age of the "new corporatism"' (1999) 23 *Melbourne University Law Review* 749, 756.
35 See transcript <www.austlii.edu.au/au/other/HCATrans/1996/294.html>.
36 *Levy v Victoria* (1997) 189 CLR 579, 644 (Kirby J).
37 *Coleman v Power* (2004) 220 CLR 1; *Unions NSW v NSW* (2013) 252 CLR 530.

a healthy and functioning democratic system of government. In fact, by the time *Levy* was decided, the Court had modified its original statements[38] in relation to representative democracy and popular sovereignty as fundamental concepts underpinning the Constitution and forming the justification for the implied freedom.[39]

I have also departed from the reasoning in the original judgments in my interpretation of the so-called legitimate end purportedly served by the Regulations: the protection of public safety. In this section of my judgment, I interpret the concept of public safety more broadly. My more expansive reading of this concept, to encompass ecological integrity and the interdependence of humanity on ecosystems and other species, is anchored in contemporary understandings of the importance of biodiversity conservation, but today would incorporate the added insights contributed by climate change science and the now prevalent realisation that the future of humanity is inseparable from the future of the planet.[40] As Pablo Solon has recently written:

> Why do we call the person who kills his neighbor a criminal, but not he who extinguishes a species or contaminates a river? Why do we judge the life of human beings with parameters different from those that guide the life of the system as a whole if all of us, absolutely all of us, rely on the life of the Earth System?[41]

Although the goal of public safety is usually perceived as an anthropocentric one, it is possible, as I have sought to do in this rewriting, to read public safety to encompass the wellbeing of non-human entities.

---

38 See, for instance, *Nationwide News v Wills* (1992) 177 CLR 1, 47–8 (Brennan J), 70 (Deane and Toohey JJ); *Australian Capital Television v Commonwealth* (1992) 177 CLR 106, 137 (Mason CJ), 149, 156 (Brennan J), 210–11 (Gaudron J); *Theophanous v Herald and Weekly Times Ltd* (1994) 182 CLR 104, 123 (Mason CJ, Toohey and Gaudron JJ), 180 (Deane J).

39 In *Lange v Australian Broadcasting Corporation* (1997) 189 CLR 520, 557, the Court referred to the specific sections in the Constitution which established representative and responsible government, rather than to general principles of representative democracy and popular sovereignty, as the constitutional basis for the implied freedom.

40 See, for instance, B Kershaw '"This is the way the world ends, not . . .?" On performance compulsion and climate change' (2012) 17 *Performance Research* 5.

41 P Solon, 'At the crossroads between green economy and rights of nature' in T B K Goldtooth and S Biggs (eds), *Rights of Nature and Mother Earth. Sowing Seeds of Resistance, Love and Change* (Movement Rights, Indigenous Environmental Network and Global Exchange, 2015) 11 <http://movementrights.org/resources/RONME-SowingSeeds.pdf>.

HIGH COURT OF AUSTRALIA
Rogers J
LAURENCE NATHAN LEVY PLAINTIFF
AND
THE STATE OF VICTORIA AND ORS DEFENDANTS
ORDER

1 Defendants' demurrer disallowed.
2 The defendants pay the plaintiff's costs.
31 July 1997

Lawrence Levy is well known as an ardent and committed animal rights activist. He is the campaign director of the Coalition against Duck Shooting. Since 1986, he has attended the hunting grounds during the opening weekend of the annual Victorian duck shooting season, retrieved and assisted wounded birds, collected deceased birds and arranged their corpses outside the Victorian Parliament house, and resolutely confronted shooters. The wounded and deceased birds include members of protected species. Levy's actions have undoubtedly raised public awareness in relation to the needless suffering and murder of wild ducks. Shooters regard him as an irritant but others uphold him as a hero of the Australian environmental movement and applaud his fearless and determined attempts to end the barbaric practice of recreational shooting of wild ducks.

On 7 June 1994, Levy was charged with three offences under Regulation 5 ('the Regulation') of the *Wildlife (Game) (Hunting Season) Regulations 1994* (Vic) ('the Regulations'). The Regulation prohibits entry by all except licenced duck shooters to the waters of any permitted hunting areas during the opening weekend of the duck shooting season in 1994. Regulation 6 imposes further constraints in that it prevents any person from coming within five metres of a duck shooter. The Regulations were made pursuant to the *Wildlife Act 1975* (Vic) and the *Conservation, Forests and Lands Act 1987* (Vic).

Levy commenced proceedings in this court, arguing that the Regulations as a whole were invalid in that they infringed the implied freedom of political communication, as found in the Australian Constitution and the Victorian Constitution.[42] He argued in his amended Statement of Claim that Regulation 5, rather than the Regulations in their entirety, impeded his right to protest effectively against the licenced practice of recreational shooting of wild birds in Victoria. The defendants demurred to the amended Statement

42 *Constitution Act 1975* (Vic).

of Claim on four grounds; of these, the argument that the Regulations did not impair implied freedoms contained in either Constitution is the most pertinent. Before I turn to this critical issue, I shall consider the objection put forward by the defendants in relation to the application of the implied freedom of political communication, as derived from the Commonwealth Constitution,[43] to the law making power of the Victorian Parliament.

A: Applicability of implied freedom to the legislative power of the Victorian Parliament

This Court has previously held that the implied freedom derived from the Commonwealth Constitution confines the legislative powers of the State Parliaments. It is clear from earlier cases[44] that, as Chief Justice Mason stated in the *Australian Capital Television* case, 'public affairs and political discussion are indivisible and cannot be subdivided into compartments that correspond with, or relate to, the various tiers of government in Australia'.[45] This is certainly the case with the controversial issues of duck shooting specifically and animal welfare more broadly. Furthermore, the application of the implied freedom to State legislatures was confirmed in the *Stephens* case.[46]

I am also persuaded by the plaintiff's argument that an implied freedom of political communication can be found in the Victorian Constitution, in that that Constitution similarly establishes a system of representative and responsible government.[47] For these reasons, it is clear that the implied freedom of political communication does act as a constraint upon the legislative powers of the Victorian Parliament.

I shall turn now to the plaintiff's argument that the Regulation is invalid because it infringes the implied freedom of political communication. The question of whether the Regulation impairs the implied freedom of political

43 *Australian Capital Television v Commonwealth* (1992) 177 CLR 106; *Nationwide News v Wills* (1992) 177 CLR 1.
44 *Australian Capital Television v Commonwealth* (1992) 177 CLR 106, 142 (Mason CJ), 168–9 (Deane and Toohey JJ), 217 (Gaudron J); *Nationwide News v Wills* (1992) 177 CLR 1, 75–6 (Deane and Toohey JJ); *Theophanous v Herald and Weekly Times Ltd* (1994) 182 CLR 104, 122 (Mason CJ, Toohey and Gaudron JJ); *Stephens v West Australian Newspapers* (1994) 182 CLR 211, 232 (Mason CJ, Toohey and Gaudron JJ), 257 (Deane J).
45 *Australian Capital Television v Commonwealth* (1992) 177 CLR 106, 142 (Mason CJ).
46 *Stephens v West Australian Newspapers* (1994) 182 CLR 211.
47 *Constitution Act 1975* (Victoria) ss 16, 26 and 34.

communication requires application of the two step test recently enunciated by this Court in the *Lange* case.[48] The Court said:

> First, does the law effectively burden freedom of communication about government or political matters either in its terms, operation or effect? Second, if the law effectively burdens that freedom, is the law reasonably appropriate and adapted to serve a legitimate end, the fulfilment of which is compatible with the maintenance of the constitutionally prescribed system of representative and responsible government and the procedure prescribed by s 128 for submitting a proposed amendment of the Constitution to the informed decision of the people (hereafter collectively 'the system of government prescribed by the Constitution'). If the first question is answered 'yes' and the second is answered 'no', the law is invalid.[49]

Thus the test involves, firstly, consideration of whether the Regulation imposes a burden upon political communication.

B: *Burden upon freedom of political communication*

The plaintiff has argued that the Regulation has prevented his engagement in acts of political protest in a manner calculated to generate the maximum publicity in relation to the issue of recreational duck shooting and its violent impact on game birds and threatened species. The televised images of dead birds, retrieved by the plaintiff and his supporters from Victorian wetlands during the opening weekend of the shooting season and presented in rows outside the Victorian Parliament House, are intended to reach as large an audience as possible in order to generate public support for legislative change in relation to duck shooting. The Regulation prevents such footage from being obtained and aired.

There is no question that imagery and spectacle have provided one of the most potent forms of political communication for protest movements in the latter half of the twentieth century. It is rare to find, in any modern Western nation, an adult who has not been exposed to such iconic images as those of peace activists offering flowers to military police and national guardsmen. In the context of environmental protest, Greenpeace in particular has pioneered the use of dramatic spectacle in promulgating its message of environmental protection. In Australia, the rhetoric of activists, politicians and environmental organisations has been rendered vastly more persuasive by

---

48 *Lange v Australian Broadcasting Corporation* (1997) 189 CLR 520.
49 *Lange v Australian Broadcasting Corporation* (1997) 189 CLR 520, 567.

accompanying images of protesters in rubber duckies bravely confronting barges on the Franklin River, images of protesters atop tripods in old growth forests, images of protesters chained to bulldozers, and images of protesters on kayaks clinging to the sides of nuclear warships. Television has greatly added to the political impact of such images by bringing them on a nightly basis into family homes.

The plaintiff is part of a long and honourable tradition of political protest, a tradition which should be not only tolerated within any representative democracy, but recognised as essential to the proper functioning of such a democracy. Members of this Court[50] have acknowledged the importance of political communication in a representative democracy in which ultimate sovereignty resides with the people, but this form of political communication has particular significance. Non-violent protest and even civil disobedience frequently lead to important legislative and social reforms and provide a mechanism by which the people can have a voice and directly contribute to legal and political developments. One important example can be found in the protest activities and civil disobedience of the British suffragettes, whose actions eventually led to votes for women in Britain. Protest and various forms of civil disobedience undoubtedly constitute political communication but the synergy between such activities and a healthy and responsive representative democratic system of government should also be acknowledged.

The first limb of the Lange test is therefore satisfied and the Regulation, in prohibiting access to the wetlands during the critical period of the opening weekend and thus constraining the activities of protesters and media coverage of such activities, imposes a burden upon the implied freedom of political communication. However, in the evolving jurisprudence in this area, the Court has always emphasised that the implied freedom is not absolute. In applying the second limb of this *Lange* test, I must also consider whether the Regulation is reasonably appropriate and adapted to achieving a legitimate end conducive to the maintenance of the constitutionally prescribed system of representative and responsible government and is therefore valid irrespective of its impact on political communication.

---

50 See, for instance, *Nationwide News v Wills* (1992) 177 CLR 1, 47–8 (Brennan J), 72 (Deane and Toohey JJ); *Australian Capital Television v Commonwealth* (1992) 177 CLR 106, 138–40 (Mason CJ), 149 (Brennan J), 210–11 (Gaudron J); *Theophanous v Herald and Weekly Times Ltd* (1994) 182 CLR 104, 121–2 (Mason CJ, Toohey and Gaudron JJ), 180 (Deane J).

C: *Reasonably appropriate and adapted to serving a legitimate end*

According to Regulation 1, the purpose of the Regulations is to 'ensure a greater degree of safety of persons in hunting areas during the open season for duck in 1994'. In the same way that the Commonwealth Parliament cannot 'conclusively "recite itself" into power',[51] a State Parliament cannot avoid an implied constraint upon its powers by an assertion in legislation or regulation that the purpose of the enactment is protection of public safety. Such legislative statements cannot automatically confer validity. They are open to judicial scrutiny and oversight.

Mr Castan QC, counsel for the plaintiff, has drawn our attention to a number of considerations which suggest that the true purpose of Regulation 5 is not protection of public safety. There is no evidence or suggestion that people have been shot or are likely to be shot. Shooters shoot ducks in flight, rather than at ground level, which makes it unlikely that protesters retrieving wounded birds from the lake would be accidentally shot. Mr Castan has pointed out that shooters 'never shoot at a duck on the water because there are other shooters'. It is, in any event, difficult to comprehend how a regulatory regime ostensibly designed to protect the safety of persons in fact authorises recreational shooting, particularly in wake of the recent tragic events at Port Arthur which starkly highlighted the threat to public safety posed by gun ownership, and led to national gun control reform. It seems somewhat incongruous, at best, and misleading and deceptive at worst, that a regulatory regime which protects the rights of shooters, and seeks to deter those who publicise the bloody consequences of their activities, is purportedly made to ensure the safety of persons.

Nevertheless, the plaintiff has refrained from arguing that the Regulations, and specifically Regulation 5, were designed to achieve ends other than public safety. His arguments are confined to the appropriateness of the means used to achieve this purported end. I shall, therefore, address the question of whether or not the Regulation is appropriate and adapted to achieving this stated goal of the safety of persons. In my view, it is not.

I am sympathetic to the plaintiff's argument that a demurrer is not the appropriate means to consider whether the Regulation is appropriate and adapted to the end of the safety of persons, and that factual evidence needs to be adduced to in relation to this issue. However, a more contextualised reading of the concept of safety of persons, or human safety, permits me to bypass this difficulty. Admittedly, such a reading is not encompassed within the view of public safety traditionally adopted by the executive and accepted

---

51 *Australian Communist Party v the Commonwealth* (1951) 83 CLR 1, 205–6 (McTiernan J).

by the courts. This extended reading is, however, required if one adopts a planetary and ecocentric perspective, and the serious ecological challenges we face now and in the future, as a species, necessitate the adoption of such a perspective. From such a perspective, we must interpret the safety of persons and human safety in the broader context of the ongoing survival of the human species and preservation of ecological integrity rather than confine its application to the prevention of danger to individual human beings in hunting areas.

Shooters indiscriminately target both endangered and non-endangered species. Despite the introduction of the government's compulsory Waterfowl Identification Test in 1990, the bodies of threatened waterfowl such as freckled ducks are often retrieved by rescuers. Levy's activities were in part directed towards demonstrating that endangered and protected species of ducks were and are being shot in Victorian wetlands, thus posing a threat to ecological integrity. It is a well-established scientific fact that the wellbeing and survival of humanity is dependent upon existing ecosystems. As far back as 1982, in the *World Charter for Nature*,[52] the General Assembly of the United Nations acknowledged that 'mankind is a part of nature and life depends on the uninterrupted functioning of natural systems'. The connection between biodiversity conservation and the ongoing wellbeing of human beings and human communities is clearly outlined in the *Convention on Biodiversity*,[53] ratified by the Australian government in 1992, in the preamble to which is recognised the importance of biodiversity conservation 'for maintaining life sustaining systems of the biosphere'. The safety of persons is dependent upon the ongoing protection of ecosystems and conservation of biodiversity.

Various individuals throughout history have felt compelled to take risks and jeopardise their own safety, and indeed their own lives, in order to save the lives of others. In John 15:13, in the King James Bible, we are told that 'Greater love hath no man than this, that a man lay down his life for his friends'. These words appear on war memorials throughout Australia. Employees in organisations such as the Red Cross International and Human Rights Watch have undertaken such risks in order to bear witness to acts of injustice and tyranny and to render assistance to victims of massacres and genocide. Few would hesitate to acknowledge the role of these organisations in promoting public safety globally. Such acts of selfless heroism and

---

52 *World Charter for Nature*, GA Res 37/7, UN GAOR, 48th plen mtg, UN Doc A/51 (adopted 28 October 1982).
53 *Convention on Biological Diversity*, opened for signature 5 June 1992, 1760 UNTS 79; 31 ILM 818 (entered into force 29 December 1993).

altruism, whether carried out on behalf of these organisations or by individuals, are highly regarded both in our legal system and by human society generally.

Other species can and do frequently become victims in mass murders and massacres and some of us feel compelled to render similar forms of assistance to non-human victims and bear similar forms of witness to such acts. The appropriateness or indeed proportionality of the restrictions on political communication placed by the Regulation must be evaluated in light of such considerations. Such deeds are no less heroic if carried out on behalf of individuals of other species, rather than on behalf of human beings. Nor are such acts any less important to the protection of the safety of persons, when we acknowledge that the safety of persons requires the preservation of ecological integrity as well as the protection of human individuals from risk.

The Regulation is invalid because it effectively burdens the freedom of political communication and is not reasonably appropriate and adapted to serve the legitimate end of protecting the safety of persons in the hunting areas or public safety generally. The defendants have failed to establish the grounds set out in the demurrer and I would disallow the demurrer.

Part VI

Looking ahead

Chapter 22

Information environmentalism and biological data

A thought experiment

Robert Cunningham

Commentary*

Australasian Information Environmentalism Alliance v Commonwealth of Australia & Ors [2037] HC 12 (hereafter *AIEA* case) is concerned with the ongoing tension between private rights and public rights flowing from 'biological data' inherent within National Parks and other nature reserves.[1] This fictitious case seeks to explore possibilities of how this tension might manifest in the twenty-first century. It is set in the future: the year 2037.

The *AIEA* case is built upon two fictional legislative initiatives developed in the 2020s: namely, the *Biological Data Mining Act 2025* (Cth) (hereafter BDMA) and the *Information Commons Rights Act 2029* (Cth) (hereafter ICRA).[2] From a common law perspective, the case also draws upon the public trust doctrine, which was reinvigorated in the Australian common law context after the imaginary judgment of *Greenspace Ltd v Commonwealth of Australia* [2035] HC 7 (hereafter *Greenspace* case).

---

* The commentary and hypothetical judgment are built upon work found in R Cunningham, *Information Environmentalism: A Governance Framework for Intellectual Property Rights* (Edward Elgar, 2014).
1 Biological data is 'the biological information inherent within the physical environment of a given location'.
2 The *Biological Data Mining Act 2025* (Cth) is aimed at facilitating economic development by allowing private companies to own and otherwise exploit biological data within the physical environment, particularly National Parks and other nature reserves. The *Information Commons Rights Act 2029* (Cth) seeks to reconcile private interests and public interests in the information environment domain; and also to establish Information Commons Rights (ICRs) in juxtaposition to Intellectual Property Rights (IPRs). The *Native Information Title Act 2028* (Cth) is also relevant, albeit the discussion of this legislation is limited to the extent there is another relevant case pending in this Court (See Annex I). This legislation is aimed at establishing *sui generis* rights for Indigenous custodians in relation to traditional knowledge, customs and practices (among other things).

The 'ideas-battle' between private and public concerns, inherent within the *AIEA* case, is not new. In Roman law terms, the struggle can be framed with reference to *res nullius* and *res communis*. *Res nullius* was built on the idea that resources belong to no one and are therefore unclaimed and for the taking.[3] *Res communis*, on the other hand, was developed on the assumption that resources belong to everyone and therefore any use of such resources is required to be for the benefit of the public at large. In general, *res nullius* benefits the strong, whereas *res communis* favours the weak.

In the twenty-first century, the overriding question is not that different from a prevailing issue of Ancient Rome: how do we strike an appropriate balance between private and public interests? Given we reside in the information age, an important consideration when answering this question is to determine who has access to and use of information – particularly information that holds the potential to secure positive environmental outcomes. The information that underpins renewable energy technology is a case in point. Similar claims could be made in relation to 3D printing, biotechnology, nanotechnology, and the list continues.

In essence, there is a symbiotic relationship between looking after the health of the information environment and securing a sustainable physical environment.[4] In particular, a sustainable future will require access to and use of an information commons, which embeds useful information that can in turn be accessed or used to secure positive environmental outcomes.[5] The concept of information here is broad, encompassing a wide range of human activities embracing everything from culture to innovation. The main qualification here, as alluded to in the *AIEA* case, is that Indigenous custodians maintain *sui generis* rights over their cultural customs, practices, language and lore.[6]

It is the symbiotic relationship between the information environment and the physical environment that ultimately led to the emergence of the

---

3 In the historical *Mabo* decision, *Mabo v Queensland (No 2)* (1992) 175 CLR 1, we saw this doctrine was largely undermined by Native Title claims.

4 The 'information environment' can be thought of as those processes and procedures relating to the production, consumption, communication, distribution and diffusion of information. The 'physical environment' can be simply thought of as the surroundings or conditions in which an animal, plant or human lives.

5 The 'information commons' embodies information that can be accessed and used by the public. For a nuanced perspective of the information commons, see R Cunningham, 'The Tragedy of (Ignoring) the Information Semicommons: a Cultural Environmental Perspective' (2010) 4 *Akron Intellectual Property Journal* 1.

6 *Sui generis* means 'one of a kind'. The term is often used to describe intellectual property right protection flowing from a new type of technology (e.g., computer software).

'information environmentalism' movement during the 2000s.[7] As observed in the *AIEA* case, this political-economic movement provides the foundation for many of AIEA's activities.[8]

In many ways, information environmentalism evolved as a counter-reaction to the broadening and deepening of information ownership, which transpired partly as a consequence of the WTO TRIPS Agreement in 1995 (and subsequent developments). The activities of Biodata Extraction and Excavation Ltd (hereafter BEEL), discussed within the *AIEA* case, symbolise this 'maximalist dynamic' of information ownership.[9] But information environmentalism has never just been about lamenting the dark by underscoring maximalist information ownership; it is also about lighting candles by building a positive programme to address underlying anxieties.[10] Open source software initiatives,[11] along with Creative Commons and other such projects, symbolise this positive approach.[12]

7 Initially, the theoretical foundations of the information environmentalism movement, and the accompanying development of Information Commons Rights, were built on scholarship that emerged from the 1980s onwards. For example, David Lange submitted in a seminal article that 'recognition of new intellectual property interests should be offset today by equally deliberate recognition of individual rights in the public domain'; see D Lange, 'Recognising the Public Domain' (1981) 24 *Law and Contemporary Problems* 147. This work has been built upon in R Cunningham, *Information Environmentalism: A Governance Framework for Intellectual Property Rights* (Edward Elgar, 2014).

8 AIEA is a not-for-profit organisation established to leverage the information commons in order to conserve the physical environment.

9 BEEL is an Australian Stock Exchange listed company, which seeks to profit from the private ownership of biological data by selling this data to pharmaceutical and biotechnology companies.

10 Along with claiming 'deliberate recognition of individual rights in the public domain', Lange also suggested the public domain should be a 'valued sanctuary for unprotected materials', inferring a positive articulation of the public domain provides a powerful antidote to the *res nullius* IPR ideology. See Lange, above n 7 and D Lange, 'Reimagining the Public Domain' (2003) 66 *Law and Contemporary Problems* 463, 470.

11 Open source software is a type of computer coding based on common-based social production methods. See R Cunningham, 'The Social Ecology of Information Ecosystems: Computer Code, Copyleft and the General Public Licence' (2006) 24(4) *Copyright Reporter* 208.

12 Lange has not been alone in advocating for a positive view of the information commons. Peter Drahos, for instance, discussed the significance of 'preservationist duties' with respect to the intellectual commons, suggesting 'duties of nurture' can be transposed; P Drahos, *Philosophy of Intellectual Property* (Dartmouth, 1996) 64. Furthermore, seemingly stemming from the information environmental work of James Boyle and others, Kahle purportedly posed the following question in a 2006 workshop discussion: 'Is there room to establish the Yosemite Park of information in Europe?'; see R Melzer and L Guibault, 'Workshop Discussions' in L Guibault and P B Hugenholtz (eds) (2006), *The Future of the Public Domain: Identifying the Commons in Information Law* (Kluwer Law International, 2006) 347, 351.

On the theoretical front, information environmentalism has sought to establish a 'governance framework', which builds upon past lessons from contemporary environmentalism. This work highlights that the physical environment and the information environment are both 'public interest' emblems. Detrimental action towards either of these environments suggests a tendency towards violating the public interest. In other words, a similar dynamic is at play when injury is imposed on the physical environment and the information environment. So how might we arrest this process?

Of course, political action is often required to protect the public interest. Yet action alone is not enough. *Appropriate* action is required. Theoretical governance frameworks can assist in this respect by setting the right direction upon which to travel. While the environmental movement has not been completely successful, it has had some meaningful gains. And these gains have mostly been built upon theoretical foundations, either consciously or unconsciously.

There are many ways of representing the contemporary environmental movement. But there are four theoretical frameworks that broadly represent the movement: welfare economics, the commons, ecology, and public choice theory. Applying each of these theories to the information environment assists in diagnosing and resolving issues of the digital age; and importantly, from a wild law perspective, will serve to create a more sustainable future.

Again, the physical environment and the information environment are entirely inter-relational, and will become more so as the twenty-first century evolves.

The lessons of applying each of the four analytical frameworks to the information environment have been dealt with elsewhere and so will not be rehearsed here. However, the following lessons are worth underscoring:

- Propertisation tends to conceal ethical considerations (welfare economics).[13]
- Before something can be protected it must be defined (the commons).[14]
- Decentralised and deliberative decision-making leads to positive outcomes (ecology).[15]
- In particular circumstances, social-based production can be more efficient than the State the market and/or the firm in producing goods and services (public choice theory).[16]

---

13 Cunningham, above n 7, pt I.
14 Ibid., pt II.
15 Ibid., pt III.
16 R Cunningham, 'The Separation of (Economic) Power: A Cultural Environmental Perspective of Social Production and the Networked Public Sphere' (2010) 11 *Journal of High Technology Law* 1.

Underpinning all of these lessons is the importance of protecting, nurturing and developing the information commons. This is at the heart of AIEA's activities. And it is also in contra-distinction to the activities of BEEL. In short, AIEA adopts a *res communis* point of view, whereas BEEL operates in accordance with *res nullius*.

The core claim of AIEA, and the information environmentalism movement more generally, is thus: in order for the symbiotic relationship between the physical environment and the information environment to flourish, it is important to protect, nurture and develop the information commons in line with the principles of *res communis* – a resource that belongs to everyone must be used for the public benefit. Of course, a key question that flows from this perspective is how to delineate between 'a resource that belongs to everyone' and a resource that is otherwise? The *AIEA* case goes some way to answering this question.

By granting the orders sought by AIEA, the High Court has got it right in the *AIEA* case. Those orders are: a revocation of the Biological Data Mining Licence granted by the first defendant, the Commonwealth of Australia; and an 'enduring injunction' against all biological data mining activities of the second defendant, BEEL, that conflicts with the *Information Commons Rights Act 2029* (Cth).[17]

HIGH COURT OF AUSTRALIA

Plaintiff: Australasian Information Environmentalism Alliance

v

First Defendant: Commonwealth of Australia (Minister of the Information Environment)

Second Defendant: Biodata Extraction and Excavation Ltd
[2037] HC 12 (1 February 2037)

I: Introduction

This matter was initially brought before the Full Federal Court of Australia in 2035. The Australian Information Environmentalism Alliance (hereafter AIEA) appealed the decision after a successful special leave application to this Court.

The appeal ultimately flows from the interpretation of two significant legislative initiatives developed in the 2020s: the *Biological Data Mining*

---

17 For the purpose of the hypothetical judgment, an 'enduring injunction' is a quasi-fictional injunction that persists provided the initial rationale of the injunction continues.

Act 2025 (Cth) (hereafter BDMA) and the *Information Commons Rights Act 2029* (Cth) (hereafter ICRA).[18]

According to the Explanatory Memorandum, the BDMA:

> aims to facilitate economic development by allowing the private sector, upon the governmental issuance of a Biological Data Mining Licence, to collect, own and sell biological data within the physical environment, particularly National Parks and other nature reserves.

Importantly, the BDMA explicitly over-turned the long-standing 'products of nature doctrine' inherent within traditional Intellectual Property Right (IPR) frameworks.[19]

The ICRA was part of a broad regulatory package that sought to reconcile private and public interests in the information environment. The initial impetus of the regulatory package was the *Global Information Commons Treaty 2027*, which was signed and subsequently ratified by most of the World Trade Organisation Member States in the late 2020s.[20]

The *Global Information Commons Treaty* was founded on a perception that the international Intellectual Property Right [hereafter IPR] framework had stretched too far and too deep, particularly in relation to biological data concerning the physical environment.[21] The Australian BDMA was singled out during the Treaty deliberations as exemplifying this dynamic, which is often referred to as 'IPR maximalism'.[22]

The ICRA gives 'the public' broad access and use rights over information that resides in the 'information commons'. Information Commons Rights

---

18 The *Native Information Title Act 2028* (Cth) is also relevant; albeit the discussion of this legislation is limited to the extent there is another relevant case pending in this Court. See Annex I.

19 The 'product of nature' doctrine has traditionally limited the ability to acquire an Intellectual Property Right over elements of nature that are discovered rather than invented. While the products of nature doctrine had already been largely diminished prior to enactment of the BDMA (as a result of common law such as *Diamond v Chakrabarty* (1980) 447 US 303), the legislative intent to eradicate the document is made abundantly clear within the BDMA.

20 The *Global Information Commons Treaty 2027* is a global treaty established by the World Intellectual Property Organisation, which seeks to redress heightened concerns about the breadth and depth of Intellectual Property Rights.

21 Intellectual Property Rights are a type of State sponsored property and/or monopoly right commonly divided into the categories of copyright, patents, design rights, trademarks, geographical indications and know-how.

22 'IPR maximalism' is a type of theoretical discourse that seeks the continued entrenchment and expansion of Intellectual Property Rights. In contrast, 'IPR minimalism' seeks to challenge the continued entrenchment and expansion of IPRs and to underscore the benefits of the information commons.

(hereafter ICRs) operate akin to open source computer software licences, which gained popularity at the end of the twentieth century. In short, ICRs allow for access and use of biological data under certain privately-ordered commons-based licencing agreements, but ownership is strictly precluded.

The relevance of the *Native Information Title Act 2030* (Cth) (hereafter NITA) is relevant to the extent that all of AIEA's activities are subject to the veto power of the relevant Indigenous custodians. The NITA sets out in clear terms the *sui generis* rights that Indigenous custodians maintain in relation to their traditional knowledge, customs and practice (among other things). As a result, AIEA's activities must be built upon the endorsement of the relevant Indigenous custodians. At present, this does not apply with respect to Biodata Extraction and Excavation Ltd (hereafter BEEL) as the NITA only applies to 'public rights' and not 'private rights'. Discussion of NITA is limited in this judgment to the extent there is another relevant matter currently pending before this Court: *Indigenous Custodians v Biodata Extraction and Excavation Ltd* [2037] HC 13.

II: Background

The facts underlying the dispute are well rehearsed in the Full Federal Court of Australia judgment. A truncated summary is as follows.

In the late 2020s, to further its objectives of conserving the physical environment, AIEA commissioned a nation-wide 'interactive biological art project' called Nature-Scape, which continues to operate throughout various National Parks and other nature reserves in Australia.[23]

According to the relevant promotional material, the aim of Nature-Scape is 'to leverage "citizen science" by co-joining conservationists with technological developments in order to secure positive environmental outcomes'.[24]

Nature-Scape was initially advertised (and trademarked) as a 'fully-immersive, five senses experience using multi-dimensional hologramaphics to deepen humanities relationship with, and understanding of, nature'.[25]

---

23 Nature-Scape is an interactive biological art project that relies on the latest Human-Nature-Interface technology in order to enhance public appreciation of nature.

24 While AIEA initially resisted technological developments, it has since adopted the position of leveraging technology in order to secure positive environmental outcomes. This approach is built upon the perspective of Donna Haraway who noted in 'A Cyborg Manifesto' that 'the boundary between science fiction and social reality is an optical illusion'; see her 1985 essay 'A Cyborg Manifesto: Science, Technology and Socialist-feminism in the late 20th century' in D Haraway, *Simians, Cyborgs and Women: The Reinvention of Nature* (Routledge, 1991) 149.

25 Hologramaphics is a technological derivative of holographics, which encompasses (but moves beyond) visual sensory experience to include the amplification and enhancement of sound, smell, taste and touch.

AIEA and the Nature-Scape project seek to enhance appreciation and understanding of nature by harnessing the latest Human-Nature-Interface (HNI) technology.[26] Specifically, AIEA provides participants of Nature-Scape with a particular type of HNI device called 'uber-sonic', which amplifies and enhances the sound, taste, smell, touch and sight of nature.

The uber-sonic device is typically worn on the external body as a piece of 'tech-clothing' (akin to the Apple Body Sensory Attire 4S) or worn as 'cyber-binoculars' (via the Google Human Nature Interface Specs 3.1).

Importantly, the use of a Nature-Scape uber-sonic device is fully dependent upon access to and use of the physical National Parks and other nature reserves used by AIEA from time to time, as well as the biological data inherent therein.

As Nature-Scape participants are immersed in the physical environment through bush-walking and other nature-based adventure activities, the uber-sonic device automatically collects and processes biological data of the surroundings. This information is in turn deployed by AIEA to advance conservation of the relevant area.

Prior to Nature-Scape, BEEL had been relying on its Biological Data Mining Licence, which had been lawfully granted by the Commonwealth of Australia in 2026. Among other things, this licence allowed BEEL to 'exploit biological data resources from National Parks and other nature reserves within Australia'. BEEL has subsequently profited by selling biological data usage licences, mainly to large pharmaceutical and industrial biotechnology and nanotechnology companies based in Germany and the United States of America.

The legal dispute culminated when BEEL sent a 'cease and desist' letter to AIEA. This correspondence stated that 'BEEL has an exclusive Biological Data Mining Licence over the biological data within particular National Parks and other nature reserves in Australia and therefore AIEA does not possess the access and/or use rights required to continue the Nature-Scape project'.

In response, after several failed attempts to negotiate a mutually satisfactory resolution, AIEA initiated legal proceedings in the Federal Court of Australia seeking two primary orders:

(1) A revocation of the Biological Data Mining Licence granted by the first defendant, the Commonwealth of Australia; and
(2) An 'enduring injunction' against all biological data mining activities of the second defendant, BEEL, that conflicts with the *Information Commons Rights Act 2029*.

---

26 HNI is a technological device that emerged during the 2020s which amplifies and/or enhances human perceptions of nature.

The Full Federal Court ultimately rejected both orders, and AIEA subsequently sought special leave for the matter to be heard in this Court. Special leave was granted.

III: Legal issues

Although the requested orders and the defendants are separate and distinct, many of the relevant factual and legal matters overlap. For this reason, the case was heard conjunctionally in the one sitting within the Full Federal Court.

The biological data activities of AIEA and BEEL are mutually exclusive to the extent that it is not possible for them to operate in the same place at the same time. This is because according to application of the relevant legislative frameworks – namely BDMA, ICRA and NITA – the issuance of a Biological Data Mining Licence negates any other access and/or use rights that AIEA (or the Indigenous custodians) would otherwise possess in relation to the biological data.

Resolution of this matter requires the application of the scales of justice to BEEL's private interests and AIEA's public interests, particularly as those interests relate to access, use and ownership of biological data.

Along with reconciling the inherent tension between the BDMA and ICRA, the dispute is to be resolved with reference to:

(1) the legal standing of AIEA to bring legal action to revoke the issuance by the Commonwealth of Australia of a Biological Data Mining Licence to BEEL; and
(2) an application of the public trust doctrine.

Prior to judgment of these two legal issues, it is necessary to provide some brief commentary on ICRs, particularly given that these rights represent a relatively new legislative initiative.

A: Information Commons Rights

For present purposes, an ICR facilitates members of the public, including civil society organisations such as AIEA, to access and use biological data content for 'the public benefit'. Information subject to an ICR cannot be subject to private appropriation under any circumstances, albeit all ICRs are subject to the veto power of Indigenous custodians.

ICRs operate in parallel fashion to the General Public Licence, which was popularised late in the twentieth century to facilitate open source computer software initiatives such as Mozilla Firefox Internet Browser as well as Wikipedia content. The main difference is that the General Public Licence was ultimately a privately-ordered initiative whereas ICRs are a publicly-ordered (State sponsored) initiative.

The ICRA was the first legislative initiative of the Commonwealth of Australia to implement the *Global Information Commons Treaty 2027*. Article 12 of that Treaty states:

> Public information resources, particularly those related to biological data, are the common property of all the people, including generations yet to come. As trustee of these resources, the Commonwealth shall conserve and maintain them for the benefits of all the people.[27]

The broad aim of the ICRA as explained in the Explanatory Memorandum is 'to enhance the governance framework of the information-age by establishing Information Commons Rights in juxtaposition to IPRs'.

B: Locus standi

The original threshold jurisprudential question in this matter is thus: does AIEA have legal standing to bring the action against the Commonwealth of Australia and BEEL with respect to the present dispute? Responding to this inquiry requires an initial exploration of how *locus standi* has applied to those civil society organisations concerned with the physical environment.

In the historic case of *Australian Conservation Foundation v Commonwealth* (1980) 146 CLR 493 this Court refused to provision standing to the Australian Conservation Foundation (hereafter ACF) because, according to Justice Gibbs, an 'intellectual or emotional concern' was not the type of interest that differentiated the ACF from other members of the public.[28]

Despite the actual outcome in *ACF v Commonwealth*, the notion of 'special damage' was nonetheless stretched to include 'special interest in the subject matter of the action'.[29] This elasticity was subsequently consolidated throughout the 1980s so that in 1989, when the ACF sought standing for a second time, the Federal Court was ready to grant it. By the late 1980s in Australia the Courts began to recognise the ecological expertise of civil

---

27 This constitutional initiative was inspired by Article 27 of the Pennsylvania State Constitution <http://sites.state.pa.us/PA_Constitution.html>. See also A Klass, 'Modern Public Trust Principles: Recognizing Rights and Integrating Standards' (2006) 82 *Notre Dame Law Review* 699 and M Kirsch, 'Upholding the Public Trust in State Constitutions' (1997) 46 *Duke Law Journal* 1169.

28 *Australian Conservation Foundation Inc v Commonwealth* (1980) 146 CLR 493, 530–1 (Gibbs J).

29 Ibid. The willingness to expand the notion of 'special interest' was further consolidated in a string of Australian cases in the 1980s such as *Onus v Alcoa of Australia Ltd* (1981) 149 CLR 27 and *Ogle v Strickland* [1987] 13 FCR 306.

society organisations such as ACF. For instance, Justice Davies stated as a matter of relevance in the case of *ACF v Minister for Resources* that:

> [p]ublic perception of the need for the protection and conservation of the natural environment and for the need of bodies such as the ACF to act in the public interest has noticeably increased, and is demonstrated by the growth of the ACF itself since the time of the 1980 ACF case.[30]

In parallel to the common law evolution, and largely as a consequence of ongoing agitation by civil society organisations during the 1970s and 1980s, many environmental legislative regimes in developed Western countries such as Australia began to facilitate and require input and comment from the public.

Although no longer in operation, section 123 of the *Environmental Planning and Assessment Act 1979* (NSW) was exemplary. This section provisioned justification for courts to subsume the 'rational truths' of civil society organisations into their judgments. For instance, Justice Stein opened his judgment in *Oshlack v Richmond River Council* in 1994 – a case concerning the habitat of koalas that may have detrimentally been affected by a residential subdivision – with reference to open standing provisions embedded within the statute:

> The notion of public interest litigation has been gaining ground in Australia over the last decade. The concept derives from principles of public law rather than private. The development springs from the increasing access of individual members of the public and groups to approach the courts and seek to enforce aspects of public law. This is particularly so in the area of environmental law where many New South Wales statutes include open standing provisions enabling 'any person' to seek to enforce breaches of the law.[31]

There are two notable elements that led to the judicial evolution of legal standing as it related to so-called 'third parties' and the physical environment: first, the application of the objective expert knowledge of ecologists and second, the 'rational truths' of civil society organisations evidenced through broad-based support within the polity.

---

30 *Australian Conservation Foundation Inc v Minister for Resources* (1989) 19 ALD 70, 73.
31 *Oshlack v Richmond River Council* (1994) 82 LGERA 236, 238. Justice Stein's judgment was overturned by the High Court with Justice McHugh dismissing the appeal of the 'concerned citizen' based on the rationale that the case was not in the 'public interest' and that in coming to a contrary view the trial judge was simply influenced by a 'social preference': *Oshlack v Richmond River Council* (1998) 193 CLR 72, 100 (McHugh J).

Both of these elements are also present in the context of the information environment.[32] This is especially the case given the rise of importance of 'information ecologists' during the 2020s, along with the increasing public support garnered by civil society organisations that are focused on the health and maintenance of the information environment.[33] As a result, the threshold level of information environmental technical expertise has now been traversed, and a body of rational truths established, particularly with respect to the protection and maintenance of the information commons.

Clearly, the procedural integrity of the ICRA is reliant upon public input concerning the application and operation of the legislation. Where the State fails to facilitate public engagement with legislative initiatives such as the ICRA, then this Court is empowered to strike down the law.[34] Relevantly, according to section 14 of the ICRA, the Department of the Information Environment is subject to a positive onus to invite civil society organisations, such as AIEA, to submit their concerns about the impacts on the information environment of proposed laws and the granting of biological data licences.

Thus, on the basis of the above reasoning, AIEA has legal standing to bring legal action against the Commonwealth of Australia and BEEL in accordance with the ICRA.

C: Public trust doctrine

Both the plaintiff and the defendants addressed the application of the public trust doctrine to the information environment within their respective submissions. It is therefore incumbent upon this Court to rule on the matter directly. This is particularly so given the public trust doctrine has only recently been revived in the Australian common law context as a result of *Greenspace Ltd v Commonwealth of Australia* [2035] HC 7 (hereafter *Greenspace* case).

While the public trust doctrine traditionally related to navigation, commerce and fishing, it was subsequently extended to include wildlife and

---

32 There is some evidence that the requisite 'broad based support within the polity' has been amassing for some decades as symbolised by legislative initiatives floated in the 2010s in Australia such as the *Lifting the Bar Bill* along with the Private Members Bill introduced by Melissa Parke; see S Smith, 'MP wants second take on gene patent laws' *ABC News*, 15 May 2012 <www.abc.net.au/news/2012–05–14/mp-urges-labor-to-back-gene-patenting-overhaul/4010612>.

33 An 'information ecologist' is a scientific expert in the study of the information environment and/or the information commons.

34 For an early example of a case relating to procedural fairness and the environment in the US see *DC Fedn of Civic Assns, Inc v Airis* (DC Cir 1968) 391 F 2d 478 (failure to hold public hearings).

marshlands along with various recreational activities such as 'boating, hunting, bathing, taking shellfish, gathering seaweed, . . . and passing and repassing'.[35] The question now before this Court is whether the public trust doctrine can extend to the information environment, and in particular the information commons?

Historically speaking, the public trust doctrine evolved from Roman law concepts of public property, or more specifically the Justinian Code inherent within Byzantine law, whereby rivers, air and the sea were dedicated to the public and therefore incapable of private ownership.[36] The doctrine was to some extent codified within *Magna Carta* and it was eventually adopted by most medieval European legal systems.[37] For instance, there is clear evidence of the doctrine operating within eleventh century French law.[38]

A more modern version of the public trust doctrine was embedded within the United States Supreme Court Case of *Illinois Central Railroad v Illinois* where the Court disallowed privatisation of navigable waters of Lake Michigan as these resources were regarded as being 'held in trust for the people of [Illinois]'. The Court stated:

> [The title to submerged lands] is a title different in character from that which the State holds in lands intended for sale. It is different from the title which the United States hold in the public lands which are open to preemption and sale. It is a title held in trust for the people of the State that they may enjoy the navigation of the waters, carry on commerce over them, and have liberty of fishing therein freed from the obstruction or interference of private parties.[39]

Inherent within *Illinois Central* is a basic principle – the public possesses certain rights in resources. This principle limits the power of legislative representatives to alienate such resources.

It is true that while United States courts have traditionally placed reliance upon the public trust doctrine with respect to environmental issues, this has not historically been the situation in Australia until the recent *Greenspace* case.

Greenspace served to reinvigorate common law perspectives of the public trust doctrine, and indeed counsel on each side of the Bar in this dispute relied heavily on the doctrine within their submissions.

---

35 *Town of Orange v Resnick* (1920) 94 Conn 573, 578.
36 *The Institutes of Justinian* bk 2, tit 1, pts 1–6, 65 (J Thomas trans, 1975).
37 M Bloch, *French Rural History: An Essay on its Basic Characteristics* (University of California Press, 1966) 183.
38 Ibid.
39 *Illinois Central Railroad v Illinois* (1892) 146 US 387.

Relevant to the present matter, the rights protected by the public trust doctrine are inherently collective in nature and are therefore held by persons as members of the public. In this way, the rights are both inalienable and indivisible from the public. As a civil society organisation with broad public support, AIEA adequately represents 'members of the public'.

The *Greenspace Case* held that the public trust doctrine can and should be applied to those resources that are considered 'inherent public property'.[40] This label refers to property whereby citizens possess free and unimpeded access. The information commons fits neatly under the 'inherent public property' banner, particularly where the information in question is biological data.

Greenspace also highlighted the traditional position, being that there are two dominant themes inherent within the public doctrine that are relevant when determining the doctrine's application: (i) fiduciary obligations and (ii) judicial review.

According to the fiduciary aspect of the public trust doctrine, an individual may abandon her private property but a public trustee cannot abandon public property.[41] Here it would seem that either the public itself possesses a property right in relation to public resources; or in the alternate, the State owes a duty to the public to protect public resources.[42] Justice Nannup (Bunjalung Country) in *Greenspace* cited with approval the following comments from the United States case *National Audubon Society v Superior Court*:

> The public trust is more than an affirmation of state power to use public property for public purposes. It is an affirmation of the duty of the state to protect the people's common heritage of streams, lakes, marshlands and tidelands, surrendering that right of protection only in rare cases when the abandonment of that right is consistent with the purposes of the trust.[43]

This was one example of many United States cases that appear to have been enlivened in Australia by the *Greenspace* case.

Regardless of whether AIEA possesses access/use rights in the biological data via ICRs, or in the alternate the State owes a duty to AIEA to protect public resources such as biological data, the result is the same: the

---

40 C Rose, 'The Comedy of the Commons: Customs, Commerce and Inherently Public Property' (1986) 53 *University of Chicago Law Review* 711, 713.

41 The case of *Arnold v Mundy* (1821) 6 NJL 1, 71–78 'marks the origin of the fiduciary aspect of public trust doctrine'. See also *State v Cleveland & Pittsburgh R R* (1916) 94 Ohio St. 61, 80 113 NE 677, 682.

42 *In Re Complaint of Steuart Transp Co* (1980) 495 F Supp 38, 40; *National Audubon Society v Superior Court* (Cal 1983) 33 Cal 3d 419, 441.

43 *National Audubon Society v Superior Court* (Cal 1983) 33 Cal. 3d 419, 441.

Commonwealth of Australia cannot facilitate the private appropriation of biological data that falls within the domain of the information commons.

With respect to judicial review, the *Greenspace* case has now put it beyond doubt that the public trust doctrine empowers this Court 'to mend imperfections in the legislative and administrative process'.[44] This is particularly relevant to the ICRA, which both implicitly and explicitly empowers this Court with judicial review authority to consider State decisions in allocating information resources, particularly where those resources do, or at least should, reside in the information commons.

From a judicial review perspective, *Greenspace* has also reinforced the growing tendency of this Court to view 'low visibility decision-making' in an unfavourable light.[45] Such decision-making occurs where critical resource-allocation decisions are made without adequate notice or publicity. This concern is particularly important with respect to the Commonwealth of Australia granting a Biological Data Mining Licence to BEEL. As *Greenspace* underscored, inadequate notice or publicity undermines participatory decision-making, diminishing the quality of social choices.

Moreover, the *Greenspace Case* highlighted the traditional public trust doctrine perspective, being that the doctrine operates to ensure the State does not revoke the public's collective rights by trading them away and/or transferring them to private parties. Justice Kwaymullina (Noongar Country) in *Greenspace* cited with approval *Illinois Central Road* when Her Honour stated that:

> The only proper exception to the general rule of inalienability is that the State can transfer public trust resources to private parties if they 'are used in promoting the interests of the public' or the transfer is made 'without any substantial impairment of the public interest'.[46]

In this instance, BEEL has been granted a private licence to exploit the biological data resources within various National Parks and other nature reserves. The granting of this licence does not promote the interests of the public; and indeed BEEL's activities impair the public interest to the extent that they undermine the activities of AIEA and the Indigenous custodians.

---

44 J Sax, 'The Public Trust Doctrine in Natural Resource Law: Effective Judicial Intervention' (1970) 68 *Michigan Law Review* 471, 509.

45 W Araiza, 'Democracy, Distrust, and the Public Trust: Process-based Constitutional Theory, the Public Trust Doctrine, and the Search for a Substantive Environmental Value' (1997) 45 *University of California Los Angeles Law Review* 385, 398; M Ryan, 'Cyberspace as Public Space: A Public Trust Paradigm for Copyright in a Digital World' (2000) 79 *Oregon Law Review* 647, 700.

46 *Illinois Central Railroad v Illinois* (1892) 146 US 387, 452–453.

Evidently, the public trust doctrine provides the jurisprudential foundation for this Court to act on behalf of citizens to override unreasonable privatisations of public resources carried out by government.[47] In applying this standard, it is clear that the BDMA undermines the inherent jurisdiction of this Court by privatising biological data that would otherwise remain, but for the BDMA, as a public right (subject to the usual exceptions concerning the Indigenous custodians).

It is now well-accepted that the contemporary application of the public trust doctrine is the precursor to all human beings having an entitlement to healthy natural ecosystems. In the present matter, the Court adopts a parallel view in relation to the information environment. Namely, that access to and use of (biological) data within the information commons is a fundamental human right.[48] Of course, like all rights, this right is not absolute to the extent that, for example, it is subject to the veto powers of the Indigenous custodians.

The Explanatory Memorandum of the ICRA makes it explicit that 'every human being has a stake in the information ecosystems that govern society'. In particular, the rationale for ICRs is clearly outlined. Namely, the information commons is the building block for critical social functions such as 'the creation of new knowledge, education, public health and safety, and deliberative democracy'.[49] In practice the only meaningful limitation on access and use of the information commons is the *sui generis* rights of Indigenous custodians.

A prerequisite for healthy ecosystems in the physical environment is that everyone should have equitable access to the information environment, and in particular the information commons.[50] This is because many problems that manifest in the physical environment are ultimately social problems

---

47 Ibid., 491–565.
48 See, for example, Earth Law Center website <www.earthlawcenter.org>. Also, parallel approaches can be found in L Helfer, 'Toward a Human Rights Approach Framework for Intellectual Property' (2007) 40 *University of California Davis Law Review* 6; V Vadi, 'Sapere Aude! Access to Knowledge as a Human Right and a Key Instrument of Development' (2008) 12 *International Journal of Communication Law and Policy* 345, 350 (arguing that 'access to knowledge is a basic human right, and that restrictions on access ought to be the exception, not the other way around').
49 E Wirten, *Terms of Use: Negotiating the Jungle of the Intellectual Commons* (University of Toronto Press, 2008) 6 (referring to Pamela Samuelson's 'thirteen public domains').
50 This view was consolidated by the United States Supreme Court in *Hague v Committee for Indus Organization* (1993) 301 US 496, 515–516 (Roberts J), where the public trust doctrine was invoked to ensure 'streets and parks as the public fora for free speech activities are open to all'. In that case, the Court further stated: 'Wherever the title of streets and parks may rest, they have immemorially been held in trust for the use of the public and, time out of mind, have been used for purposes of assembly, communicating thoughts between citizens, and discussing public questions. Such use of the streets and public places has, from ancient times, been a part of the privileges, immunities, rights, and liberties of citizens'.

that require social solutions. The information commons is important in this regard, and the legislature has bound the Court to this view.

The application of the public trust doctrine to the present dispute requires this Court to begin with a presumption in favour of 'public use, access and enjoyment'.[51] For this reason we are required to give sufficient consideration to public interest rather than a pure focus on individual interests, particularly with respect to the information environment and the information commons. The public dimension of the information environment should not be seen as a tack-on or somehow incidental, but rather a foundation of the general regulatory framework.

An important question for this Court is how to demarcate between 'public informational resources' and 'private informational resources'? As the ICRA (and other relevant legislation such as the NITA) is mostly silent on this issue, it is now incumbent upon this Court to synthesise the substantive and procedural values underpinning the relevant legislation.

The substantive values inherent within the ICRA relate to free (libre) access, use and flow of knowledge and information afforded to the public. The procedural values concern the democratisation of the information environment through the facilitation of further public participation engagement in shaping laws relating to the access, use and flow of the information environment, specifically the information commons.

Applying the public trust doctrine to the information environment – taking into account the substantive and procedural values inherent within the relevant legislative initiatives – empowers this Court to impose a judicial check on State power to allocate informational resources in an appropriate manner.

Consequently, the BDMA violates the inherent jurisdiction of this Court by allocating public resources to private interests. Just as the State cannot grant property rights to individuals that undermine reasonable use of certain physical resources by the public, so too the State cannot grant legal rights (for example IPRs) over informational resources purely for the benefit of private parties.[52] The public purpose of any provision of rights by the State must be paramount, and must always trump pure private interests.

For these reasons, the Federal Court of Australia judgment is overturned; and the orders sought by AIEA are granted.

---

51 In *Re Water Use Permit Applications* (2000) 94 Hawai'i 97, 142 and *In re Waiolo O Molokai Inc* (2004) 103 Hawai'i 401, 432.

52 See, for example, *Matthews v Bay Head Improvement Association* (1984) 95 NJ 306. This approach was shared in *Sony Corp of Am v Universal City Studios* (1984) 464 US 17 429 where the Court stated: 'The monopoly privileges that Congress may authorize are neither unlimited nor primarily designed to provide a special private benefit. Rather, the limited grant is a means by which an important public purpose may be achieved'.

IV: Conclusion

The Court orders:

(1) A revocation of the Biological Data Mining Licence granted by the first defendant, the Commonwealth of Australia; and
(2) An 'enduring injunction' against all biological data mining activities of the second defendant, BEEL that conflicts with the ICRA.
(3) Both defendants are jointly and severally liable for the plaintiff's costs.

Annex I: Indigenous *sui generis* claim

As is now standard practice, and in accordance with the *Native Information Title Act 2028* (Cth), the traditional Indigenous custodians maintain veto power over information that flows from physical environments subject to Native Title.

Within its submission, AIEA made it clear that any ICR that it maintained access to was to be subject to the veto power of the Indigenous custodians. A Memorandum of Understanding between AIEA and the relevant Indigenous custodians was submitted. This document outlined broad endorsement by the Indigenous custodians of AIEAs activities. However, that endorsement was subject to the *sui generis* rights of Indigenous custodians, which relates to traditional knowledge, customs and practice, as well as expanded categories of *sui generis* Indigenous rights within the NITA.

The Indigenous custodians have recently relied upon their inherent *locus standi* rights to seek an order that largely mirrors the AIEA order:

> An 'enduring injunction' against all biological data mining activities of the second defendant, BEEL that conflicts with the *Native Information Title Act 2028.*

Judgment of this order will be considered in the forthcoming decision pending before the Federal Court: *Indigenous Custodians vs Biodata Extraction and Excavation Ltd* [2037] HC 13.

Index